Edward John Chapman

Minerals and Geology of Central Canada

Comprising the provinces of Ontario and Quebec

Edward John Chapman

Minerals and Geology of Central Canada
Comprising the provinces of Ontario and Quebec

ISBN/EAN: 9783744752060

Printed in Europe, USA, Canada, Australia, Japan

Cover: Foto ©berggeist007 / pixelio.de

More available books at **www.hansebooks.com**

THE
MINERALS AND GEOLOGY

OF

CENTRAL CANADA,

COMPRISING THE

PROVINCES OF ONTARIO AND QUEBEC.

BY

E. J. CHAPMAN,
PH.D., LL.D.

PROFESSOR IN THE UNIVERSITY OF TORONTO, AND SCHOOL OF
PRACTICAL SCIENCE.

THIRD EDITION,
REVISED AND IN GREAT PART REWRITTEN.

TORONTO:
THE COPP, CLARK COMPANY, (LIMITED.)
1888.

PREFACE.

THE plan and mode of treatment adopted in the original edition of this work, are retained in the present edition; but the subject matter has been greatly extended, and, with the exception of a few pages, the work has been entirely rewritten. It aims to convey, in language as little technical as possible, a practical knowledge of the minerals and general geology of the two central provinces of the Dominion—Ontario and Quebec. In its plan, it embraces five leading sections or subdivisions. These follow each other in logical order, and are discussed throughout in an essentially explanatory form.

The first section describes briefly the more salient characters or properties by which the determination of minerals is effected; and it includes a sufficient notice of the blowpipe to enable anyone to employ that useful instrument in the practical examination and rough analysis of mineral bodies.

The second part or section contains descriptions of all the minerals hitherto recognized within the Provinces to which the work refers. In these descriptions, minute crystallographic and chemical details are purposely avoided, as unsuited to the character of the book.* The descriptive portion of the section is preceded by a couple of Determinative Tables, drawn up expressly for the present work, by the use of which, the name of any mineral occurring in central Canada, may be easily ascertained.

The two succeeding sections, PARTS III and IV, are introductory to PART V, in which the geological features of Ontario and Quebec are passed under review. Part III discusses the classification, structural characters and other technical points belonging to the study of rock masses generally, and that of mineral veins; and it includes also a brief outline of the Earth's rock-recorded history (pp. 199 to 207). Part IV comprises an epitome or systematic synopsis of Canadian Palæontology, with figures and descriptions of our more characteristic

---

* A synopsis of the crystallographic characters of the more important mineral species will be found in the *Notes* attached to the *Determinative Tables* of the author's *Blowpipe Practice.* In these *Notes*, spectroscopic characters, where readily determinable, are also given.

fossils, and many original groupings and generalizations. The figures are somewhat roughly executed, but they serve sufficiently for the identification of the forms to which they refer.

Finally, in Part V, the subdivisions, economic minerals, characteristic fossils, and general distribution of the rock formations of the two Provinces are given in systematic outline. Here, as elsewhere throughout the work, I have been careful to acknowledge my obligations where information has been specially derived from other sources. The subdivisions adopted with regard to the geological areas of the Provinces are practically the same as those given in the author's "Outline of the Geology of Canada" published in 1876; but their arrangement has been slightly altered, and the greater portion of this section has been entirely rewritten for the present work.

An Index, containing upwards of three thousand references to minerals, rock-formations, fossils, localities, etc., within the two Provinces, concludes the volume, and will add much, it is thought, to its utility.

E. J. C.

Toronto, April 30, 1888.

CONTENTS.

—o—

ADDITIONS AND CORRECTIONS.

— ·o——

P. 6, note—for "thas" read "that."

P. 17, line 8—for "pa" read "pan."

P. 22, line 11—for "fluorhydric" read "hydrofluoric."

P. 51, insert in bracket 18—BB, on charcoal, white incrustation and arsenical odour—*Mispickel* (some varieties), No. 22.

P. 57, insert at the commencement of the "not malleable" series—Tin-white or greyish. BB, arsenical odour—*Mispickel* (some varieties), No. 22.

P. 61, line 1 — for "SIMPLE SUBSTANCE" read "SIMPLE SUBSTANCES."

P. 64, line 15—add: "Gold occurs also in many of the veins around Big Stone Bay, Lake of the Woods ; and quite recently it has been discovered in the township of Dennison and elsewhere in the Huronian rocks of the Sudbury district."

P. 64, line 28—after Rivière du Loup, add (Beauce Co).

P. 71—add Sudbury to the localities of Purple Copper Pyrites.

P. 76, line 13—for "Cape Ibberwash" read "Cape Ipperwash."

P. 84, line 15—for "altered Silurian" read "crystalline."

P.P. 90, 91—The figures of crystals on these pages are placed in reversed position.

P. 90, line 8 from bottom—for "from" read "form."

P. 93, for "distinct" read "distinctive."

P. 99, line 2—for "iron is" read "iron oxides ore."

P. 104, lines 7 and 8 from bottom—erase "or Upper Laurentian strata."

P. 110—transpose lines 5 and 6.

P. 110, line 15—for "Nepheletic" read "Nephelitic."

P. 115—add Sheffield to localities of Phlogopite. A valuable deposit of this species of mica has been recently discovered in the township.

P. 126, line 6 from bottom—for "or" read "of."

P. 127, line 2—for "Bentick" read "Bentinck."

P. 128, last line—for "chlorite" read "chloritic."

P. 130, line 8 from bottom—for "from" read "form."

P. 148, line 4—for "more limited" read "a more or less limited depth."

P. 149, line 26—for "exhibit" read "exhibits."

P. 163—the cut, figure 88, is placed incorrectly. The lettering shews its proper position.

P. 169, line 8—add : a very fine-grained variety, with predominating feldspar, has been named granulite or feldspar-rock.

P. 172, line 11—for "agillite" read "argillite."

P. 172, line 24—for "Huronian rock" read "Huronian rocks."

P. 173, line 5 from bottom—for "from" read "form."

P. 189, line 17—add "at" before "Gros Cap."

P. 191, line 22—for "are not uncommon" read "have been fouud."

P. 191, line 5 from bottom—insert "lava-like products " after " but."

P. 204, line 21—for " Equiseta " read " Equiseti."

P. 205, line 10—for "Parts V. and VI." read " Part V."

P. 207, lines 6 and 7—erase " and to have been experienced also in the extreme south."

P. 210, line 26—for " literal " read " littoral."

P. 216, line 16—for "hollow-jointed" read "hollow, jointed."

P. 224, lines 19, 23, 25—for " siliceons " read " siliceous."

P. 227, line 8 from bottom—insert after "all " " Upper Cambrian or."

P. 229, line 2 from bottom—for " Hydrocorolla " read " Hydrocoralla.

P. 232, line 3—insert (7) before " Syringopora."

P. 239, line 3 from bottom—for "bifercating " read " bifurcating."

P. 240, line 22—for "Enerincus and Enerinite" read "Encrinus and Encrinite."

P. 241, line 11—for " oval " read " anal."

P. 245, line 10—for "species " read " examples."

P. 248, line 3 under Vermes—for " six " read " seven."

P. 251, note—for " Phyllorarida " read " Phyllocarida."

P. 254, line 24—for " states " read " slates."

P. 260, line 23—for " Limulas " read "Limulus."

P. 266, under Fig. 181—for " Fenstella " read " Fenestella."

P. 267, line 10—for " Brachipods " read " Brachiopods.

P. 269, lines 6 and 10 from bottom—for " Brown " read " Bronn."

P. 272, line 10 from bottom—for " Pernian " read " Permian."

P. 279, line 8—for " arrange " read " arranged."

P. 280, line 6—for " Prosobranchiala " read " Prosobranchiata."

P. 284, line 1—for " Cytolites " read " Cyrtolites."

P. 288, line 9—for " Tetrebranchiata " read " Tetrabranchiata."

P. 289, line 5 from bottom—for " resembling " read " resembles."

P. 302, at close of first paragraph—add : " Quite recently, important discoveries of Native gold and auriferous ore have been made in the vicinity of Sudbury: more especially in Dennison township."

P. 340, line 8—for "outlying " read " inlying."

A POPULAR EXPOSITION

OF THE

MINERALS AND GEOLOGY

OF

CENTRAL CANADA.

—◆—

INTRODUCTORY NOTICE.

The aim of the present work is to impart, in a simple and condensed form, a practical knowledge of Canadian minerals and rock formations, including with the latter, the various fossilized bodies which so many of these rocks contain, and by which their respective ages and positions are principally established. Geology, in the proper acceptation of the term, comprises the History of the Earth as distinct from records of human action and progress : a history revealed to us by the study of the rock masses which lie around and beneath us ; and by a comparison of the results of ancient phenomena, as exhibited in these rocks, with the forces and agencies still at work in modifying the surface of the globe. As Geology is thus essentially based on the study of rocks and their contents, and as rocks are not only made up of a certain number of simple minerals, but contain also many of these latter in veins and other more or less accidental forms of occurrence, it is advisable at the outset to obtain a certain knowledge of the distinctive characters of minerals, and of the application of these characters to the determination of mineral bodies generally. This achieved, we may proceed to the study of the more extended mineral masses, or rocks proper : their classification, structural characters, composition, modes of formation, and other related points of inquiry. The study of Organic Remains comes next in order—these bodies, the representatives of departed forms of life, occurring in great numbers in many strata. They serve not only for the practical identification of the rock groups in which they are enclosed—thus enabling us to determine, for instance, whether a

2

given bed lie above or below the great coal formation or other geological horizon—but they make known also many interesting facts with regard to the climatic relations of the Past, and serve to explain to some extent the embryology and development of existing forms. Finally, with the information obtained from these preliminary sections, the reader may turn with profit to the study of our local geology.

In accordance with these views, the subject-matter of the present treatise is discussed under the following sub-divisions :

I.—The Distinctive Characters of Minerals.

II.—The Minerals of Central Canada, or Provinces of Ontario and Quebec.

III.—Rocks and Rock-producing Agencies.

IV.—Fossilized Organic Bodies.

V.—The Geology of Central Canada—comprising the Subdivisions, Characteristic Fossils, Economic Materials, and Distribution, of the various Geological Formations occurring within the Provinces of Ontario and Quebec.

PART I.

——◆——

THE DISTINCTIVE CHARACTERS OF MINERALS.

Preliminary Remarks :—The various bodies which occur in Nature are of two general kinds—Organic and Inorganic, respectively. The former constitute Vegetables and Animals, and all bodies of vegetable or animal origin. In the living state, they possess certain structural parts or organs by which they assimilate or take into their substance external matter, and thus increase in bulk or maintain vitality. Inorganic bodies, on the other hand, are entirely destitute of functional organs of this nature. They comprise all products of chemical, electrical, and mechanical forces, acting independently of life; and thus include all metals, stones, and rocks, and also air and water.

Mineral or inorganic bodies are in themselves, also, of two general kinds. Some possess a definite composition and definite physical characters. Others are mixed bodies or compounds of more or less variable character. The former constitute simple minerals or minerals proper; the latter form rocks or rock-matters. In Parts I and II of this Treatise, minerals proper are alone considered. Rocks and rock-producing agencies, come under review in Part III and in succeeding portions of the work.

Minerals are distinguished from one another by certain characters or properties which they possess: such as form, degree of hardness, relative fusibility, &c.

Mineral characters are of two principal kinds : *physical* or *external*, and *chemical*, respectively. Physical characters comprise the various properties exhibited under ordinary conditions by mineral bodies : colour, form, &c., are examples, Chemical characters, on the other hand, comprise the properties developed in minerals by the application of heat, or by the action of acids or other re-agents, by which, in general, a certain amount of chemical decomposition is effected.

A.—PHYSICAL CHARACTERS OR PROPERTIES.

The physical properties of minerals are somewhat numerous; but many, although of the highest interest in indicating the existence of natural laws, and in their relations to physical science generally, are not readily available as a means of mineral discrimination. These, consequently, will be omitted from consideration in the following pages; and the other characters will be discussed only in so far as they admit of direct application to the end in view—namely, the practical discrimination of minerals one from another.*

The following are the characters in question:

1. ASPECT OR LUSTRE.
2. COLOUR.
3. STREAK.
4. FORM.
5. STRUCTURE.
6. HARDNESS.
7. SPECIFIC GRAVITY.
8. RELATIVE MALLEABILITY.
9. MAGNETISM.
10. TASTE.

Aspect or Lustre.—In reference to this character we have to consider first, the *kind*, and, secondly, the *degree* or *intensity* of lustre, as possessed by the mineral under examination. The kind of lustre may be either *metallic*, as that of a piece of copper, silver, &c.; or *sub-metallic*, as that of most kinds of anthracite coal; or *non-metallic*, as that of stones in general. Of the non-metallic there are several varieties, as, more especially: the *adamantine* lustre or that of the

---

* Viewed collectively, the Physical Characters of Minerals may be arranged for the purposes of study, under six groups, as follows:

FIRST GROUP:—*Morphological Characters:*—1, Form. 2, Surface-condition. 3, Structure. 4, Cleavage. 5, Fracture.

SECOND GROUP:—*Optical Characters:*—1, Aspect or Kind of Lustre. 2, Degree of Lustre. 3, Colour. 4, Streak. 5, Degree of Transparency. 6, Refraction. 7, Polarization.

THIRD GROUP:—*Cohesion Characters:*—1, Hardness. 2, Tenacity. 3, Malleability. 4, Expansibility.

FOURTH GROUP:—*Sensationary Characters:*—1, Weight (Specific Gravity). 2, Feel. 3, Taste, 4, Odour. 5, Sound.

FIFTH GROUP:—*Physical Characters, proper:*—1, Magnetism. 2, Electricity. 3, Phosphorescence.

SIXTH GROUP:—*Epigenic Characters:*—1, Tarnish. 2, Ordinary Disintegration and Decomposition. 3, Efflorescence. 4, Deliquescence.

diamond, carbonate of lead, &c.; the *vitreous* or glassy lustre—example: rock crystal; the *resinous* lustre—ex.: native sulphur; the *pearly* lustre—ex.: talc; the *silky* lustre (usually accompanying a fibrous structure)—ex.; fibrous gypsum; the *stony* aspect; the *earthy* aspect, &c. These terms sufficiently explain themselves. Occasionally, two kinds of non-metallic lustre are simultaneously present —either blended, as seen in obsidian, which exhibits a "resino-vitreous" aspect; or distinct as regards different crystal faces or external and internal surfaces. Many of the so-called Zeolites, for example, present a pearly lustre on the surfaces produced by cleavage (see beyond), whilst the external lustre is vitreous. In Apophyllite, the basal or terminal crystal-plane is pearly, the others vitreous. Micas, and some few other minerals, present a *pseudo-metallic* lustre. This may be distinguished from the metallic lustre properly so-called, by being accompanied by a degree of translucency, or by the powder of the mineral being white or faintly coloured: minerals of a true metallic aspect being always opaque, whilst their powder is either black or dictinctly coloured. Very few minerals exhibit (in their different varieties) more than one general kind of lustre: metallic or non-metallic. Thus, galena (the common ore of lead), copper pyrites, &c., always present a metallic lustre; whilst, on the other hand, quartz, feldspar, calc-spar, gypsum, &c., are never metallic in aspect. Hence, by means of this easily-recognized character, we may divide all minerals into two broad groups; and thus, if we pick up a specimen, and wish to ascertain its name, we need only look for it among the minerals of that group with which it agrees in lustre. The first step towards the determinataion of the substance will in this way be effected.

The *degree* of lustre may be either splendent, shining, glistening, glimmering or dull; but the character is one of comparatively little importance.

Colour.—When combined with a metallic aspect, colour becomes a definite character, and is thus of much value in the determination of minerals. As regards a substance of metallic aspect, for example, specimens brought from different localities, or occurring under different conditions, rarely vary in colour beyond a slight difference of depth or shade. Thus, galena the common ore of lead is always lead-gray; copper pyrites, always brass-yellow; native gold, always gold-yellow; and so forth. When accompanied, however, by a vitreous or other non-metallic lustre, colour becomes a character of no practical value, as a mineral of non-metallic aspect may present, in its different varieties, every variety of colour. Thus, we have colourless quartz, amethystine or violet quartz, red quartz, yellow quartz, &c. Also, feldspars, fluorspars, and other minerals of variable colour: just as in the Vegetable Kingdom, we have red, white and yellow roses, and dahlias, &c., of almost every hue. The more common shades of metallic colour are as follows.

White	Silver-white	ex. Native silver.
	Tin-white	ex. Pure tin; cobalt ore.
Grey	Lead-grey	ex. Galena.
	Steel-grey	ex. Specular iron ore.
Black	Iron-black (usually with sub-metallic lustre)	ex. Magnetic iron ore.
Yellow	Gold-yellow	ex. Native gold.
	Brass-yellow	ex. Copper pyrites.
	Bronze-yellow (a brownish-yellow)	ex. Magnetic pyrites.
Red	Copper-red	ex. Native copper.

These metallic colours are often more or less obscured by a black, brownish, purple, or iridescent *surface-tarnish*. In noting the colour of a mineral, this must be constantly borne in mind, and if possible a newly-fractured surface should be observed. The non-metallic colours comprise, white, grey, black, blue, green, red, yellow, and brown, with their various shades and intermixtures: as orange-yellow, straw-yellow, reddish-brown, greenish-black, &c. In minerals of a non-metallic aspect, the colour is sometimes uniform; and at other times, two or more colours are present together in spots, bands, &c., as in the varieties of quartz called agate, blood-stone, jasper, and so forth· In most varieties of Labradorite, or Labrador Feldspar, a beautiful play or change of colour is observable in certain directions. The finer varieties of Opal also exhibit a beautiful and well known iridescence.

Streak.—Under this technical term is comprised the appearance or colour of the scratch, produced by drawing or "streaking" a mineral across a file or piece of unglazed porcelain. The character is a valuable one on account of its uniformity: as no matter how varied the colour of a mineral may be in different specimens, the streak will remain of one and the same colour throughout. Thus, blue, green, yellow, red, violet, and other specimens of fluor spar, quartz, &c., exhibit equally a white or "uncoloured" streak. The streak is sometimes "unchanged," or of the same tint as the external colour of the mineral; but far more frequently it presents a different colour. Thus, whilst Cinnabar, the ore of mercury, has a red colour and red streak, Realgar, red sulphide of arsenic, has a red colour and orange-yellow streak; Copper Pyrites, a brass-yellow colour, and greenish-black streak; and so forth. In certain malleable and sectile minerals, the scratched surface presents an increase of lustre. The streak is then said to be "shining." Finally, it should be remarked, that in trying the streak of very hard minerals, we must crush a small fragment to powder, in place of using the file; because otherwise, a greyish-black streak, arising from the abrasion of the file, might very possibly be obtained, and so conduce to error.

Form.—The forms assumed by natural bodies are of two general kinds: (1) *Accidental* or *Irregular*, depending rather on external conditions than on the actual nature of the body: and (2), *Essential* or *Regular*. Accidental forms occur only as monstrosities in Organic Nature. Amongst minerals, on the other hand, they are of frequent occurrence; but the Mineral Kingdom possesses also its definite or essential forms. These, whether transparent or opaque, are termed *crystals*. This term was first applied to transparent vitreous specimens of quartz or rock-crystal, from the resemblance of these to ice; but as it was subsequently found that many opaque specimens of quartz present exactly similar forms, and that opaque as well as transparent forms of other minerals occur, the term, in scientific language, gradually lost its original signification, and came to be applied to all the geometrical or regular forms of minerals and other inorganic bodies, whether transparent, translucent, or opaque. As already remarked, minerals of a metallic lustre are always opaque; and many of these, galena, iron-pyrites, arsenical-pyrites, &c., occur frequently in very regular and symmetrical crystals.

As regards the regular or essential forms of Nature, two distinct and in a measure antagonistic form-producing powers—*Vitality* and *Crystallization*—thus appear to exist. Forms which arise from a development of the vital force, exhibit rounded and confluent outlines; whilst those produced by crystallization are made up of plain surfaces, meeting, in sharp edges, under definite and constant angles.*

---

* This law is affected within slight limits by isomorphous replacements, and also by changes of temperature. The law itself appears to have been discovered by Nicolaus Steno (a naturalized Florentine) as early as 1669, but its true importance was not appreciated until the

Crystals originate in almost all cases in which matter passes from a gaseous or liquid into a solid state; but if the process take place too quickly, or the matter solidify without free space for expansion, crystalline masses, in place of regular crystals, will result. If a small fragment of arsenical pyrites, or native arsenic, be heated at one end of an open and narrow glass tube, the arsenic, in volatilizing, will combine with oxygen from the atmosphere, and form arsenious acid, which will be deposited at the other end of the tube, in the form of minute octahedrons (Fig. 3, below). In like manner, if a few particles of common salt be dissolved in a small quantity of water, and a drop of the solution be evaporated gently (or be left to evaporate spontaneously) on a piece of glass, numerous little cubes and hopper-shaped cubical aggregations will result. Boiling water, again, saturated with common alum, will deposit octahedral crystals on cooling: the cooled water not being able to retain in solution the full amount of alum dissolved by the hot water. Finally, it may be observed that bismuth, antimony, and many other bodies, crystallize by slow cooling from the molten state. Although, as explained above, crystals usually originate when matter passes slowly from the gaseous or liquid condition into the solid state, crystallization and solidification are not actually identical. Various substances, such as silica in certain conditions, its hydrate (constituting the different opals), gums, many resins, &c., appear to resist altogether the action of crystallization.

The crystal forms and combinations met with in Nature, exclusive of those produced by the chemist in his laboratory, are exceedingly numerous, many thousands being known to exist. By the help of certain laws, however, and, more especially, by the aid of one, termed "the Law of Symmetry," we are enabled to resolve these multitudinous combinations into six groups or systems. The forms of the same group combine together, and may be deduced mathematically from each other; whilst those of distinct groups are unrelated. Thus, although the cube, the rhombic dodecahedron, and the regular octahedron (Figs. 1, 2 & 3) appear at first sight to be unconnected forms, their co-relations may

---

re-announcement, or rather re-discovery of the law in 1772, by the French crystallographer, Romé de l'Isle. Many of the contemporaries of the latter—amongst others the celebrated Buffon—attempted to deny its existence; but being susceptible of practical proof, its truth was soon established.

be readily shown by the Law of Symmetry. This law, for instance, exacts one of three things, of which the most important is to this effect, *viz.*, that if an edge or angle of a crystal be modified in any way, all the similar edges or angles in the crystal must be modified in a similar manner. Now the cube has twelve similar edges and eight similar angles. Consequently, if one edge or one angle be truncated, or, to use a term more in conformity with the actual operations of Nature, if one of these be *suppressed* during the formation of the crystal, the other edges (or angles) must be suppressed also; and if the new planes, which thus arise, be extended until they meet, the rhombic dodecahedron on the one hand, and the regular octahedron on the other, will result.* These forms, moreover, as well as their intermediate oscillations, frequently occur in the same substancee : red oxide of copper may be cited as an example. But between the cube, a square prism, a regular hexagonal prism, and a rhombic prism, no relations of this kind exist. Neither are these solids related physically : their optical, thermal, and other physical characters are equally distinct. By considerations of this sort, therefore, we are able to establish six (or really seven) distinct Crystal Systems. These (named chiefly in accordance with the relations of their axes, or certain right lines assumed to pass through the centre of each crystal, and terminate in opposite planes, edges, or angles) are enumerated in the annexed tabular view :

Crystal-axes of one length. Refraction, single	*The Regular* or *Isopolar System* (including the cube, rhombic dodecahedron, octahedron, &c., with their various combinations.)
Crystal-axes of two lengths. Refraction, double, with one neutral line or optical axis	*The Tetragonal System* (including square-based prisms and pyramids with their various combinations.)
	The Hexagonal System (including regular hexagonal prisms and pyramids, rhombohedrons, &c., with their combinations.)

---

* The Law of Symmetry, in its exact acception, may be thus expressed :

(1.) If an edge or angle of a crystal be modified, *all* the *similar* edges or angles will exhibit a similar modification.

Or (2.) *One-half* or *one-mth* of the corresponding angles or edges, in alternate positions, will be equally modified. *Example.*—Cube and Tetrahedron (Boracite ; Arseniate of Iron.)

Or (3.) *All* the similar edges or angles will be modified by *one-half* or *one-mth* the normal or regular number of planes. *Example.*—Cube and Pentagonal Dodecahedron (Iron Pyrites, Cobaltine.)

Conditions 2 and 3 produce *hemihedrons* or *part-forms.*

Crystal-axes of three lengths Refraction, double, with two neutral lines or optical axes	Axes at right-angles.	The *Rhombic* or *Ortho-Rhombic System* (including right rectangular prisms and pyramids, rhombic prisms and pyramids, and combinations of these).
	One axis oblique.	The *Clino-Rhombic* or *Monoclinic System* (including oblique rectangular and rhombic combinations).
	All the axes oblique.	The *Triclinic* or *Anorthic System* (including doubly-oblique combinations).

The study of these Crystal groups, and that of crystal forms and combinations generally, constitutes the science of Crystallography. To enter into the details of this science would extend our present discussion much beyond its proposed limits and object, the simple determination of commonly occurring minerals; but it will be advisable for the student to impress upon his memory the names of the groups in question, with the general aspect of their more common forms and combinations, as given in the following enumeration.

The Isopolar or Regular System.—This group includes the cube (Fig. 1), the rhombic dodecahedron (Fig. 2), the regular octahedron (Fig. 3), trapezohedrons (Fig. 4), pentagonal dodecahedrons (Fig. 5), &c. Figs. 6 and 6* are combinations of the cube and octahedron;

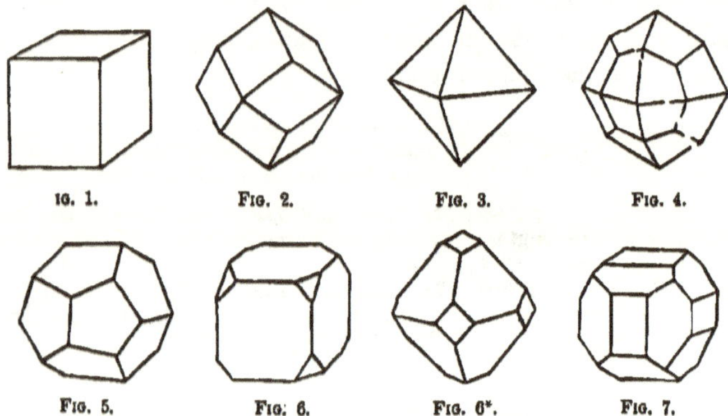

No. 7, a combination of the cube and pentagonal dodecahedron. Native gold, silver, copper, iron pyrites, galena, zinc blende, grey copper ore, red copper ore, magnetic iron ore, spinel, garnet, fluor spar, rock salt, and numerous other minerals, crystallize in this system.

The Tetragonal System—This includes, principally, square-based prisms and pyramids, and their combinations. Figures 8 to 9 are ex-

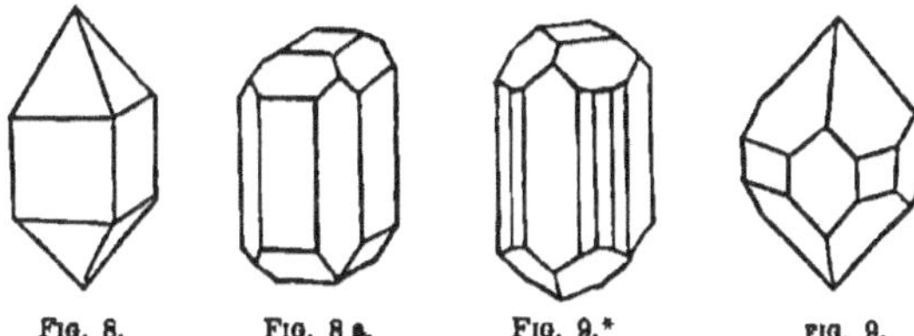

Fig. 8.   Fig. 8 a.   Fig. 9.*   Fig. 9.

amples of Tetragonal crystals. Amongst minerals, Copper Pyrites, Tinstone, Rutile, Anatase, Zircon, Idocrase, Scapolite &c., may be cited as belonging to the group.

The Hexagonal System.—Regular six-sibed prisms (Fig. 10, and pyramids (Fig. 11), combinations of these (Fig. 12), three-sided prisms, rhombohedrons (Figs. 13 and 14), and scalenohedrons (Fig. 15); are included under this system. Graphite, Red Silver Ores, Cinnabar, Specular Iron Ore, Corundum, Quartz, Beryl, Tourmaline, Apatite or Phosphate of Lime, Phosphate and Arseniate of Lead, Calcite,

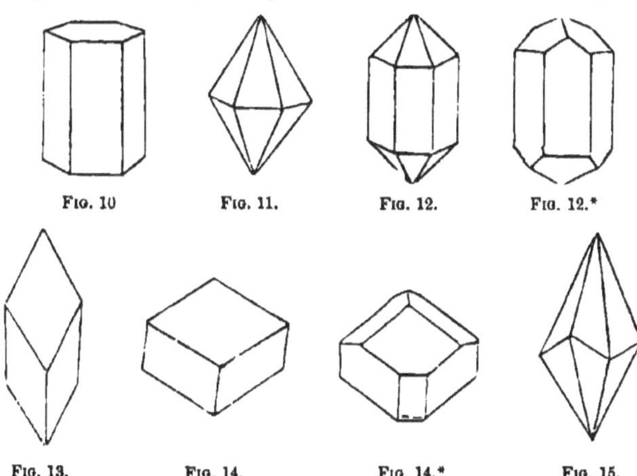

Fig. 10   Fig. 11.   Fig. 12.   Fig. 12.*

Fig. 13.   Fig. 14.   Fig. 14.*   Fig. 15.

Dolomite, and Carbonate of Iron, are some of the principle minerals which belong to it.

The Rhombic System—This system includes right-rhombic prisms, rectangular prisms, rhombic octahedrons, &c., and their combinations. Fig. 16 is a rhombic prism; Figs. 17 to 21 represent other crystals of this system.   Prismatic Iron-pyrites, Mispickel or Arsenical-pyrites, Native Sulphur, Topaz, Staurolite, Arragonite, Heavy spar, Celestine,

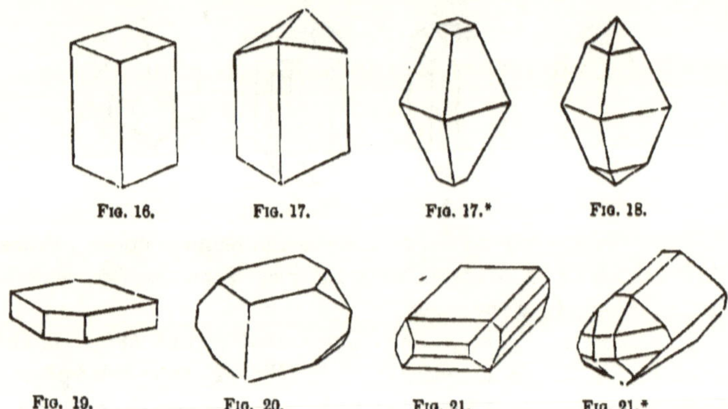

FIG. 16.   FIG. 17.   FIG. 17.*   FIG. 18.

FIG. 19.   FIG. 20.   FIG. 21.   FIG. 21.*

and Epsom salt, are some of the principal minerals which belong to the Rhombic group.

The Monoclinic or Clino-Rhombic System.—Rhombic prisms and pyramids, and rectangular prisms and pyramids, with *oblique or sloping base,* belong to this system.   Figs. 22 to 24 are Monoclinic combinations.   Characteristic minerals comprise: Augite, Hornblende, Epidote, Sphene, Orthoclase or Potash Feldspar, Gypsum, and Iron Vitriol or Sulphate of Iron.

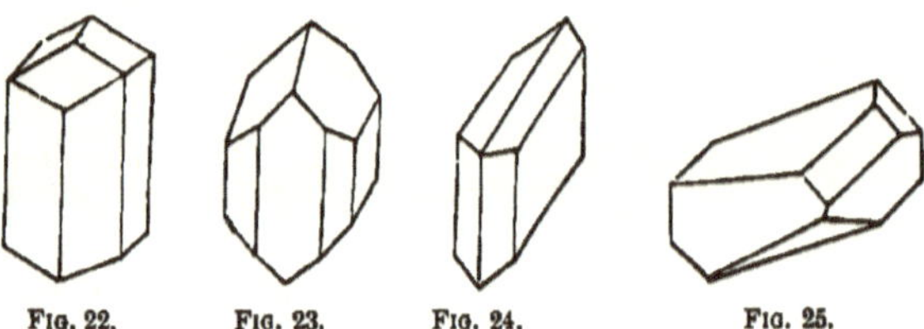

FIG. 22.   FIG. 23.   FIG. 24.   FIG. 25.

The Triclinic or Anorthic System—The forms of this system are oblique in two directions.   The crystals in general are more or less flat and unsymmetrical in appearance.   No two planes meet at right angles; and there are never more than two similar planes present in any crystal belonging to the group.   Axinite, Albite or Soda-Feldspar, and Sulphate of Copper, Fig. 25, are examples of Triclinic minerals.

The *Irregular Forms* assumed by minerals are of very subordinate importance. The following are some of the more common:—*Globular* or *nodular*, ex. quartz, iron pyrites; *reniform* or *kidney-shaped*, ex. quartz, &c.; *botryoidal* or *mammillated:* a form made up of a series of rounded elevations and depressions, or otherwise exhibiting a surface of this character, ex. red and brown iron ore, calcedony, &c.; *stalactitic*, ex. calc. spar, &c.; *coralliform*, resembling certain branching corals, ex. arragonite; *acicular*, in minute needle-like forms, ex. millerite; *dendritic* or *arborescent*, a branching form, often made up of small aggregated crystals, ex. native silver, native copper, &c.; *filiform* or *wire-like*, ex. native silver. When a mineral presents a perfectly indefinite shape, it is said to be *massive*. Other terms used in connection with the irregular forms of minerals, such as *incrusting*, *disseminated*, &c., explain themselves. The term *amorphous* is applied to obsidian, opal, and other minerals in which crystalline structure and cleavage planes are altogether wanting.

Structure and Cleavage.—In the majority of minerals a certain kind of structure, or, in other words, the shape as well as the mode of aggregation of the smaller masses of which they are composed, is always observable. Structure in minerals may be either *lamellar*, *laminar* or *foliated*, *prismatic*, *fibrous*, *granular* or *compact*. When the mineral, as in most varieties of calc-spar, heavy-spar, feldspar, and gypsum, for example, is made up of broad, tabular masses producing a more or less stratified appearance, the structure is said to be lamellar. When the tabular masses (whether straight, wavy or curved) become extremely thin or leafy, as in mica more especially, the structure is said to be laminar, or foliated, or sometimes micaceous. The scaly structure is a variety of this, in which the laminæ are of small size. When the component masses are much longer than broad or deep, as in many specimens of tourmaline, beryl, calc-spar, &c., the structure is said to be prismatic or columnar. When the prismatic concretions become very narrow, the fibrous structure originates. Fibrous minerals may have either: a straight or parallel-fibrous structure, as in many specimens of gypsum, calc-spar, &c.; an irregularly-fibrous structure, as in many specimens of augite and hornblende; or a radiated-fibrous structure, as in the radiated varieties of iron pyrites, natrolite, wavellite, and many other minerals,—the fibres radiating from one or more central points. Minerals made up

of small grains or granular masses are said to have a granular structure; ex. granular or saccharoidal limestone, granular gypsum, &c.
Finally, when the component particles are not apparent, the mineral
is said to have a compact structure, as in the native malleable metals,
obsidian, and most varieties of quartz. Hard and vitreous minerals
of a compact structure (ex. obsidian) generally show, when broken, a
conchoidal fracture, or a series of circular markings resembling the
lines of growth on the external surface of a bivalve shell.

Almost all minerals, especially those of a lamellar structure, break
or separate more readily in certain directions than in others. This
peculiarity is called *cleavage*. When cleavage takes place in more
than one direction, the resulting fragments have often a perfectly
regular or definite form. Thus the purer specimens 'of calc-spar, no
matter what their external form, break very readily into rhombohedrons, which measure 105°5' over their obtuse edges. Galena, the
common ore of lead, yields rectangular or cubical cleavage forms;
whilst the cubes of fluor-spar break off most readily at the corners or
angles, and yield regular octahedrons (Figs. 6 and 3).

Hardness.—The hardness of a mineral is its relative power of
resisting abrasion, not that of resisting blows, as many of the hardest
minerals are exceedingly brittle. Practically, the character is of
great importance. By its aid, gypsum may be distinguished in a
moment from calc-spar or ordinary limestone, calc-spar from feldspar,
and copper pyrites from iron pyrites, not to mention other examples.*
The degree of hardness in minerals is conventionally assumed to
vary from 1 to 10 (1 being the lowest), as in the following scale,
devised by the German mineralogist, *Möhs*, and now generally
adopted :

1. Foliated TALC.
2. ROCK SALT, a transparent cleavable variety.
3 CALCAREOUS SPAR, a transparent variety.
4. FLUOR SPAR.
5. APATITE.

* Gypsum may be scratched by the finger-nail; Calc-spar and copper pyrites are scratched
easily by a knife; whilst feldspar and iron pyrites are hard enough to scratch window-glass.
Some years ago, as mentioned by Sir William Logan, a farmer in the Ottawa district was put to
much expense and annoyance by mistaking feldspar for crystalline limestone, and attempting
to burn it into lime. On a late visit to the township of Marmora, we found, near a deserted
kiln, a large heap of quartz fragments, on which n similar attempt had evidently been made.

6. FELDSPAR.
7. ROCK CRYSTAL.
8. TOPAZ.
9. CORUNDUM.
10. THE DIAMOND.

In order to ascertain the hardness of a mineral by means of this scale, we attempt to scratch the substance, under examination, by the different specimens belonging to the scale, beginning with the hardest, in order not to expose the specimens to unnecessary wear. Or, proceeding in another manner, we take a fine file, and compare the hardness of the mineral with that of the individual members of the scale, by drawing the file quickly across them. The comparative hardness is estimated by the resistance offered to the file; by the noise occasioned by the file in passing across the specimens; and by the amount of powder so produced. The degree of hardness of the mineral is then said to be equal to that of the member of the scale with which it agrees the nearest. Thus, if the mineral agree in hardness with fluor-spar, we say, in its description, H (or hardness) $= 4$. If, on the other hand, it be somewhat softer than fluor-spar, but harder than calcareous spar, we say, $H = 3.5$. Finally, if, as frequently happens, the hardness of a mineral vary slightly in different specimens, the limits of the hardness are always stated. Thus, if in some specimens, a mineral agree in hardness with calc-spar, and in others with fluor-spar, we say, $H = 3$ to 4; or, more commonly, $H = 3 - 4$. If the hardness be very rigorously tested, it will frequently be found to differ slightly on different faces of a crystallized specimen, or on the broad faces and the edges of the laminæ of foliated specimens; but this, so far as regards the simple determination of minerals, is practically of little moment.

As the minerals of which the scale of Möhs consists are not in all places obtainable, or always at hand when required, the author of this work devised, many years ago, a scale of hardness so contrived as to agree closely enough for practical purposes with that of Möhs, whilst exacting for its application only such objects as are always to be met with. The following is the scale in question: its use explains itself.

Chapman's Convenieat Scale of Hardness, to correspond with that of Möhs.

1. Yields easily to the nail.
2. Does not yield to the nail. Does not scratch a copper coin.*
3. Scratches a copper coin, but is also scratched by one, being of about the same degree of hardness.
4. Not scratched by a copper coin. Does not scratch glass (ordinary window-glass).
5. Scratches glass very feebly. Yields easily to the knife.
6. Scratches glass easily
7. Yields with difficulty to the edge of a file.
8, 9, 10. Harder than flint or rock-crystal.

Convenient objects for the estimation of degrees of hardness above No. 7 cannot be easily obtained ; but that is of little consequence, as there are but few minerals which exhibit a higher degree, and these are readily distinguished by other characters.

Specific Gravity.—This is also a character of great value in the determination of minerals. The specific gravity of a body is its weight compared with the weight of an equal bulk of pure water. In order to ascertain the specific gravity of a mineral, we weigh the specimen first in a' · and then in water. The loss of weight in the latter case exactly equals the weight of the displaced water, or, in other words, of a volume of water equal to the volume of the mineral. The specific gravity of pure water, at a temperature of about 62°, being assumed to equal 1, or unity, it follows that the specific gravity of a mineral is obtained by dividing the weight of the latter in air by its loss of weight in water. Thus, if a = the weight in air, and w = the weight in water, G, or *sp. gr.* $= \dfrac{a}{a-w}.$

Example.—A piece of calcareous spar weighs 66 grs. in air, and 42 grs. when immersed in rain or distilled water. Hence its sp. gr.
$= \dfrac{66}{66 - 42} = \dfrac{66}{24} = 2\cdot75.$

The weight of the mineral may be ascertained most conveniently, and with sufficient exactness for general purposes, by a pair of small scales such as are commonly called " apothecaries' scales." These may

* Thas is, an old-fashioned copper coin, not a modern bronze coin. The scale was published in 1843.

be purchased for a couple of dollars or even less. A small hole must be made in the centre of one of the pans for the passage of a horse-hair or silken thread (about four inches in length) furnished at its free end with a "slip-knot" or running noose to hold the specimen whilst this is being weighed in water. The strings of the perforated pan may also be somewhat shortened, but the balance must in this case be brought into equilibrium by a few strokes of a file on the under side of the other pan, or by attaching thinner strings to it.

As an application of specific gravity, apart from the employment of the character in the determination of minerals, it may be observed that the weight of masses of rock, heaps of ore, etc., may be readily ascertained by reference to this property. The length, breadth, and depth of the body being taken in feet and decimal parts of a foot, and these dimensions being multiplied together, we get the contents of the body in cubic feet. This value is then multiplied by 62.32, the weight in lbs. of a cubic foot of water. This gives the weight of an equal bulk of water, which must finally be multiplied by the average sp. gr. of the body. The weight of the latter is thus obtained in lbs. This weight divided by 2,000 gives the weight in American or Canadian tons; and by dividing it by 2,240, we get the weight in British tons.

Relative Malleability.—Some few minerals, as native gold, native silver, sulphide of silver, native copper, &c., are *malleable* or *ductile*, flattening out when struck, instead of breaking. A few other minerals, as talc, serpentine, &c., are *sectile*, or admit of being cut by a knife; whilst the majority of minerals are *brittle*, or incapable of being cut or beaten out without breaking. In testing the relative malleability of a mineral, a small fragment should be placed on a little anvil, or block of steel polished on one of its faces, and struck once or twice by a light hammer. To prevent the fragment from flying off when struck, it may be covered by a strip of thin paper, held down by the forefinger and thumb of the left hand. Thus treated, malleable bodies flatten into discs or spangles, whilst brittle substances break into powder.

Magnetism.—Few minerals attract the magnet in their natural condition, although many do so after exposure to the blowpipe. (See below.) In trying if a mineral be magnetic, we chip off a small fragment, and apply to it a little horse-shoe magnet, such as may be

3

purchased anywhere for a quarter of a dollar; or otherwise we apply the specimen to a properly suspended magnetic needle. In this manner many of the black granular masses which occur so frequently in our Gneissoid or Laurentian rocks, and in the boulders derived from these, may easily be recognised as magnetic iron ore.* Most specimens of this mineral (and also of magnetic pyrites) exhibit "polarity," or attract, from a given point, one end of the needle, and repel the other.

Taste.—This is a very characteristic although limited property, being, of course, exhibited only by soluble minerals. In these, the taste may be saline, as in Rock Salt; or bitter, as in Epsom Salt; or metallic, as in Sulphate of Iron, and so forth.

B. CHEMICAL CHARACTERS.

The chemical characters of principal use in the determination of minerals comprise the phenomena developed by the action of acids, and those produced by the application of the blowpipe. Before referring to these characters, the reader should be familiar with certain chemical terms of common employment in Mineralogy.

A substance of any kind, whether of natural or artificial formation, must be either a *simple* or a *compound* substance. If the former, it cannot be decomposed or subdivided into more simple bodies by any process of art. If *compound*, on the other hand, a decomposition of this kind may be more or less readily effected. Thus, whilst from a piece of sulphur, copper, or iron, if pure, nothing but sulphur, copper or iron respectively, can be extracted, a piece of copper pyrites will yield all three of these substances—each, as before, resisting further subdivision. Hence sulphur, copper, and iron are regarded as simple substances, whilst copper pyrites is a compound body. These so-called "simple" substances, it must be understood, may not be, and probably are not, absolutely simple; but they are simple, *id est*, undecomposable, in the present state of science. They are often known as Elements. Up to the present time between sixty and seventy have been recognized, but many occur only in a few rare minerals. Some—oxygen, nitrogen, chlorine, fluorine, hydrogen—exist in the free state as gases; two at ordinary temperatures are liquid; the rest

---

* The other dark-coloured cleavable masses, in these rocks, consist mostly of mica, ho n-blende, or tourmaline.

are solid. Some few occur naturally, at times, in the free or simple state. These form the so-called "native substances" (as Native Sulphur, Native Platinum, Native Gold, &c.,) of Mineralogists. Others occur only in combination. Some have a remarkable tendency to attack and combine with other bodies. Oxygen, chlorine, fluorine, sulphur and arsenic, in reference to natural compounds, may be especially cited in this respect. The binary compounds formed by these elements may be more or less passive bodies or *bases*, or active bodies or *acids*, although in some cases a strict line of demarcation cannot be drawn between the two. The bases have their generic name always terminated by the monosyllable "ide." Thus oxygen, in forming a compound of this kind, produces an oxide; chlorine, a chloride; fluorine, a fluoride; sulphur, a sulphide; and arsenic, an arsenide. Sulphur and arsenic compounds of this sort were formerly known as "sulphurets" and "arseniurets," but these terms have now passed out of use. When more than one compound of the above kind occurs, a distinctive prefix is added to the term. Thus red oxide of copper is often known as the "sub-oxide," whilst the black oxide of copper is known as "oxide" simply. The red oxide contains two parts (combining weights,) of copper to one of oxygen; the black oxide, equal parts (combining weights,) of each element. In like manner the oxide or protoxide of iron consists of equal combining weights of iron and oxygen, whilst the the sesquioxide or peroxide has one-and-a-half parts of oxygen to one of iron—or, as more commonly given, three combining weights of the former to two of the latter element. [By some authors these compounds are known, respectively, as cuprous and cupric oxide, ferrous and ferric oxide—the termination "ic" denoting the presence of the larger amount of oxygen.] Active or "mineralizing" compounds into which the above and other elements enter, are still commonly known as "acids," although many of these, it must be remembered, are insoluble compounds, and hence have no acid properties in the common acceptation of the term. By many modern chemists they are designated as "anhydrides." For present purposes we need only refer to oxygen compounds of this class. In these, the combining weights of oxygen are always greater than in bases or oxides proper. Some elements form, with oxygen, several acids. Where two exist, the one with least oxygen has its generic name terminating in the monosyllable " ous "—as sulphurous acid, arsenious acid, &c. ; whilst the monosyllable " ic " terminates the generic name of the more highly oxidized compound, as sulphuric acid, arsenic acid, &c. As regards minerals or natural inorganic substances, "ous" acids are all but unknown. The more common acids of the Mineral Kingdom—some of which occur both in the free state and in combination with bases, others in the latter condition only—comprise Silicic acid, (conventionally known as Silica), Carbonic acid, Sulphuric acid, Phosphoric acid, Arsenic acid, &c. These acids have a great tendency to combine with bases. The generic name of the compounds which thus result, terminates in either the syllable *ite* or *ate*. *Ous* acids give *ite* compounds with bases, and *ic* acids give *ate* compounds. Except in a few rare instances the latter only are met with among natural bodies. Silicic acid, in combining with a base or with several bases, produces a *silicate ;* carbonic acid, in like manner, produces a *carbonate ;* sulphuric acid, a *sulphate ;* arsenic acid, an *arseniate*, and so forth. Many of these compounds yield water when ignited : they are then known as hydrous or hydrated silicates, sulphates, &c.

In these oxidized compounds, it will be observed, three elements are present. Thus, the mineral cyanite (a silicate,) contains aluminium, silicon and oxygen ; and carbonate of iron contains iron, carbon and oxygen. If these minerals be chemically decomposed, they separate into an oxidized base on the one hand, and into an oxidized acid on the other. In other words, the silicate cyanite yields alumina, or oxide of aluminium, and anhydrous silicic acid ; whilst from carbonate of iron, oxide of iron and carbonic acid are obtained. Calcite or calcareous spar, in like manner, may be formed from, and decomposed into, lime or oxide of calcium and carbonic acid. If the mineral be exposed to a red heat, carbonic acid is expelled in the form of an invisible gas, and lime remains behind ; and if this lime be exposed to the atmosphere it will gradually absorb carbonic acid from the latter, and the original compound will again result.

Action of Acids.—As a general rule, the use of acids may be dispensed with in the ordinary determination of minerals, or resorted to only as a confirmatory test, when the name of the substance has been ascertained by other means. A drop of acid serves, however, very conveniently, to distinguish *carbonates* from most other bodies, by the effervescence which is produced by the liberation of carbonic acid from these salts. The test acids chiefly used, are nitric acid and hydrochloric acid. These must be kept in well-stoppered glass bottles provided with glass caps, as their fumes soon destroy cork, and are otherwise highly corrosive and deleterious. For geological purposes (testing calcareous rocks, &c.) strong hydrochloric acid diluted with about an equal volume of pure water, is principally used. The small bottle in which this is kept, may have a long stopper extending into the acid ; and a little nest or wicker-work pocket may be provided for its reception near the upper edge of the specimen basket. In examining a mineral with an acid, the substance should be reduced, in ordinary cases, to a fine powder, and covered in a test-tube or small porcelain capsule with a few drops of the acid, the latter being subsequently warmed or brought to the boiling point over the flame of a small spirit lamp. The following are some of the principal effects produced by this treatment :

(*a.*) Simple solution :—Example, gypsum, &c.

(*b.*) Solution with effervescence and simultaneous evolution of a colorless inodorous gas :—Ex. carbonates generally. Some of these, as calc spar, malachite, &c., dissolve with effervescence in cold and more or less dilute acid ; but others, as dolomite or bitter spar, and carbonate of iron, only effervesce, as a rule, in heated acid. Either acid may be used, except in the case of carbonate of baryta or strontia ; as with these minerals, strong hydrochloric acid forms an insoluble coating of chloride of barium or strontium, by which the further action of the acid is entirely prevented. If the acid be used in a diluted state, however, this effect is prevented, chlorides of barium and strontium being readily soluble in water.

(*c.*) Partial solution, with separation of a gelatinous residuum :—Ex. various silicates : these are said to "gelatinize in acids." Boiling hydrochloric acid is generally required to produce the effect. The gelatinous matter consists of silicic acid or silica. Some silicates (Vesuvian, Epidote, &c.), which do not gelatinize under ordinary conditions, exhibit the effect after fusion or strong ignition.

(*d.*) Partial solution. with separation of granular silica, ex. harmotome, labradorite, &c. Boiling hydrochloric acid must be used, and the mineral should be finely pulverized.

(*e.*) Oxidation and solution, or partial solution, with evolution of sulphuretted hydrogen, known by its fetid odour. Example—Most Sulphides. The effect is most readily produced by boiling the mineral in powder with hydrochloric acid.

(*f.*) Oxidation and solution, or partial solution, without odour of sulphuretted hydrogen. Ex. red copper ore, native copper, native silver, and some other minerals, when treated with hot nitric acid. The acid gives up part of its oxygen to the dissolving mineral, and the portion of the acid, thus altered, escapes in ruddy fumes. Sulphides and other non-oxidized bodies also cause the evolution of these coloured fumes. Acid cupreous solutions are green o greenish-blue in colour. A piece of polished steel or iron immersed in a diluted solution of this kind, becomes coated with metallic copper.

(*g.*) Solution, or partial solution, and production, with hydrochloric acid, of chlorine fumes. Ex. pyrolusite or black manganese ore, &c. The chlorine is, of course, derived from the decomposition of the acid. Care must be taken not to inhale its fumes.

(*h.*) Solution, or partial solution, with production of fluohydric acid in corrosive fumes. Example—Fluor spar, in powder, with hot sulphuric acid. The evolved fumes corrode glass. The experiment should be performed in a platinum or lead vessel. If a piece of glass coated on its under side with a thin layer of wax through which a pattern has been traced, be laid over the vessel for a few minutes, and then removed and washed in warm water, the lines of the pattern will be found more or less deeply etched on the surface of the glass. Great care must be taken to prevent the fumes from being inhaled.

(*i.*) The substance may remain undissolved and unattacked. Example— Quartz, orthoclase, zircon, &c.

Application of the Blowpipe :—The blowpipe in its simplest form is merely a narrow tube of brass or other metal, bent round at one extremity, and terminating, at that end, in a point with a very fine orifice (fig. 26). If we place the pointed end of this instrument just within the flame (and a little above the wick) of a lamp or common candle, and then blow gently down the tube, the flame will be deflected to one side in the form of a long narrow cone, and its heating power will be wonderfully increased. Many minerals, when held in the form of a thin splinter at the point of a flame thus acted upon, melt with the greatest ease ; and some are either

Fig. 26.

wholly or partially volatilized. Other minerals, on the contrary, remain unaltered. Two or more substances, therefore, of similar appearance, may often be separated and distinguished in a moment, by the aid of the blowpipe.

The blowpipe (in its scientific use) has, strictly, a three-fold application. It may be employed, as just pointed out, to distinguish

minerals from one another : some of these being fusible, whilst others
are infusible ; some attracting the magnet after exposure to the blowpipe, whilst others do not exhibit that reaction ; some imparting a
colour to the flame, others volatilizing, and so forth. Secondly, the
blowpipe may be employed to ascertain the general composition of a
mineral ; or the presence or absence of some particular substance, as
copper, lead, iron, cobalt, manganese, sulphur, arsenic, and the like.
Thirdly, it may be used to determine in certain special cases, the
actual amount of a metallic or other ingredient previously ascertained
to be present in the substance under examination.

In using the blowpipe, the mouth is filled with air, and this is
forced gently but continuously down the tube by the compression of
the muscles of the cheeks and lips, breathing being carried on simultaneously by the nostrils. By a little practice this operation becomes exceedingly easy, especially in ordinary experiments, in which
the blast is rarely required to be kept up for more than fifteen or
twenty seconds at a time. The beginner will find it advisable to
restrict himself at first to the production of a steady continuous
flame, without seeking to direct this on any object. Holding the
blowpipe in his right hand, (with thumb and two outside fingers
below, and the index and middle finger above the tube,) near the
lower extremity, he should let the inner part of his arm, between the
wrist and the elbow, rest against the edge of the table at which he
operates. The jet or point of the blow-pipe is turned to the left, and
inserted either into or against the edge of the flame, according to the
nature of the operation, as explained below. After a few trials, when
sufficient skill to keep up a steady flame has been acquired, the point
of the flame may be directed upon a small splinter of some easily
fusible material, such as natrolite or lepidolite, held in a pair of forceps with platinum tips.* Some little difficulty will probably be
experienced at first in keeping the test-fragment exactly at the flame's
point ; but this, arising partly from irregular blowing and partly from
the beginner being constrained to look at the jet of the blow-pipe
and the object simultaneously, is easily overcome by half-an-hour's
practice. A small cutting of metallic lead or particle of grey antimony ore supported on a piece of well burnt soft-wood charcoal can

---

* If forceps of this kind cannot be procured, a pair of steel forceps with fine points, such as
watchmakers use, may serve as a substitute. It will be advisable to twist some silk thread or
fine twine round the lower part of these in order to protect the fingers. The points must be
kept clean by a file.

be examined in a similiar manner. In these experiments the beginner must be careful not to operate on fragments of too large a bulk. The smaller and more pointed the subject submitted to the flame, the easier and and more certain will be the experiment.

In out-of-the-way places the common form of blowpipe described above is frequently the only kind that can be obtained. It answers well enough for ordinary experiments, but the moisture which collects in it by condensation from the vapour of the breath is apt to be blown into the flame. This iuconvenience is remedied by the form of construction shown in the annexed figures, in which the instrument consists of two principal portions, a main stem closed at one end, and a short tube fitting into this at right angles near the closed

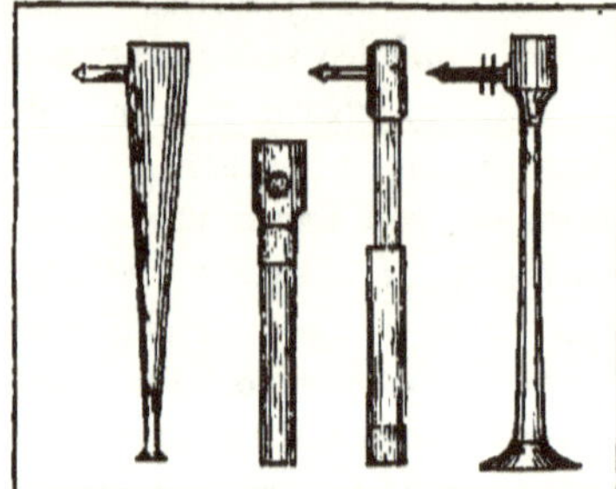

FIG. 27. FIG. 28. FIG. 29. FIG. 30.

extremity. The short tube is also commonly provided with a separate jet or nozzle of platinum. In this case, the jet can cleaned by simple ignition before the blow-pipe flame, or over the flame of the spirit lamp. In the variety of blow-pipe known as "Black's Blowpipe," FIG. 27, the main tube is usually constructed of japanned tin-plate, and the instrument is thus sold at a cheap rate. Mitscherlich's Blowpipe, FIG. 29, consists of three seperate pieces which fit together, when not in use, as shewn in FIG. 28. This renders it as portable as an ordinary pencil-case. FIG. 30 represents Gahn's or Berzelius's Blowpipe, with a trumpet shaped mouth-piece of horn or ivory as devised by Plattner. This mouth-piece is placed, of course, on the outside of the lips. It is preferable to the ordinary mouth-piece, but is not readily used by the beginner. In length, the blowpipe varies from about seven-and-a-half to nine inches, according to the eye-sight of the operator.

In addition to the blowpipe itself, and the forceps described above, a few other instruments and appliances are required in blowpipe operations.* The principal of these comprise : a few pieces of plati-

---

*It will of course be understood, that merely a slight sketch of the application of the blowpipe is given in these pages. Hence only the more necessary operations, instruments, &c. are alluded to. For fuller information, the author's BLOWPIPE PRACTICE (Copp and Co, Toronto) may be consulted.

num wire, three or four inches in length, of about the thickness of
thin twine, to serve as a support in fusions with borax, &c. (see
below) ; two or three small glass flasks, or, in default, a narrow test
tube or two, used chiefly for the detection of water in minerals (see
below) ; a few pieces of narrow glass tubing, in lengths of four or five
inches, open at both ends ; a small hammer and anvil, or piece of
hard steel, half-an-inch thick, polished on one of its faces ; a bar or
horse-shoe magnet ; a pen-knife or small steel spatula ; a small agate
pestle and mortar ; spirit-lamp, &c.; and a few wooden boxes or small
stoppered bottles to hold the blowpipe reagents. These latter are
employed for the greater part in the solid state, a condition which
adds much to their portability, and renders a small quantity sufficient
for a great number of experiments. The principal comprise : Carbonate of soda (abbreviated into *carb. soda*, in the following pages), used
largely for the reduction of metallic oxides, and in testing for sulphur
and sulphuric acid, manganese, &c., as explained below ; Biborate of
soda, or Borax, used principally for fusions on the platinum wire,
many substances communicating peculiar colours to the glass thus
formed ; and Phosphate of soda and ammonia, commonly known as
microcosmic salt or phosphor salt, used for the same purpose as borax,
and also for the detection of silicates and chlorides, as explained
further on. Re-agents of less common use comprise : nitrate of
cobalt (in solution) ; bisulphate of potash ; black oxide of copper ;
chloride of barium ; metallic tin ; and a few other substances of
special employment.

The effects produced by the blowpipe cannot be properly understood without a preliminary knowledge of the general composition
and structural parts of Flame. If the flame of a lamp or
candle, standing in a place free from draughts, be carefully
examined, it will be seen to consist of four more or less
distinct parts, as shown in in the annexed diagram, Fig.
31. A dark cone, a, will be seen in the centre of the flame.
This consists of gases, compounds of carbon and hydrogen,
which issue from the wick, but which cannot burn as they
are cut off from contact with the atmosphere. A bright
luminous cone b, surrounds this dark central portion,
except at its extreme base. In this bright cone the carbon, or a portion of it, separates from the hydrogen of

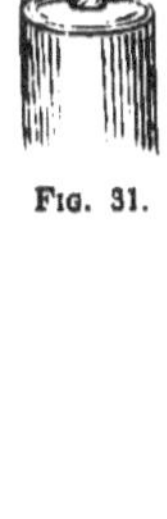

Fig. 31.

the gaseous compounds pumped up by the wick. The carbon becomes ignited in the form of minute particles, and these, with the liberated hydrogen and undecomposed gas, are driven partly outwards, and partly downwards, or into the blue cup-shaped portion *c*, which lies at the base of the flame. At the latter spot, the carbon, meeting with a certain supply of oxygen, is converted into carbonic oxide, a compound of equal combining weights of carbon and oxygen. Finally, in the flame-border or outer envelope, *d*, of a pale pinkish colour, only discernible on close inspection, complete combustion, *i. e.*, union with oxygen, of both gases, carbon and hydrogen, takes place. The carbon burns into carbonic acid, a compound of two combining weights of oxygen with one of carbon (hence, now commonly known as carbon dioxide); and the hydrogen, uniting with oxygen, forms aqueous vapour. If a cold and polished body, for example, be brought in contact with the edge of a flame of any kind, its surface will exhibit a streak or line of moisture.

Now these different parts of flame, possess, to some extent, different properties. The dark inner cone is entirely neutral or inert. Bodies placed in it, become covered with soot or unburnt carbon. The luminous or yellow cone possesses *reducing powers*. Its component gases, requiring oxygen for their combustion, are ready to take this from oxidized bodies placed in contact with them. The luminous cone, however, in its normal state, has not a sufficiently high temperature to decompose oxidized bodies, except in a few special cases; but its temperature, and consequently its decomposing or deoxidizing power, becomes much increased by the action of the blowpipe, as shown below. The blue portion of flame possesses also reducing powers, but of comparatively feeble intensity, as the carbon is there able to obtain from the atmosphere a partial supply of oxygen. Finally, in the outer or feebly luminous envelope, in which complete combustion takes place, the flame attains its highest temperature; and, having all the oxygen it requires from the surrounding atmosphere, it exerts an oxidizing influence on bodies placed in contact with it, since most bodies absorb oxygen when ignited in the free air.

In subjecting a body to the action of the blowpipe, we seek, (1) either to raise its temperature to as high a degree as possible, so as to test the relative fusibility of the substance; or (2) to oxidize it, or cause it, if an oxide, to combine with a larger amount of oxygen ; or (3) to reduce it, either to the metallic state, or to a lower degree of oxidation. The first and second of these effects may be produced by the same kind of flame, known as an oxidating flame (or O. F), the position of the substance being slightly different : whilst the third effect is obtained by a so-called reducing flame (or R. F.), in which the yellow portion is developed as much as possible, and the substance kept within it, so as to be off from contact with the atmosphere.

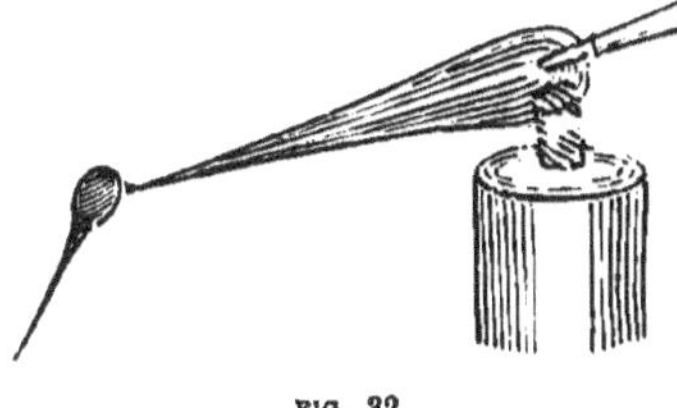

An oxidating and fusion flame is thus produced. The point of the blowpipe is inserted well into the flame of the lamp or candle under use, so as almost to touch the surface of the wick. The deflected flame is thus well supplied with

FIG. 32.

oxygen, and its reducing or yellow portion becomes obliterated. It forms a long narrow blue cone, surrounded by its feebly luminous mantle. The body to be oxidized should be held a short distance beyond the point of the cone, as in FIG. 32 ; but to test its fusion, it must be held in contact with this, or even a little within the flame. In thia position many substances, as those which contain lithia, stronia, baryta, copper, &c., impart a crimson, green, or other colour to outer or feebly luminous cone

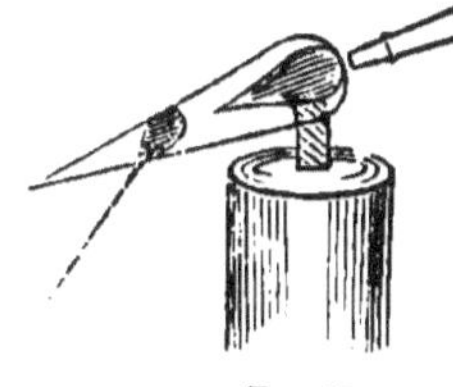

For the production of a reducing flame the orifice of the blowpipe must not be too large. The point is held just on the out-side of the flame, a little above the level of the wick, as shown in FIG. 33. The flame in its deflected state, then retains the whole or a large portion of its yellow

FIG. 33.

cone. The substance under treatment must be held within this (although towards its pointed extremity) so as to be entirely excluded from the atmosphere ; whilst at the same time, the temperature is raised sufficiently high to promote reduction. As a general rule, bodies subjected to a reducing treatment should be supported on charcoal.

For ordinary experiments, such as testing the relative fusibility, &c., of minerals, the blowpipe may be used with the flame of a common candle. The wick of the candle should be kept rather short (but not so as to weaken the flame), and it should be turned slightly to the left, or away from the point of the blowpipe, the stream of air being blown along its surface. A lamp flame, or that of coal gas, gives a higher temperature, and is in many respects preferable. The wick-holder (or jet, if gas be used) should be of a rectangular form, with its upper surface sloping towards the left at a slight angle. Either good oil, or, better, a mixture of about 1 part of spirit of turpentine, or benzine, with 7 or 8 parts of strong alcohol, may be used with the lamp. If the latter mixture is used, equal volumes of the two ingredients must be first well shaken up together, and then the rest of the alcohol added. If the wick crust rapidly, the turpentine will be in excess, in which case another volume of alcohol may be added to the mixture. For stationary use, a Bunsen burner provided with an additional tube to slip into the main tube, and so cut off the supply of air, is still more convenient. The projecting top of this additional tube should have the sloping form of the rectangular wick-holder described above.

The following are some of the more general operations required in the examination of minerals by the blowpipe. A few others of special employment are referred to under the Reactions given on a subsequent page.

(1) *The Fusion Trial:*—In order to ascertain the relative fusibility of a substance, we chip off a small particle, by the hammer or cutting pliers, and expose it, either in the platinum-tipped forceps or on charcoal, to the point of the blue flame (Fig. 26, above). If the substance be easily reduced to metal, or if it contain arsenic, it must be supported on charcoal (in a small cavity made by the knife-point for its reception), as substances of this kind attack platinum.* In other cases, a thin and sharply pointed splinter may be taken up by the forceps, and exposed for about half-a-minute to the action of the flame. It ought not to exceed, in any case, the size of a small carraway seed—and if smaller than this, so much the better. If fusible, its point or edge (or on charcoal, the entire mass) will become rounded into a bead or globule in the course of ten or twenty seconds.

---

* In order to prevent any risk of injury to the platinum forceps, it is advisable to use charcoal as a support for all bodies of a metallic aspect, as well as for those which exhibit a distinctly coloured streak or high specific gravity.

Difficultly fusible substances become vitrified only on the surface, or rounded on the extreme edges; whilst infusible bodies, though often changing colour, or exhibiting other reactions, preserve the sharpness of their point and edges intact.

The more characteristic phenomena exhibited by mineral bodies when exposed to this treatment, are enumerated in the following table :—

(a) The test-fragment may "decrepitate" or fly to pieces. Example: most specimens of galena, heavy spar, &c. In this case a larger fragment must be heated in a test-tube over a small spirit lamp, and after decrepitation has taken place, one of the resulting fragments can be exposed to the blowpipe flame as directed above. Decrepitation may sometimes be prevented if the operator expose the test-fragment cautiously and gradually to the full action of the flame.

(b) The test-fragment may change colour (with or without fusing) and become attractable by the magnet. Example, carbonate of iron. This becomes first red, then black, and attracts the magnet, but does not fuse. Iron pyrites, on the other hand, becomes black and magnetic, but fuses also.

(c) The test-fragment may colour the flame. Thus, most copper compounds impart a rich green colour to the flame; compounds containing baryta and many phosphates and borates, with the mineral molybdenite, colour the flame pale green; sulphur, selenium, lead and chloride of copper colour the flame blue of different degrees of intensity; compounds containing strontia and lithia impart a crimson colour to the flame; some lime compounds impart to it a pale red colour; soda compounds, a deep yellow colour; and potash compounds, a violet tint.

(d) The test-fragment may become caustic. Example, carbonate of lime. The carbonic acid is burned off, and caustic lime remains. This restores the blue colour of reddened litmus paper. It also imparts, if moistened, a burning sensation to the back of the hand.

(e) The test-fragment may take fire and burn. Example, native sulphur; common bituminous coal, &c.

(f) The test-fragment may be volatilized or dissipated in fumes, either wholly or partially, and with or without an accompanying odor. Thus, grey antimony ore volatilizes with dense white fumes; arsenical pyrites volatilizes in part, with a strong odor of garlic; common iron pyrites yields an odor of brimstone, and so forth. In many cases the volatilized matter becomes in great part deposited in an oxidized condition on the charcoal. Antimonial minerals form a white deposit or incrustation of this kind. Zinc compounds, a deposit which is lemon-yellow whilst hot, and white when cold. Lead and bismuth are indicated by sulphur-yellow or orange-yellow deposits. Cadmium by a reddish-brown incrustation.

(g) The test-fragment may fuse, either wholly or only at the point and edges ; and the fusion may take place quietly, or with bubbling, and with or without a previous "intumescence" or expansion of the fragment into a cauli-flower-like mass. Most of the so-called Zeolites, for example, (minerals abundant in Trap rocks) swell or curl up on exposure to the blowpipe, and then fuse quietly ; but some, as Prehnite, melt with more or less bubbling.

(h) The test-fragment may remain unchanged. Example, Quartz and various other infusible minerals.

(2) *Treatment in the Flask or Bulb-tube (The Water Test) :—* Minerals are frequently subjected to a kind of distillatory process by ignition in small glass tubes closed at one end. These tubes are of two general kinds. One kind has the form of a small flask, and is commonly known as a " bulb-tube." Where it cannot be procured, a small-sized test-tube may supply its place. It is used principally in testing minerals for water. The other kinds consist simply of narrow pieces of glass tubing, closed and sometimes drawn out to a point at one extremity. They are chiefly employed in testing for mercury and arsenic (see below). Our present description refers solely to the use of the bulb-tube. Many minerals contain a considerable amount of water, or the elements of water, in some unknown physical condition. Gypsum, for example, yields nearly 21 per cent. of water. As the presence of this substance is very easily ascertained, the water test is frequently resorted to, in practice, for the formation of determinative groups, or separation of hydrous from anhydrous minerals. The operation is thus performed. The glass is first warmed gently over the flame of a small spirit-lamp to ensure the absence of moisture, and is then set aside for a few moments to cool. This effected, a piece of

the substance under examination, of about the size of a small pea, is placed in it and ignited over the spirit-lamp, as shown in the annexed figure. If water be present in the mineral, a thin film, condensing rapidly into little drops, will be deposited on the neck or upper part of the tube. As soon as the moisture begins to show itself, the tube must be held in a more horizontal position, otherwise a fracture may be occasioned by the water flowing down and coming in contact with the hot

Fig. 34.

part of the glass. A small spirit-lamp may be made by passing a piece of glass tubing of about an inch in length, to serve as a wick-holder, through an orifice in the cork of a short, flat bottle. When the lamp is not is use the wick should be covered with a glass or other cap to prevent the evaporation of the spirit. A mineral may also be examined for water, though less conveniently, by igniting it before the

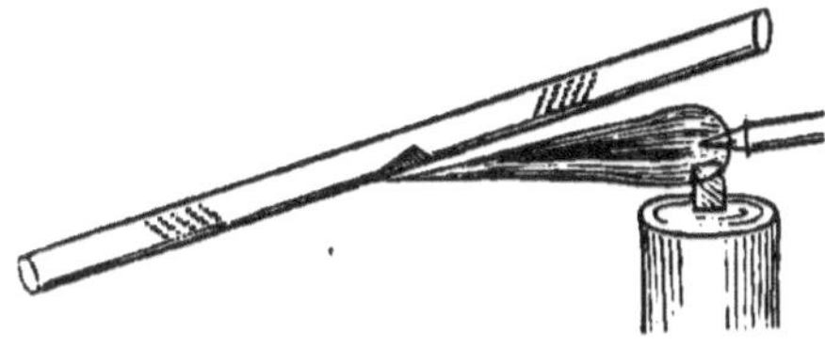

blowpipe-flame in a piece of open tubing, as shown in FIG. 35. To prevent the tube softening or melting, a strip of platinum foil may be folded round it where the test-

FIG. 35.

fragment rests. The latter is pushed into its place by a thin iron wire. The moisture condenses on each side of the test-matter.

(3) *Treatment with Nitrate of Cobalt:*—This operation serves, in certain cases, for the detection of alumina, magnesia, oxide of zinc, and some few other substances; but it is not applicable to deeply coloured or easily fusible bodies, nor to such as possess a metallic lustre or coloured streak. A fragment of the substance, under treatment, is reduced by the hammer and anvil, and afterwards by the use of the agate mortar, to a fine powder. This is moistened with a drop of the cobalt solution (nitrate of cobalt dissolved in water), and the resulting paste is strongly ignited on charcoal by being held about an inch before the point of the flame, fusion being carefully avoided. Thus treated, alumina assumes on cooling a fine blue colour; magnesia (and the comparatively rare tantalic acid), a flesh-red tint; baryta, a dull brownish-red colour ; oxide of zinc, bin-oxide of tin, antimony oxides, a green colour. With other substances, a grey, blueish-grey, brownish-black, or other indefinite coloration is produced, unless fusion take place, in which case a glass may be obtained, coloured blue by the dissolved oxide of cobalt.

(4) *Roasting :*—Tne principal object of this operation is the elimination of sulphur, arsenic, and certain other volatile bodies, from the mineral under examination, as these prevent the reduction of many substances to the metallic state, and also mask, to some extent, their other characteristic reactions. By roasting, the substance is not only deprived of sulphur, &c., but is also converted into an oxidized con-

dition.   The operation is most readily performed as follows.   A small fragment of the mineral is reduced to powder.   Some of this is made into a paste by moistening with a drop of water, and is spread over the surface of a piece of charcoal, or broken fragment of a porcelain evaporating-dish or thin crucible.   It is then ignited before the point of an oxidating flame (Fig. 32), the heat being kept low, at first, to prevent fusion.   It is sometimes necessary to remove the ignited paste to the mortar, and to break it up again with a fine steel spatula (the end of a flattened wire, or knife-point), and renew the operation.   When the roasting is terminated, the powder will present a dull earthy aspect, and cease to omit fumes or odour.   It is then ready for operations 5 and 6, described below.   By reducing the substance to powder before roasting, the risk of decrepitation and fusion is prevented, and the process itself is more efficiently performed.

Roasting is sometimes effected in a piece of open glass tubing as in Fig. 35—only the test object is placed near one end of the tube, and the tube itself is held in a more inclined position.   Sulphur eliminated from bodies by this treatment, is converted into sulphurous acid (a compound of sulphur and oxygen, the latter taken up from the atmosphere); and arsenic forms arsenious acid, which deposits itself in the shape of numerous microscopic octahedrons on the cool sides of the glass near the upper part of the tube.   Sulphurous acid in escaping from the open end of the tube is easily recognized by its odour (identical with that emitted by an ignited match), as well as by its property of changing the blue colour of a slip of moistened limus paper to red.   Antimonial compounds form a dense white uncrystalline sublimate.

(5.) *Formation of glasses on platinum wire or charcoal :*—This operation is one of constant utility in the determination of the constituents of minerals.   The glasses, in question, are formed by the fusion of small portions of borax, phosphor salt, or carbonate of soda : the latter reagent, however, being only occasionally used.   Most substances, dissolve in one or the other of these glasses before the blowpipe, and many communicate to them peculiar colours by which the nature of the test-matter is made known.   If the matter to be tested contain sulphur or arsenic, it should be roasted before being subjected to the action of these fluxes.   Metals and metallic alloys, as well as metallic oxides, chlorides, &c , of very easy reduction, must be examined on charcoal, but in other cases it is more convenient to employ a piece of platinum wire as a support. One end of the wire may be inserted into a cork or special handle, or, if the wire be from $2\frac{1}{2}$ to 3 inches in length, it may be held in the naked fingers, as platinum conducts heat very slowly. The other end is bent into a small loop or ear. This, when borax or phosphor-salt is used, is ignited by the blow-pipe flame, and plunged into the flux, the adhering portion of the latter being then fused into a glass. If a sufficient portion to fill the loop be not taken up at first, the process must be repeated. With beginners, the fused glass is often brownish or discoloured by smoke, but it may be rendered clear and transparent by being kept in ignition for a few moments before the extreme point of the flame, the carbonaceous matter becoming oxidized and expelled by this treatment. When carbonate of soda is used, a small portion of the flux must be moistened and kneaded in the palm of the left hand, by a knife-point or a small spatula, into a slightly cohering paste, which is placed on the loop of the wire, and fused into a bead. Whilst hot, the bead is transparent, but it becomes opaque on cooling. The portion of test-matter added to a glass or bead, formed by these reagents, must be exceedingly small, otherwise the glass may become so deeply coloured as to appear quite black. In this case, the colour may be observed by pinching the bead flat between a pair of forceps, before it has time to cool. It is always advisable, however, in the first instance, to take up merely a minute particle or two of the test-substance, and then to add more if no characteristic action be obtained. The glass, in all cases, must be examined first before an oxidating flame, and its colour observed both whilst the flux is hot, and when it has become cold ; and, secondly, it must be kept for a somewhat longer interval in a good reducing flame (Fig. 33), and its appearance noted as before.* With certain substances (lime, magnesia, &c.) the borax and phosphor-salt glasses become milky and opaque when saturated, or when subjected to the intermittent action of the flame—the latter being urged upon them in short puffs, or the glass being moved slowly in and out of the flame—a process technically known as *flaming*.

---

* The colour of the glass ought not of course, to be examined by the *transmitted* light of the lamp or candle flame. Strictly, it should be observed by daylfght.

4

The colours, &c., communicated to these glasses by the more commonly occurring constituent bodies, are shown in the annexed tabular view.

BORAX.

Colour of Bead after exposure to an Oxidating Flame.	Compounds of :	Colour of Bead after exposure to a Reducing Flame.
Violet or Amethystine.........	Manganese....	Colorless, if quickly cooled. Violet-red if quickly cooled.
Violet-brown (whilst hot).... Clear-brown (when) cold.....	Nickel	Grey and opaque.
Blue (very intense)...........	Cobalt	Blue (very deep).
Green (whilst hot .... Blue or greenish-blue (cold)..	Copper........	More or less colorless or indistinctly colored whilst hot ; brownish-red and opaque on cooling.
Green or blueish-green	Cobalt + Iron....	Green or blueish-green.
Green (dark)	Copper + Nickel Copper + Iron..	Brownish-red, opaque on cooling.
Yellowish or reddish (hot)... Yellowish-green (when cold..	Chromium.......	Emerald-green.
Yellow (whilst hot).......... Greenish-yellow (cold).......	Vanadium.....	Brownish (whilst hot) Emerald-green (when cold).
Yellowish or reddish..........	Iron	Bottle-green.
Yellowish or reddish........ Enamelled by flaming.......	Uranium	Green (black by flaming).
Yellow (whilst hot) .... Pale yellowish (cold)........ Enamelled by flaming.......	Cerium	Colorless or yellowish. Opaque-white, if saturated.
Yellow (hot)................ Colorless (cold) .... Enamelled by flaming........	Titanium.......	Yellow or yellowish-brown. Enamelled light-blue by flaming. [low. V. under phosphor-salt be-
Yellow (hot) .... Colorless (cold) .... Enamelled by flaming..	Tungstenum....	Yellow or yellowish-brown. Enamelled by flaming. V. under phospher salt below.
Yellow (hot)................ Colorless or yellowish (cold).. Greyish and opaque by flaming	Molybdenum ..	Brown or grey, semi-opaque, often with separation of black specks. [low. V. under phosphor-salt be-
Yellow or yellowish-red (hot) Yellowish or colorless, and often opaline, when cold ..	Lead .... Bismuth.... ... Silver.......... Antimony......	Grey and opaque on cooling but after continued subjection to the flame, the glass becomes clear: the reduced metallic particles either collecting together or volatilizing.
Yellowish (hot)............. Colorless (cold) .... Opaque-white when saturated.	Cadmium	Colorless—the reduced metal being volatilized.

Colorless (permanently clear) Slowly dissolved ....

- Aluminium .....
- Silicon ....
- Tin............

Colorless: permanently clear. (Tin compounds dissolve in small quantity only. On charcoal, they become reduced to metal, especially if a little carb. soda be added to the glass.)

Colorless. When saturated, opaque-white on cooling or by flaming ... .

- Tantalum .....
- Zirconium......
- Glucinum ......
- Yttrium, &c....
- Thorium ....
- Magnesium ....
- Calcium........
- Strontium......
- Barium ....
- Lithium ....
- Natrium ....
- Kalium ....

Colorless. When saturated, opaque-white on cooling or by flaming.

See Reactions, below.

PHOSPHOR-SALT.

The glasses produced by the fusion of constituent bodies with this reagent are for the greater part identical with those obtained by the use of borax, although somewhat less deeply coloured as a general rule. The principal exceptions are the glasses formed in a reducing flame with compounds of Molybdenum, Tungstenum, and Titanium, respectively. The molybdenum glass presents, when cold, a fine green colour, and the tungstenum glass becomes greenish-blue. If the latter contain iron, the colour of the glass is changed to blood-red or brownish red. Titanium in the presence of iron gives a similar reaction : but when free from iron, the glass is yellow whilst hot, and violet-coloured when cold. Phosphor salt is an important reagent for the detection of silica in silicates, as the silica remains for the greater part undissolved in the glass, in the form of a translucent flocculent mass technically known as a "silica skeleton," the associated constituents being gradually taken up by the flux. A small amount of silica is also generally dissolved, but this is precipitated as the bead cools, rendering it semi-transparent or opaline. Phosphor-salt is likewise employed for the detection of chlorine, &c. (See Experiment 3, page 43.

CARBONATE OF SODA.

This reagent is principally used to promote the reduction of oxidized and other bodies to the metallic state, as explained under the description of that process. (Operation 6, below.) It is also of frequent employment as a test for sulphur in sulphides and oxidized bodies. (See under Reactions.) It is rarely used, on the other hand, for the formation of glasses on platinum wire, except as a test for the presence of manganese ; although when employed, in this manner, it serves to distinguish salts of the alkalies, and those of strontia and baryta from all other salts : the alkalies, with baryta and strontia, dissolving completely and rapidly in the bead, whereas lime, magnesia, alumina, and other bases, remain unattacked.   Manganese compounds form by oxidizing fusion with this reagent a green glass, which becomes blue or bluish-green and opaque on cooling.   A very minute amount of manganese may be thus detected.   The delicacy of the test is increased by the addition of a small quantity of nitre, as this promotes oxidation ; and if the 'substance contain much lime, magnesia, iron oxides, or other bodies more or less insoluble in carb. soda, it is advisable to add a little borax to the test-mixture.   The blue or bluish-green bead thus produced, is technically known as a "turquoise enamel."   Chromium compounds produce a somewhat similar reaction ; but if the bead be saturated with silica or boracic acid, it will remain green in the latter case.   If the green colour result from the presence of manganese, on the other hand, a violet or amethystine glass will be obtained.   Some other applications of carbonate of soda as a blowpipe re-agent will be found under the head of REACTIONS.

6. *Reduction :*—This term denotes the process by which an oxidized or other compound is converted into the metallic state.   Some 'compounds become reduced by simple ignition ; others require for their reduction the addition of certain reagents ; and some, again, resist reduction altogether.   The reduced metal is in some cases so highly volatile that it connot be obtained except by a kind of distillatory process.   In other cases, one or more fusible globules, or a number of minute infusible grains, are obtained in blowpipe operations.   Reducible metals may be thus distributed into three groups, as shown (with omission of a few metals of rare occurrence) in the annexed Table :—

A. *Yielding metallic globules :*—Gold, Silver, Copper, Tin, Lead, Bismuth, Antimony.

B. *Yielding infusible metallic grains :*—Platinum, Iron, Nickel, Cobalt, Molybdenum, Tungstenum.

C. *Yielding metallic vapours only, when treated on charcoal* :—Mercury, Arsenic, Cadmium, Zinc.

A metal of the first group may be obtained, unless present in very small quantity, by a simple fusion of the previously roasted test-substance, with some carbonate of soda, on charcoal, in a good reducing flame (Fig. 33 above).   In ordinary cases, metallic globules are rapidly produced by this treatment.   By a little management the globules may be brought together so as to form a single large globule.   This must be tested on the anvil as regards its relative malleability, &c.   Gold, silver, copper, tin and lead are malleable ; bismuth and antimony, more or less brittle.   Gold and silver (if pure) retain a bright surface after subjection to an oxidating flame. Copper becomes covered with a black film, and tin with a white crust. Lead and bismuth volatilize more or less readily, and deposit on the charcoal a yellow coating of oxide. Antimony is rapidly volatilized with deposition of a dense white incrustation on the charcoal. It is not, of course always necessary to subject the test-substance to a previous roasting (Operation 4, above) but it is always safer to do so. Sulphur in most, and arsenic in all cases, must be driven off by this preliminary treatment before the actual process of reduction is attempted.

When the metal to be reduced belongs to the second group, or if the amount of fusible metal in the test-substance do not exceed 4 or 5 *per cent.*, the operation is performed as follows. A small portion of the substance in powder—subjected previously to the roasting process, if it contain sulphur or arsenic—is mixed with 3 or 4 volumes of carbonate of soda (or neutral oxalate of potash, or a mixture of about equal parts of carb-soda and cyanide of potassium—the latter, it must be remembered, a highly poisonous substance), and the mixture is exposed on charcoal to a good reducing flame, until all the alkaline salt has become absorbed. Some more of the flux is then added and the operation is repeated until the whole or the greater part of the test-matter is also absorbed. The charcoal at this spot is finally separated by a sharp knife-point and carefully ground to powder in a small agate mortar or porcelain capsule, whilst a fine stream of water is protected upon it from time to time, until all the carbonaceous and other non-metallic particles are gradually washed away. For this purpose, the mortar or capsule may be placed in the centre of an ordinary plate ; and if the operator be not provided with a chemical washing-bottle, he may use a small syringe, or, still more economically, a simple piece of glass tubing, five or six inches in length and about the fourth of an inch in diameter, drawn out at one end to a point. This is filled by suction, and the water expelled, with the necessary force, by blowing down the tube. The metallic grains or spangles obtained by this process must be examined by the magnet. Those of iron, nickel and cobalt are magnetic. Sometimes, however, when but a trace or very small percentage of reducible metal is contained in the test-substance, its presence is only indicated by a few metallic streaks on the sides and bottom of the mortar. Metallic markings of this kind can be removed by a piece of pumice.

Metallic compounds referable to the third group, yield no metal on charcoal, or by other treatment in open contact with the atmosphere. The presence of arsenic, however, is easily made known by the garlic-like odour evolved during fusion with reducing agents (or alone) on charcoal. Cadmium and zinc may also be recognized by the oxidized sublimates which they deposit on the charcoal. The Cadmium subli-mate is reddish-brown; the zinc sublimate, lemon-yellow and phos-phorescent whilst hot, and white when cold. Mercury forms no incrustation on charcoal; but its presence in any compound may be determined by reduction with carbonate of soda or iron-fillings in a glass tube of narrow diameter. A small test-tube or piece of glass tubing closed at one end before the blow-pipe, may be used for the experiment. The test-substance, in powder, mixed with 3 or 4 vols. of perfectly dry carb. soda, is inserted into the tube by means of a narrow strip of glazed writing-paper bent into the form of a trough, so as to prevent the sides of the glass from being soiled, and the mix-ture is strongly ignited by the spirit-lamp or by the blowpipe-flame. If mercury be present, a grey metallic sublimate will be formed near the upper part of the tube. By friction with an iron wire, or the narrow end of a quill-pen, &c., the sublimate may be brought into the form of fluid globules, which can be poured out of the tube, and are thus easily recognized as metallic mercury.

7. *Cupellation:*—Gold and silver are separated by this process from other metals. The test-metal is fused with several times its weight of pure lead. The button, thus obtained, is exposed to an oxidating fusion on a porous support of bone-ash, known as a cupel. The lead and other so-called base metals become oxidized by this treat-ment, and are partly volatilized, and partly absorbed by the bone-ash, a globule of gold or silver (or the two combined) being finally left on the surface of the cupel. For blowpipe operations, cupels are gen-erally made by pressing a small quantity of dry boneash into a cir-cular iron mould, the latter being fixed, when presented to the flame, in a special support, consisting essentially of a wooden foot and pillar with three or four short cross-wires (between which the cupel-mould rests) at the top of the latter. Instruments of this kind cannot be obtained in remote places, but the process may be performed equally well by the use of a small iron spoon, of about half-an-inch in dia-meter. Enough bone-ash to fill this, is taken up in it, and warmed

over the spirit-lamp or by the blowpipe flame. The spoon is then placed on the blowpipe anvil, and, whilst the smooth or unused end of the agate pestle (or other similar object, a glass button cemented to a cork, for example) is pressed firmly on the surface of the bone-ash, the handle of the spoon is moved three or four times from side to side. The surface of the cupel thus formed is then exposed for a few moments to the point of the flame, so as to render the bone-ash thoroughly dry ; and if its smooth condition be in any way affected by this treatment, the pressure with the pestle is repeated. Another equally good, if not better support consists of a cylindrical piece of pumice or well-baked clay with a small saucer-shaped depression for the bone-ash at its transverse end. The substance to be cupelled must be in the metallic state ; if not in this condition, therefore, it must first be subjected to the reducing process described above. In actual assaying or quantitative operations, this process is modified in various ways, but in the present work, in which merely a brief out-line of the use of the blowpipe is attempted, it would be out of place to enter into these details. The piece of test-metal, which may weigh about a couple of grains (or from 100 to 200 milligrammes) is wrapped in a piece of pure lead-foil of three or four times its weight, and the whole is exposed on the surface of the cupel to the extreme point of a clear oxidating flame. If the substance consist of argentiferous lead, as obtained from galena, &c., the addition of the lead-foil is of course unnecessary.* Six or seven grains (or from 400 to 500 milli-grammes) may be taken for the experiment : a beginner, at least, will not find it advisable to operate on a larger quantity at one time. As

---

* "In reducing galena, with a view to test the reduced lead for silver by cupellation, the reduc-tion may be conveniently performed as follows : a small portion of the galena, crushed to powder, is mixed with about twice its volume of carb. soda to which a little borax has been added. This is made into a stiff paste by the moistened knife-blade or blowpipe spatula, and a short piece of thin iron wire is stuck through it. The whole is then placed in a tolerably deep cavity scooped in a good piece of charcoal, and is exposed for a couple of minutes to the action of a reducing flame. By a little management, the minute globules of lead, which first result, can easily be made to run into a single globule. The iron assists in taking up sulphur from the galena. When sufficiently cool the fused mass is removed by a sharp knife-point, and flattened (under a strip of paper) on the anvil. The disc of reduced lead, thus separated from the slag, is then ready for cupellation." CHAPMAN'S BLOWPIPE PRACTICE.

In the case of ordinary ores or matters suspected to contain gold or silver, the roasted test-substance must be mixed with about an equal quantity of pure litharge or granulated lead and a proper amount of flux, and subjected to fusion in a charcoal cavity. The lead or reduced litharge takes up any gold or silver that may be in the ore, and this is set free by subsequent cupellation as described above.

soon as fusion takes place, the cupel must be moved somewhat farther from the flame, so as to allow merely the outer envelope of the latter, or the warm air which surrounds this, to play over the surface of the globule. By this treatment, the lead will become gradually converted into a fusible and crystalline slag. When this collects in large quantity, the position of the cupel must be slightly altered, so as to cause the globule to flow towards its edge, the surface of the lead being thus kept free for continued oxidation. When the globule becomes reduced to about a fourth or fifth of its original bulk, the process is discontinued, and the cupel set on the anvil to cool.* This is the first or concentration stage of the process. Another cupel is then prepared and dried ; and the concentrated globule (after being carefully separated from the slag in which it is imbedded) is placed on this, and again subjected to the oxidizing influence of the flame. During this second part of the process, the flame is made to play more on the surface of the cupel around the lead button, than on the button itself, by which a complete absorption of the oxidized lead is effected. The flame should be sharp and finely pointed, and urged down on the cupel at an angle of forty or forty-five degrees. Finally, if the test-metal contain gold or silver, a minute globule of one (or both) of these metals will be left on the surface of the bone-ash. By concentrating several portions of a test-substance, melting the concentrated globules together, again concentrating, and finally completing the cupellation, as small an amount as half-an-ounce of gold or silver in a ton of ore—or in round numbers, about one part in sixty-thousand ---may be readily detected by the blowpipe.

During cupellation, the process sometimes becomes suddenly arrested. This may arise from the temperature being too low, in which case the point of the blue flame must be brought for an instant on the surface of the globule, until complete fusion again ensue. Or the hindrance may arise from the boneash becoming saturated, when a fresh cupel must be taken. Or it may be occasioned, especially if much copper or nickel be present, by an insufficient quantity of lead. In this latter case, a piece of pure lead must be placed in contact with the globule, and the two fused together ; the cupel being then

---

* This is not always necessary, as in many cases the entire cupellation may be effected without interruption on the same cupel.

moved backward from the flame, and the oxidizing process again established.

Reactions :—Certain reactions of the more commonly occurring constituents of mineral bodies have already been mentioned in illustration of the various operations given above. In the present place a few additional reactions are described, and the whole are arranged in systematic form.*

A.—DETERMINATION OF THE CHEMICAL GROUP TO WHICH A MINEARL MAY BELONG.

In the examination of a mineral by the blowpipe, it is advisable to look first to its general chemical nature—or, in other words, to determine the chemical group to whicn it belongs—and afterwards, to seek for the base or bases which it may contain. The more important chemical groups of natural occurrence, comprise: Sulphides, Arsenides, Chlorides, Fluorides, Oxides, Sulphates, Silicates, Carbonates, Borates, Nitrates, Phosphates, and Arseniates. The group of simple Oxides can only be determined by negative characters, but the other groups are easly recognized by a few simple experiments.

Experiment 1. Fuse the substance, in powder, with 2 or 3 vols. of carb. soda and a little borax, in a good reducing flame, on charcoal.

This experiment serves directly for the detection of *Sulphides, Sulphates, Arsenides,* and *Arseniates.*

a. A strong odour of garlic is emitted :—*Arsenides* and *Arseniates.* The former possess a metallic aspect, and emit the garlic odour when ignited *per se.* The latter never exhibit a metallic aspect. As occuring in nature, arseniates are mostly of a green, blue, or red colour, depending on the nature of the base.

b. A reddish or dark mass is produced. This, when moistened and placed on a bright silver coin or on a glazed visiting card, forms a dark stain. The moistened mass smells also of sulphuretted hydrogen : *Sulphides* and *Sulphates.* The former possess a metallic aspect, or, if the lustre be non-metallic, the streak is always distinctly coloured.† With few exceptions, they omit an odour of burning brimstone (sulphurous acid) when ignited *per se ;* and in the open tube,

---

* A more complete method for the rapid determination of the chemical nature and composition of mineral bodies will be found in the author's BLOWPIPE PRACTICE, pages 60-65.

† Certain specimens of Zinc Blende are the only exceptions to this, so far at least as regards naturally occurring minerals, to which alone the statements of the text apply.

the evolved acid reddens moistened litmus-paper. (See Operation 4, above.) The natural sulphates do not possess a metallic aspect, and the streak is either colourless or pale green or blue. They do not omit the smell of brimstone when heated,

Other results, if exhibited, may be noted down for after reference.

Remarks :—Reactions *a* and *b* are sometimes produced by the same mineral, from the simultaneous presence of sulphur and arsenic, (Arsenical Pyrites, Realgar, Orpiment, &c.) Reaction *b* is also produced by *Selenides* and *Seleniates*, but these are of exceedingly rare occurrence, and they evolve at the same time a strong odour of cabbage-water or decomposing vegetable matter. The rare *tellurides* also exhibit the reaction.

Experiment 2. Fuse a solid particle of the test-mineral with phosphor-salt on platinum wire.

This experiment serves directly for the detection of *Carbonates* and *Silicates*.

a. The substance dissolves rapidly and with marked effervescence : —*Carbonates* (essentially).*

Note :—Sulphates, Phosphates, and various other compounds, also dissolve readily by fusion with phosphor-salt, but produce no effervescence.

b. The substance dissolves in part only, the undissolved portion retaining the original form of the test-fragment but becoming more or less translucent.   (On cooling, the glass often becomes opalescent) :—*Silicates* (see under " Phosphor-salt," page 35, above). Free silica, or quartz, melts into a clear glass with carb. soda, in expelling, with effervescence, the carbonic acid from the latter. Some silicates produce the same reaction. The test-substance should be added little by little. If the soda be in excess, the glass remains opaque, and with too much silica it becomes infusible.

Note :—Other reactions that may ensue from this experiment, such as the coloration of the glass, &c., may serve to detect the base or bases in combination with the carbonic or silicic acid. These reactions, therefore, should be noted down for after reference.

---

*Nitrates and certain bodies (Pyrolusite or Black Manganese Ore, &c.) which evolve oxygen on ignition, also dissolve in phosphor-salt with effervescence before the blowpipe ; but these bodies are of comparatively exceptional occurrence. To avoid risk of error, however, the substance may be warmed in a test tube with a few drops of hydrochloric acid. Thus treated, all carbonates dissolve with marked effervesence, and evolve a *colorless, inodorous* gas.

Experiment 3. Dissolve a few particles of black oxide of copper in phosphor-salt on platinum wire, so as to form a strongly-coloured glass. (Or simply melt some of the salt in a loop of thin copper-wire.) To this, add the test-substance, in powder, and expose the whole to the point of the blue cone.

This experiment serves directly for the detection of *chlorides.*

a. The fused bead is surrounded by a bright azure-blue flame. *Note :*—The coloration is produced by the volatilization of chloride of copper. It ceases therefore, after a time, but may be renewed by more of the test-substance being fused into the bead. The rare *Bromides* and *Iodides* can also be distinguished by this experiment. The former produce a blue flame with green streaks and edges, the latter a bright emerald-green coloration.

Experiment 4. Moisten the substance, in powder, with a drop of sulphuric acid, and expose on platinum wire to the point of the blue flame.

This experiment serves for the detection of *Phosphates* and *Borates.* as these bodies impart, when thus treated, a clearly marked green colour to the flame-border. The borates communicate also a green colour—after previous treatment with a few drops of sulphuric acid— to the flame of alcohol. The phosphates and borates of natural occurrence are without metallic aspect. All dissolve readily in borax and phosphor-salt before the blowpipe. Many communicate a green colour to the point of the flame when strongly ignited, *per se.* It must not be forgotten, however, that certain other bodies, oxide of copper, baryta, &c., also colour the flame green. Phosphates may also be detected as follows :—Melt some of the substance in fine powder with about 3 vols. of carb. soda, on platinum wire, or in a small platinum spoon. Treat the fused mass with a few drops of boiling water (in a test-tube, or, better, in a small porcelain or platinum capsule, over the spirit lamp), decant the clear solution from the insoluble residuum, add some nitric acid, and place in the solution a fragment of ammonium molybdate. This forms a canary-yellow precipitate with solutions of phosphates. In most cases the mineral may be treated directly (in powder) with nitric acid, and the diluted solution tested with ammonium molybdate. The yellow precipitate rapidly forms on the solution being warmed. It is readily soluble in ammonia.

Experiment 5.   Heat a small portion of the substance, in powder, at the bottom of a test-tube, with a few drops of strong sulphuric acid.

This experiment serves for the detection of *Fluorides* and *Nitrates*.

a.   The inside of the tube is more or less corroded, and also covered, where damp, with a deposit of silica :—*Fluorides.*   The results are best seen by washing out the tube, and then drying thoroughly in the flame of the spirit lamp.   The corrosion arises from the formation of a compound of fluorine and hydrogen which readily attacks silica, producing a volatile compound of fluorine and silicon.   This is decomdosed by water, with deposition of silica.   The latter re-action may be seen on the damp sides of the glass, and still more distinctly if a piece of narrow tubing with a drop of water at the end (kept there by the pressure of the finger at the other extremity) be brought within the mouth of the test-tube.   The deposit of silica adheres to the glass with great tenacity.

b.   Brownish or orange-coloured fumes are evolved :—*Nitrates.* The fumes possess the peculiar sweetish smell of nitrous acid.   All nitrates of natural occurrence are more or less soluble in water. They deflagrate when ignited on charcoal or in contact with other organic bodies.

B.—REACTIONS OF THE MORE COMMON MINERAL BASES.

In many minerals, the so-called base—lead, for example, in sulphide of lead (galena), copper in red or black oxide of copper, baryta in carbonate of baryta, and so forth—may be easily recognized by the use of the blowpipe.   This is especially the case, when the base consists of a single and easily reducible metal or metallic oxide, such as silver, lead, copper, tin, &c., or where it imparts a colour to borax or other reagents, as in the case of copper, iron, cobalt, nickel, manganese, &c.; or forms a deposit on charcoal, communicates a colour to the flame, or exhibits other characteristic reactions.   Even when several bodies of this kind are present together in the base, their recognition, as a general rule, is easily effected.   Earthy and alkaline bases, when in the form of carbonates, sulphates, phosphates, fluorides, &c., can also be made out, in general, without difficulty, unless several happen to be present together, in which case it is not always possible, by the simple aid of the blowpipe, to distinguish them individually.   When these bases are combined with silica, on the other hand, the blowpipe alone is rarely sufficient for their detection. This however, so far as practical purposes are concerned, is of little consequence, as no economic value in silicates of this kind is dependent on the base.

A complete scheme for the detection of mineral bases by the blowpipe, does not fall within the province of the present work, but an arrangement of the more important of these bodies, in groups, founded on blowpipe characters, is given below. Before referring to these groups, the unpractised operator is recommended to subject the specimen under examination to three or four simple experiments, and to note down the results. These experiments comprise :—1, Ignition in the bulb-tube, for detection of water. (This experiment may be omitted as a general rule, if the substance possess a metallic lustre.) 2, Treatment *per se* on charcoal or in the foreceps (see Operation 1, page 30 above), the characters more especially to be looked for, being coloration of the flame, formation of a coating on the charcoal, assumption of magnetism, &c. 3, Treatment (after previous roasting [Operation 4], if sulphur, &c., be present) with borax, phosphor-salt, and carb. soda, respectively : observing if the glass be coloured, if the substance dissolve entirely in it, if a reduction to metal take place, and so forth (Operations 5 and 6, above). These experiments will in general be sufficient to determine the nature of the base ; but occasionally, certain special operations may be required in addition, such as testing with nitrate of cobalt, or examination for mercury in the closed tube, as described on a preceding page (Operations 3 and 6).

SECTION 1.—GIVING *per se*, OR WITH CARB. SODA, ON CHARCOAL, METALLIC GLOBULES OR METALLIC GRAINS.

Group 1. *Yielding malleable metallic globules, without deposit on the charcoal.*

Gold. Silver. Copper.

Gold is insoluble in the fluxes. *Silver* is not oxidized *per se*, but retains a bright surface after exposure to an oxidating flame. *Copper* becomes encrusted on cooling with a black coating. It imparts a green colour to the flame-border ; and forms strongly coloured glasses with borax and phosphor-salt : (green (hot) blue (cold) in O F ; redbrown, opaque, in R F). Gold and Silver may be separated from copper, &c., by fusion with lead, and subsequent cupellation (Operation 7). If gold and silver be present together, the bead is generally more or less white. By fusing it in a small platinum spoon with bisulphate of potash, the silver dissolves, and the surface of the globule becomes yellow. If the globule be flattened out into a disc on the anvil, before treatment with bisulphate of potash, the silver is more rapidly extracted. The sulphate of silver must be removed by treating the spoon, in a porcelain or platinum capsule, with a small quantity of water, over the spirit lamp. By evaporation, and fusion of the residuum with carb. soda on charcoal, metallic silver can be again obtained.

Group 2. *Yielding infusible metallic grains, without deposit on the charcoal:*

Platinum. Iron. Nickel. Cobalt.

Platinum is not attacked by the blowpipe fluxes. *Iron, Nickel,* and *Cobalt,* are readily dissolved by fusion with borax or phosphor-salt, producing a coloured glass (see under Borax, page 34, above.) These metals are also magnetic. As a general rule, if a substance become attractable by the magnet after exposure to the blowpipe, the presence of iron may be inferred, cobalt and nickel compounds being comparatively rare. The presence of cobalt is readily detected by the rich blue colour of the borax and phosphor-salt glasses, in both an oxidating and reducing flame; but if much iron be present also, the glass is blueish-green. With borax in the R. F., nickel compounds give reduced metal, and the glass becomes grey and troubled. It is also attracted by the magnet.

Group 3. *Yielding metallic globules, with white or yellow deposit on the charcoal.*

Tin. Lead. Bismuth. Antimony.

Tin and *Lead* give malleable globules. The sublimate formed by tin, is white, small in quantity, and deposited on, and immediately around, the globule. The lead sublimate is yellow, and more or less copious. *Bismuth* and *Antimony* give brittle globules. The bismuth sublimate is dark yellow; the antimony sublimate, white, and very abundant. Lead imparts a clear blue colour to the flame-border; Antimony, a greenish tint. As a general rule, a yellow deposit on the charcoal may be regarded as indicative of the presence of lead ;*

---

* Some lead compounds *per se* give a white or greyish sublimate; but if the test-substance be mixed with carb. soda, the sublimate is always yellow.

whilst the emission of copious fumes, and deposition of a white coating on the charcoal, may be safely considered to indicate antimony. The coating or sublimate formed by Zinc (see below), although white when cold, is lemon-yellow whilst hot. Bismuth compounds if fused in powder with a mixture of sulphur and iodide of potassium produce on charcoal a vivid scarlet incrustation (Von Kobell). If metallic tin and lead, in about equal proportions, be fused together, the resulting globule immediately oxidises, and on removal from the flame continues to push out white and yellow excrescences (Chapman's Blowpipe Practice, p. 92).

SECTION 2.—REDUCIBLE ; BUT YIELDING NO METAL ON CHARCOAL. (This arises from the rapid volatilization of the reduced metal.)

Group 1. *Volatilizing without odour, and without formation of deposit on the charcoal.*

Mercury.

For the proper detection of this metal, a small portion of the test-substance in powder must be mixed with some previously dried carb. soda, and the mixture strongly ignited at the bottom of a small tube or narrow flask. If mercury be present, a grey sublimate will be formed. By friction with a wire, &c., this runs into small metallic globules which may be poured out of the tube.

Group 2. *Volatilizing without odour, but forming a deposit on the charcoal.*

Cadmium. Zinc.

The deposit produced by cadmium is dark brown or reddish-brown. That produced by zinc is lemon-yellow and phosphorescent whilst hot, and white when cold. If moistened with a drop of nitrate of cobalt and ignited, it becomes bright green.

Group 3. *Volatilizing with strong odour of garlic.*

Arsenic.

See additional reaction under Operation 4, page 32, above.

SECTION 3.—NOT REDUCIBLE BEFORE THE BLOWPIPE.

Group 1. *Imparting a colour to borax.*

Manganese. Chromium.

Manganese compounds impart, before an oxidating flame, a violet colour to borax ; *Chromium compounds*, a clear green colour. See also under " Carbonate of Soda " page 36, above.

The rare metals cerium, uranium, &c., belong also to this group.
Reference should also be made to iron, nickel, cobalt and copper, as
the oxides of these metals, if in small quantity, might escape detection
by the reducing process.

Group 2. *Imparting no colour to the fluxes. Slowly dissolved by
borax, the glass remaining permanently clear* :
Alumina
Moistened with nitrate of cobalt and then ignited, this base assumes
on cooling a fine blue color.

Group 3. *Imparting no colour to the fluxes. Rapidly dissolved
by borax, the glass becoming opaque on cooling or when flamed* :
Magnesia. Lime.
Moistened with nitrate of cobalt, and ignited, *Magnesia* becomes
pale-red in colour ; Lime, dark-grey.

Group 4. *Entirely dissolved by fusion with carb-soda.*
Baryta. Strontia. Lithia. Soda. Potash.
Baryta compounds impart a distinct green colour to the point and
border of the flame. *Strontia* and *Lithia* colour the flame deep carmine-red. The crimson coloration is destroyed in the case of strontia
if the substance be fused with chloride of barium. *Soda* colours the
flame strongly yellow. *Potash* communicates to it a violet tint ; but
this colour is completely masked by the presence of soda, unless the
flame be examined through a deep blue glass.*

---

* The presence of alkalies or alkaline earths (magnesia excepted) is most readily ascertained
in minerals by the use of a small pocket spectroscope. See the author's *Blowpipe Practice*.

PART II.

THE MINERALS OF CENTRAL CANADA.

The preceding sub-division of this work is of a purely introductory
character, explanatory of the more common properties possessed by
minerals in general, and of certain technical terms employed in mineralogical definitions.   In the present Part, the Minerals of Central
Canada, comprising the Provinces of Ontario and Quebec, are classified, and described.   In these descriptions, in accordance with the
stated plan of the work, minute chemical and crystallographic details
are purposely omitted : details of this kind being obviously out of
place in a work intended for general use.   Localities also except in a
few instances, are only stated generally, *i. e.* without precise reference
to lots and concessions ; but an attempt is made in all cases to give
the localities in systematic order, based, as much as possible, on geological relations.

The classification, adopted in the work, is founded essentially on
composition, as being the most convenient for practical reference.   It
is preceded, however, by an Analytical Key, by means of which
the name and place of any mineral described under the Classification
proper, may be easily arrived at ; and a Simplified Key or Tabular
Arrangement, including minerals of common occurence only, is also
given for the same purpose.   The method of application is explained
fully at the end of the principal key : a certain knowledge of technical terms, and of the more common properties of minerals as explained
in the preceding division of the work, being of course supposed on
the part of the reader.

ANALYTICAL KEY,

By which the name of any Canadian Mineral may be easily ascertained. *

Note—In this Key, minerals of common or extensive occurrence are denoted by the name
being printed in large capitals, and minerals of tolerably common occurence, by the use of

---

* As regards the determination of minerals generally, the Reader may consult the *Mineral
Tables* attached to the author's *Blowpipe Practice.* Also the anthor's *Mineral Indicator*
Copp, Clark & Co., Toronto).

5

small capitals. Names in ordinary type, refer to minerals of rare occurrence, or obscure character : so far, at least, as regards the presence of these minerals in Canada. The initials BB, signify "before the blowpipe." The number placed within brackets after the name of a mineral, refers to the position of the substance in the classification proper, in which its description is given, at the end of the key

1 { Aspect metallic or sub-metellic 2
  { Aspect non-metallic (*i. e.* vitreous, stony, &c.) 35

2 { Occurring in detached grains or scales 3
  { Occurring under other conditions 6

3 { Soiling, or marking on paper .... GRAPHITE No. 1.)
  { Not marking or soiling 4

4 { Yielding by trituration a white or light-grey powder.. MICA (Nos. 77-78.)
  { Not yielding a white powder by trituration 5

  (Colour, yellow. Fusible.......................... ....Gold (No. 3.)
5 { Colour, tin or greyish white. Infusible .... Platinum (No. 4.)
  (Colour, black ; magnetic.......... MAGNETIC IRON SAND (No. 31.)
                                          [Also Iserine (No. 32)

6 { Hardness sufficient to scratch glass.. 7
  { Hardness insufficient to scratch glass......... 15

7 { BB, emitting fumes, or odour of garlic or brimstone 8
  { BB, no fumes or odour 10

8 { Colour, light brass-yellow.. 9
  { Colour, tin-white or greyish.............. ARSENICAL PYRITES (No. 22.)

  (In cubes or other Monometric Crystals (p. 14), or massive
  (IRON PYRITES (No. 20.)
9 { In pointed, Prismatic Crystals of the Rhombic System (p. 16), mostly
  (arranged in curved rows..... Marcasite or Prismatic Pyrites* (No. 21.)

10 { BB, easily fusible ....Wolfram (No. 39.)
   { BB, infusible, or nearly so 11

11 { Streak-powder, dull-red............ SPECULAR IRON ORE (No. 29.)
   { Streak powder, black or brown 12

12 { Strongly magnetic.................. MAGNETIC IRON ORE (No. 31.)
   { Not (or very feebly) magnetic 13

13 { Streak, black, brown, reddish-brown, or greenish 14
   { Streak, brownish-yellow. Yielding water in the bulb-tube...
                                          BROWN IRON ORE (No. 34.)

   (Black, sub-metallic. BB, with phosphor-salt in R. F., a fine green glass
14 {   CHROMIC IRON ORE (No. 33.)
   (Black, sub-metallic. BB, with phosphor-salt in R. F., a red-brown glass
                                          TITANIFEROUS IRON ORE (No. 30.)

15 { More or less distinctly malleable................................... 16
   { Not malleable... 20

16 { BB, no fumes, or deposit on charcoal................................ 17
   { BB, copious fumes, or incrustation on charcoal.................... 18

---

* Iron Pyrites and Marcasite have exactly the same composition (Sulphur 53.3, Iron 46.7) but their crystal forms are quite distinct. Iron Pyrites is very abundant : Marcasite, in Canada, comparatively rare. Marcasite is especially prone to decomposition ; specimens are thus often coated wit a greenish-white efflorescence, or minute hair-like crystals, of sulphate of iron.

17 {
Colour, yellow (soft) ....: ....GOLD (No. 3.)
C. silver-white (soft)....................................SILVER (No. 5.)
C. copper-red (soft)................................. ....COPPER (No. 6.)
C. steel-grey ; magnetic (hard)..................Meteoric Iron (No. 10.)
}

18 {
BB, on charcoal, a copious yellow incrustation 19
BB, on charcoal, no incrustation. Colour, black..SILVER GLANCE (No. 11.)
}

19 {
Colour, lead-grey. Perfectly malleable ... .Lead (No. 7.)
Colour, tin-white. Slightly malleable ....,.Bismuth (No. 8.)
}

20 {
Structure distinctly scaly or micaceous, the substance admitting of sepa-
  ration into thin leaves, plates, or scales.......................... 21
Structure not micaceous or scaly { Marking or soiling............... 21
                                  { Not marking or soiling........... 23
}

21 {
Marking on paper. Streak, black 22
Not marking. Streak, white or greyish (*Mica*) .... 21 *bis.*
}

21 *bis* {
Not attacked by acids...MUSCOVITE or POTASH-MICA (No. 77.)
Decomposed (in powder) by sulphuric acid.... .PHLOGOPITE or
                                      MAGNESIA MICA (No. 78.)
}

22 {
Colour, black. BB. not dissolved by fluxes........GRAPHITE (No. 1.)
Colour, lead-grey, BB, giving sulphur-reaction (see p. 44) with carb.
  soda and borax.............. ....MOLYBDENITE (No. 23.)
}

23 {
Attracting the magnetic needle. Colour, brownish-yellow....MAGNETIC
                                      PYRITES (No. 19.)
Not affecting the magnetic needle............................... 24
}

24 {
BB, easily fusible (with or without previous decrepitation) 25
BB, infusible, or nearly so 34
}

25 {
BB, a magnetic globule.. 26
Fusion-globule not magnetic....................... 28
}

26 {
Colour dark-lead grey..............................Tennantite (No
Colour metallic-yellow or red .... 26 *bis*
}

26 *bis* {
In acicular form only'.................... .. ..Millerite (No. 18.)
Not acicular.................................. 27
}

27 {
Colour brass yellow (sometimes with iridescent tarnish......COPPER
                                      PYRITES (No. 16.)
Colour reddish, but with purple tarnish..........BORNITE, (Horse-
                                      flesh Ore, No. 15.)
Colour pale-red or yellowish. BB, yielding arsenical fumes......Nickeline
                                      some examples, (No. 17.)
}

28 {
Colour pale-red or yellowish..........Nickeline (most examples, No. 17.)
Colour, metallic white or grey 29
}

29 {
BB, no coating on charcoal........ ... .COPPER GLANCE (No. 14.)
BB, a white or yellow coating on charcoal.......................... 30
}

30 {
BB, a white coating on charcoal................................. 31
BB, a yellow incrustation on charcoal.. 32
}

31 {
Lamellar or fine-granular in structure..........Native Antimony (No. 9.)
Fibrous or bladed....................... Antimony Glance (No. 25.)
}

32 {
BB, with mixture of potassium iodide and sulphur forming on charcoal
  a vivid scarlet coating..... 33
BB, as above, no scarlet coating............................. 33 *bis*
}

33 { Sp gr. over 9, reddish tin white............. ..Native Bismuth, (No. 8.)
{ Sp. gr. under 7, light lead-grey, often iridescent............... Bismuth
                    Glance (No. 24.)

33 bis { Breaking into rectangular fragments............GALENA (No. 12.)
{ BB, yielding antimonial fumes... .Menighinite and Plumbigerous
              Antimony Ore, (Nos. 25 and 25 bis.)

34 { Lustre distinctly metallic; streak greyish-black; mostly fibrous or
{ acicular.. ...Manganite (No. 35.)
{ Lustre sub-metallic; streak, mostly pale brown; BB, sulphur-reaction,
{ p. 44 ....ZINC BLENDE (No. 13.)

35 { Soluble or partially soluble in water. Taste bitter or metallic. Occur-
{ ring chiefly as an efflorescence or incrustation...... 36
{ Insoluble { Occurring in earthly masses or crusts (which soil or mark
{   { more or less).... 39
{   { Occurring under other conditions........................ 48

36 { BB, with borax, a coloured glass or bead......................... 37
{ BB, with borax, a white or colourless beard........................ 38

37 { Solution giving a deep blue precipitate with red or yellow "prussiate of
{ potash."*...................Green Vitriol (Sulphate of Iron) (No. 100.)
{ Solution giving a greenish-white precipitate with "yellow prussiate."
            Sulphate of Nickel (No. 101.)

38 { BB, with nitrate of cobalt, a blue mass after ignition (see p. 34.).......
{   Alum (No. 102.)
{ BB, with nitrate of cobalt, a pale-red mass...........Epsomite (No. 99.)

39 { Colour, yellow or yellowish-brown................................. 40
{ Colour, red, black, brownish-black, blue, or green.................. 43

40 { BB, taking fire and burning with blue flame. ....Sulphur (No. 2.)
{ BB, not inflammable.. 41

41 { BB, becoming black and magnetic................................... 42
{ BB, not rendered magnetic........, ...Uran Oehre (No. 37.)

42 { Occurring in thin crusts on bituminous shale.....Humboldtine (No. 108.)
{ Occurring under other conditions....... ......YELLOW OCHRE (No. 34.)

43 { Colour, red. BB, becoming magnetic..RED OCHRE and SCALY RED
{   IRON ORE (No. 29.)
{ C. black, dark-brown, blue, or green................... 44

44 { C. black or dark-brown 45
{ C. blue or green.. 46

45 { BB, inflammable....................................Asphalt (No. 110.)
{ BB, not inflammable. Forming with carb-soda a "turquoise enamel,"
{ p. 39............. ....EARTHY MANGANESE ORE (No. 36.)

46 { Colour, blue.. 47
{ Colour, green. Effervescing in acids.. ...Malachite (Green Carbonate
             of Copper) (No. 95.)

---

* As the iron is always partly peroxidized, a blue precipitate is produced by either of these reagents.

47 { Effervescing in acids ; BB, reactions of Copper (p. 37.)..Blue Carbouate of Copper (No. 95.)
BB, rendered magnetic.........Vivianite (Phosphate of Iron) (No. 104.)

48 { Hardness sufficient to scratch window-glass distinctly............... 49
Hardness insufficient to scratch glass distinctly........ 69

49 { Fusible or partially fusible, *per se**.................. 50
Infusible, *per se*.. 61

50 { Sp. gr. = 3.0 or *less*. (Colour, mostly pale.)..................... . 51
Sp. gr. over 3.0.................... 56

51 { Yielding water by ignition in bulb-tube (see p. 34).................. 52
Not yielding water, or yielding traees only, on ignition.............. 53

52 { Fusible on thin edges, only. C. dark-green.........Chloritoid (No. 81.)
Easily fusible. C. light-green, greenish-white....PREHNITE (No. 67.)
Easily fusible. C. peach-blossom red....Wilsonite: var. of Scapolite (No. 63.)

53 { Easily fusible............... ....SCAPOLITE OR WERNERITE (No. 63.)
Fusible on edges only, unless in thin splinters...................... 54

54 { White, red, &c. In masses with smooth rectangular cleavage......... ORTHOCLASE (No. 57.)
Cleavage not rectangular. Cleavage planes faintly striated.......... 55

55 { White, reddish, &c. BB, imparting a yellow colour to the flame...... ALBITE (No. 58.)
Grey, often with coloured reflections........LABRADORITE (No. 60.)

56 { In rhombic dodecahedrons or trapezohedrous (p. 14), or in imbedded granular masses mostly of a red colour............GARNET (No. 47.)
In fibrous masses or prismatic crystals............................ 57

57 { In black, brown, or green triangular prisms (often broken and disjointed or in fibres with triangular cross fracture ....TOURMALINE (No. 46.)
In other forms..................................... 58

58 { In tetragonal (square prismatic) crystals (p. 15). Sp. gr. 3.5 or more.. Idocrase (No. 48.)
In other forms....................................... 59

59 { Fusible into a globule or rounded mass.. ....AMPHIBOLE (No. 52.)
PYROXENE (No. 53.)
Fusible on the surface, only, into a dull slag or scoria............... 60

60 { In flat wedge-like crystals, mostly dark-brown or yellowish......... ...
SPHENE (No. 51.)
In green fibrous masses and long prismatic crystals.....EPIDOTE (No. 49.)

61 { Streak-powder, white, or greyish in dark specimens. 62
Streak-powder, black, brown, greenish or red....................... 66

---

* In testing this character, only a thin, pointed splinter of the mineral must be taken ; otherwise the substance, although really fusible or fusible on the edges, may be regarded by beginners as infusible.

62 { Sp. gr. under 2.8 ; H = 7.0 ; vitreous ; fusible with carb-soda into a clear glass...QUARTZ (No. 43.)
Sp. gr. over 3.0... 63

63 { Harder than quartz 64
Less hard than quartz... 65

64 { Crystallization, Hexagonal ; H = 9.0 ; sp. gr. 3.8 — 4.1.....Corundum (No. 41.)
Crystallization, Octahedral (Regular System) ; H = 8.0 ; sp. gr... . 3.5 — 4.5..Spinel (No. 42.)
Cystallization, Square-pyramidal (Tetragonal System) ; H=7.5 ; sp. gr. 4.0 — 4.7..Zircon (No. 44.)
Crystallization, Rectangular-prismatic ; H = 7.0 — 7.5 ; sp. gr. 3.1 — 3.2...Andalusite (No. 45.)

65 { Red or orange ; Lustre inclined to semi-metallic ; sp. gr. 4.1 — 4.3..... Rutile (No. 40.)
Yellow ; in small granular masses (mostly with graphite in crystalline limestone); sp. gr. 3.1 — 3.2.....................Condrodite (No. 56.)
Green, brownish-yellow ; in crystalline grains in eruptive rocks ; sp. gr. 3.3 — 3.5 ....Olivine (No. 55.)

66 { Strongly magnetic................MAGNETIC IRON ORE (No. 31.)
Feebly (or non-) magnetic... 67

67 { Streak-powder, black or brown 68
Streak-powder, pull-red ....RED IRON ORE (No. 29.)

68 { BB, with borax, an emerald-green glass....CHROMIC IRON ORE (No. 33.)
BB, with borax, a dingy-green glass,..TITANIFEROUS IRON ORE (No. 30.)

69 { BB, fusible, or imparting distinct colour to the flame, or both........ 70
BB, infusible (or fusible only at the external point)................ 77

70 { BB, easily dissolved by borax or phosphor-salt, the saturated glass becoming opaque on cooling or when flamed (p. 37)................ 71
BB, slowly and incompletely dissolved by borax or phosphor-salt, a "silica skeleton" (p. 39) separating in the latter reagent........... 74

71 { BB, yielding sulphur-reaction with carb-soda and silver foil (p. 44).... 72
BB, no sulphur-reaction with carb-soda, &c. Mostly in cubical crystals.. FLUOR SPAR (No. 106.)

72 { Yielding a large amount of water by ignition in bulb-tube............. GYPSUM (No. 98.)
No water on ignition... 73

73 { BB, imparting an apple-green tint to the flame-border. Fusible with . difficulty. BARYTINE (No. 96.)
BB, imparting a carmine-red colour to the flame-border............. CELESTINE (No. 96.)

74 { BB, imparting a green tint to point of flame.........Datolite (No, 68.)
BB, imparting a yellowish or indistinct colour to the flame.......... 75

75 { BB, fusible quietly ....Analcime (No. 75.)
BB, intumescing...76

APPLICATION OF THE ANALYTICAL KEY.

The method of employing the above Key is shewn in the following
example.   Let the reader be supposed to have a massive piece of
magnetic pyrites, of the name and nature of which he is ignorant.
Turning to the first bracket of the Key, he finds:

1 { Aspect metallic or sub-metallic........................... 2
   { Aspect non-metallic (i. e., vitreous, stony, etc)............ 35

As the substance possesses a metallic aspect or lustre, he turns to bracket 2.　There he finds :

$$2 \begin{cases} \text{Occurring in detached grains or scales} \dots \dots \dots \dots \dots \dots & 3 \\ \text{Occurring under conditions} \dots \dots \quad \dots \dots \dots \dots \dots \dots \dots & 6 \end{cases}$$

As the specimen is not in the form of loose grains or scales, but in that of a solid mass, he turns to bracket 6, which reads :

$$6 \begin{cases} \text{Hardness sufficient to scratch glass} \dots \dots \dots \dots \dots \dots \dots & 7 \\ \text{Hardness insufficient to scratch glass} \dots \dots \dots \dots \dots \dots & 15 \end{cases}$$

As the mineral is not hard enough to scratch glass, bracket 15 must be referred to, which reads :

$$15 \begin{cases} \text{More or less distinctly malleable} \dots \dots \dots \dots \dots \dots \dots & 46 \\ \text{Not malleable} \dots \dots \dots \dots \dots \dots \dots \dots \dots \dots \dots & 20 \end{cases}$$

As the substance is not malleable—a small piece breaking readily into powder under the hammer—the inquirer turns to bracket 20, He there finds :

$$20 \begin{cases} \text{Structure distinctly scaly or micaceous, the substance admitting of separation into thin leaves, plates, or scales} \dots \dots \ 21 \\ \text{Structure not micaceous or scaly} \dots \begin{cases} \text{Marking or soiling} \dots \dots \ 21 \\ \text{Not marking or soiling.} \ 23 \end{cases} \end{cases}$$

As the mineral under investigation does not present a scaly or micaceous structure, and does not soil the hands or leave a mark on paper, reference is made to bracket 23.　This reads :

$$23 \begin{cases} \text{Affecting the magnetic needle ; colour, browish-yellow} \dots \dots \\ \qquad\qquad\qquad\qquad\qquad \text{MAGNETIC PYRITES (No. 19.)} \\ \text{Not magnetic} \dots \dots \dots \dots \dots \dots \dots \dots \dots \dots \dots \dots \dots & 24 \end{cases}$$

A small particle or two being chipped off the specimen, and tried by a common magnet—or the entire specimen being held near a magnetic needle—attraction is found to ensue; hence the substance is shewn to be *Magnetic Pyrites*, No. 19 of the classified series described in the following pages.　By reference to the description there given, the various physical and chemical characters of the substance, its percentage composition, localities, etc., may at once be ascertained. In using the Key, care must be taken to pass regularly from one indicated bracket to the other, without attempting, on account of foregone conclusions respecting the nature of the substance, to jump over any of them, or to refer to others than those actually indicated. If this be not attended to, errors and confusion may easily arise.

As the above Key contains a good many minerals of rare or comparatively exceptional occurrence, the beginner may frequently avoid unnecessary trouble, in making out the name of an unknown substance, by consulting in the first instance the annexed simplified Key, in which Canadian minerals of common occurrence are alone included. Reference should then be made, for confirmatory proofs, to the complete description of the species indicated by the Key.

A TABULAR GROUPING OF CANADIAN MINERALS OF COMPARATIVELY FREQUENT OCCURRENCE.

* *Aspect Metallic or Sub-Metallic.*

  ** *Hard enough to scratch glass distinctly. Not scratched, or very slightly scratched, by the point of a knife. (Streak distinctly coloured.)*

    (*a*) Pale brass-yellow (Often in cubes) :—*Iron Pyrites* (No. 20).

    (*b*) Tin-white, or between silver-white and pale-grey (Emitting a garlic-like odour on ignition) :—*Arsenical Pyrites* (No. 22).

    (*c*) Steel-grey ; powder, dull-red :—*Specular Iron Ore* (No. 29).

    (*d*) Iron-black ; powder, black ; strongly magnetic :—*Magnetic Iron Ore* (No. 31)

    (*e*) Iron-black ; powder, black or brown ; feebly or non-magnetic :—*Titaniferous Iron Ore* (No. 30) : also *Chromic Iron Ore* (No. 33).

  *** *Too soft to scratch glass. Easily scratched by a knife-point.*

Malleable.

    (*a*) Colour, yellow :—*Native Gold* (No. 3).

    (*b*) Colour, silver-white (but often with dark tarnish) ;—*Native Silver.*

    (*c*) Colour, black :—*Silver Glance* (No. 11).

Not Malleable.

    (*d*) Brownish-yellow ; slightly magnetic :—*Magnetic Pyrites* (No. 19.)

    (*e*) Brass-yellow (often with varigated tarnish) ; streak, greenish black :—*Copper Pyrites* (No. 16.)

    (*f*) Reddish, with purple tarnish ; streak, greyish-black :—*Purple Copper Pyrites* (No. 15.)

    (*g*) Dark-grey (ofthe with blue or green tarnish) ; cleavage indistinct :—*Copper Glance* (No. 14.)

    (*h*) Lead-grey ; breaking readily, with rectangular cleavage, into cubical fragments ; very heavy :—*Galena* (No. 12.)

    (*i*) Light lead-grey ; in soft scaly masses ; marking :—*Molybdenite* (No. 23.)

    (*k*) Black ; soft, mostly in scaly or leafy masses ; marking and soiling,—*Graphite* (No. 1.)

    (*l*) Lustre, metallic-pearly : brown, black, silvery-white, &c. In foliated or scaly masses with white or light streak ; easily separated into thin leaves :—*Mica*, including chiefly *Muscovite* (No. 77) and *Phlogopite* (No. 78.)

† *Aspect : vitreous, stony, or earthy.*

  †† *Hard enough to scratch glass distinctly. Not scratched by a knife-point.*

    (*a*) Vitreous : colourless, amethystine, brownish, &c. Mostly in hexagonal prism-pyramids, or in groups of sharply-pointed crystals; otherwise massive. No lamellar structure. (Infusible):—*Crystalline Quartz*, including *Rock Crystal, Amethyst, Smoky Quartz,* &c. (No. 43).

    (*b*) Vitreous or stony. In nodular masses of grey, red, bluish, and other colours, two or more tints being often present together in spots or bands. (Infusible) :—*Calcedonic Quartz*, including the various *Agates*, &c. (No. 43).

(*c*) Stony or pearly. Vitreous. White, grey, red, green, &c. Mostly in lamellar masses, which cleave easily in several directions, presenting smooth and somewhat pearly cleavage-planes. Fusible, but as a general rule not very easily: the point of a thin splinter is soon rounded or vitrified, however, in a properly sustained flame :— The various *Feldspars*, including more especially *Orthoclase* or *Potash-Feldspar* (No. 57); *Albite* or *Soda-Feldspar* (No. 58); and *Labradorite* or *Lime-Feldspar* (No. 60).

(*d*) Vitreous. Greenish-white or pale-green. Mostly in botryoidal masses with crystalline surface. Easily fusible. Yielding a little water in the bulb-tube :—*Prehnite* (No· 67).

(*e*) Dark or bright-red, brown, &c. Mostly in rhombic dodecahedrons or in small rounded masses. Fusible. (Sp. gr. over 3.4) :—*Garnet* (No. 47).

(*f*) Black or dark-brown. Mostly in triangular (and often broken) prisms, or in acicular or fibrous groups. Easily fusible :—*Tourmaline* or *Schorl* (No. 46).

(*g*) Black, brown, green, greenish-white or colourless. In small crystals (mostly imbedded in crystalline limestone, or otherwise in trap rocks), and also in cleavable and granular masses. Fusible :—*Pyroxene* (including *Augite*, &c.) No. 53; and also *Amphibole* (including *Hornblende*, &c.) No. 52.

††† *Too soft to scratch glass. Easily scratched, or cut, by the point of a knife.*

Streak, white, or very faintly coloured.

Not yielding water by ignition in the bulb-tube.

(*a*) White, grey, &c. Effervescing in cold acids. Infusible :—*Calcite* (No. 88).

(*b*) White, brownish, &c. Effervescing only in heated acids. Infusible :—*Dolomite* (No. 90); also *Magnesite* (No. 91).

(*c*) Green, reddish-brown. Mostly in six-sided prisms with rounded edges. Infusible, or nearly so. (H = 5.0) :—*Apatite* (No. 103).

(*d*) Violet-blue, green, greyish, &c. Mostly in cubes. Fusible :—*Fluor Spar* (No. 106).

(*e*) White, yellowish, greyish, pale-red, &c. Mostly in cleavable masses. Very heavy (sp. gr. = 4.4 — 4.7. Fusible, but not easily, tinging the flame pale-green :—*Heavy Spar* (No. 96).

(*f*) White, pale-blue, reddish, &c. Mostly in cleavable masses or small crystals in limestone rocks. Fusible, tinging the flame carmine red : *Celestine* (No. 97).

Yielding water, or distinct traces of moisture, by ignition in the bulb-tube.

(*g*) White, greyish, &c. Scratched by the nail. Fusible; becoming at once opaque and dull white when held at the edge of a candle-flame; yielding a large amount of water by ignition in the bulb-tube :—*Gypsum* (No. 98).

(*h*) White, greenish, green and brown, &c; often mottled. Very sectile, and more or less soapy to the touch :—*Talc* and *Steatite* (No. 82). Also *Serpentine* (No. 83).

(*i*) Dark or light green, scaly or earthy :—*Chlorite*.

(*k*) Pearly-white, brown, black, &c. In leafy and scaly masses with more or less pseudo-metallic lustre. Splitting into thin plates :—The various *Micas*, including more especially: *Muscovite* or *Potash-Mica* (No. 77), and *Phlogopite* or *Magnesia-Mica* (No. 78).

Streak, distinctly coloured.

(*l*) Streak, pale-brown. Colour, brown, black, yellow, &c. Mostly in indistinct crystals or small cleavable masses :—*Zinc Blende* (No. 13).

(*m*) Streak, dull or bright-red. Colour, brick-red. Magnetic after ignition :—*Red Ochre* and other varieties of *Red Iron Ore* (No. 29).

(*n*) Streak, brownish-yellow. C., dark or light-brown. Magnetic after iguition, and yielding water in the bulb.tube :—*Yellow Ochre* and *Brown* or *Boy Iron Ore* (No. 34).

(*o*) Streak, pale-green ; colour, green :—*Malachite* (No. 94). Some *Chlorites* (No. 80).

(*p*) Streak, pale-blue ; colour, blue. Mostly in crusts or earthy masses: —*Blue Carbonate of Copper* (No. 95). Also *Phosphate of Iron* (No. 104.)

SYSTEMATIC ARRANGEMENT OF MINERALS.

Mineral bodies are characterized partly by composition, and partly by physical properties. Composition alone, is not sufficient in all cases to define or individualize a mineral species, as certain substances —Carbon, for instance, in Graphite and the Diamond ; Carbonate of Lime, in Calcite and and Arragonite—may occur in nature under two or more distinct physical conditions. On the other hand, a close resemblance, in general aspect and other physical characters, may be exhibited by minerals of very dissimilar composition. Minerals have thus a double nature, so to say—chemical and physical : the one frequently in apparent opposition to the other ; and in this lies the difficulty of framing a satisfactory classification of minerals. A system of arrangement based on chemical composition, although unavoidably artificial in many of its details, is especially convenient for practical reference, and on the whole is perhaps best suited to meet the requirements of the general student. A system of this kind is adopted, therefore, in the present work. It comprises five leading groups or classes. First, a group of simple or so-called Native Substances, as Native Sulphur, Native Gold, Native Silver, &c., the naturally occurring elementary bodies of chemical language (See under " Chemical Characters " in Part I). Secondly, a group of Sulphides and Arsenides, or compounds of sulphur, or of arsenic, with various metals : galena, iron pyrites, arsenical pyrites, are examples. Thirdly, a large group of oxygenized compounds, including Simple Oxides, as red iron ore, &c., and various so-called oxygen-salts, as Silicates, Carbonates, Sulphates, and the like. (See explanation of Chemical Terms in Part I. Also the explanatory remarks prefixed to the different groups and sub-divisions, in the following pages.)

Fourthly, a group of Fluorides and Chlorides, compounds of fluorine or chlorine with bases. And, finally, a small group of carbonaceous matters, usually classed as Organico-Chemical substances, and regarded commonly as products of alteration derived from Organic Nature. The sub-divisions of the system adopted, are shewn by way of index, in the annexed tabular view.

I.—SIMPLE SUBSTANCES:
    A. Native Non-Metallic Substances (1—2).
    B. Native Metals (3-10).

II.—ARSENIDES AND SULPHIDES:
    A. Sulphides of Silver, Lead, and Zinc (11-13.)
    B. Sulphides of Copper (14-16).
    C. Arsenides and Sulphides of Nickel and Iron (17-22).
    D. Sulphide of Molybdenum (23).
    E. Sulphides of Bismuth and Antimony (24-26).

III.—OXYGEN COMPOUNDS:
    A. Copper Oxides (27-28).
    B. Iron Oxides:
        (1) Hematite group of Iron Oxides (29-30).
        (2) Magnetite group of Iron Oxides (31-33).
        (3) Limonite group of Iron Oxides (34).
    C. Manganese Oxides (35-36).
    D. Uranium Oxides (37-38).
    E. Tungstenum Compounds (39).
    F. Titanium Oxides (40).
    G. Alumina and Aluminates (41-42).
    H. Silica and Silicates:
        (1) Quartz group (43).
        (2) Group of Basic Silicates (44-51).
        (3) Group of Pyroxenic Silicates (52-54).
        (4) Group of Chrysolitic Silicates (55-56).
        (5) Group of Feldspathic Silicates (57-59).
        (6) Group of Calcareo-Feldspathic Silicates (60-64).
        (7) Group of Nephelitic Silicates (65-66).
        (8) Group of Zeolitic Silicates (67-76)·
        (9) Group of Micaceous and Chloritic Silicates (77-81).
        (10) Group of Talcose Silicates (82-83).
        (11) Group of Kaolinic Silicates (84-85).
        (12) Group of Copper and Nickel Silicates (86-87).
    I. Carbonates:
        (1) Group of Anhydrous Carbonates (88-93).
        (2) Group of Hydrous Carbonates (94-95).
    K. Sulphates (96-102).
    L. Phosphates and Arseniates (103-105).

IV.—FLUORIDES AND CHLORIDES:
    A. Fluorides (106).
    B. Chlorides (107).

V.—BODIES OF ASSUMED ORGANIC ORIGIN:
    A. Oxalates (108).
    B. Carbonaceous substances (109-113).

I.—SIMPLE SUBSTANCE.

[This group includes the Native Non-Metallic Elements and Native Metals of Canadian occurrence. Three of these, *Graphite*,—often termed Plumbago or " Black Lead," but consisting essentially of carbon,—*Native Gold*, and *Native Silver*, are entitled to rank amongst the economic products of the country ; and *Native Copper* may eventually perhaps be added to the list. The rest occur in small quantities only, or under more or less obscure conditions.]

A.—NATIVE NON-METALLIC SUBSTANCES.

1. *Graphite* (Plumbago) :—Iron-black or dark steel-grey, with black lustrous streak, and metallic or sub-metallic aspect. Found occasionally in tabular hexagonal crystals, but more commonly in small scales, and in foliated and granular masses, which soil the hands, and leave a dark metallic trace on paper. Very sectile, and greasy or soapy to the touch. H = 1.0 — 2.0; sp. gr. 2.0 — 2.3 in pure specimens, but sometimes as high as 2.5. BB, quite infusible, and not dissolved by borax or ordinary fluxes. Consists essentially of carbon, with a variable amount of intimately intermixed siliceous or ferruginous matter, the so-called " ash." This, which becomes visible when the carbon is burnt off by long continued ignition, may vary from a mere trace to 40 or 50 per cent. The actual amount of ash scarcely affects the value of the plumbago. Samples holding 40 or more per cent. may possess as much marketable value as others in which no more than 8 or 10 per cent. is present. But a great deal depends on the composition of the ash, at least as regards certain uses. If the ash contain more than a very slight amount of lime or magnesia, the graphite is scarcely suitable for the manufacture of crucibles. A selected sample, from Buckingham, on the Ottawa, shewed the following composition :

Carbon 80.12	Silica............................	12.86
	Alumina.......................	4.33
Ash...........18.58	Iron Oxide	1.07
	Lime........	0.16
	Magnesia.....................	trace
Moisture...... 1.30	Loss......	0.18

Another sample yielded : moisture 1.14, ash 22.06, carbon 76.80.

In the form of small scales and flaky masses, graphite is widely disseminated throughout the area occupied by the Laurentian series of rocks (Part V.) It occurs most commonly in the beds of crystalline limestone of this series ; but sometimes also in the gneissoid strata, where it appears occasionally to replace the mica of these rocks. It occurs also in large flakes in some of the beds of iron-ore associated with the Laurentian limestones, as at Hull, on the Ottawa. In other places, graphite forms large lenticular masses, or actual beds a foot or more in thickness, in these limestones. Occasionally also, it occurs in the form of distinct veins, traversing different strata of the Laurentian series. The more important localities comprise, the townships of Buckingham, Lochabar, Petite Nation, and Grenville, on the left bank of the Ottawa, where this useful mineral occurs in comparative abundance, and is more or less largely worked. Other localities comprise, more especially, the township of Burgess in Lanark county, and Loughborough and Bedford in Frontenac ; but small quantities are met with in almost every locality in which crystalline limestone occurs. Graphite is found also in thin coatings and finely disseminated scales amongst many of the altered slates of the metamorphic region south of the St. Lawrence (See Part V.), as in Melborne, Shipton, and elsewhere, but nowhere in workable quantities. The chief employment of graphite or plumbago is in the manufacture of drawing pencils, and refractory crucibles, the common kinds and refuse being used as a polishing material for stoves, grates, &c. It is also occasionally employed to remove friction in machinery.

2. *Sulphur :*—Normally, in Ortho-Rhombic crystals (chiefly acute rhombic octahedrons), and in granular masses of a yellow or yellowish-grey colour. H = 2.5 or less ; sp. gr. 2.0. Inflammable, burning with blue flame and sulphurous odour, and melting into brownish-yellow drops which become pale-yellow on cooling.

In Canada, sulphur occurs very sparingly in the simple state : chiefly as an efflorescent crust on specimens of decomposing pyrites from Lake Superior, and elsewhere. It is also occasionally deposited as an incrustation from springs containing sulphuretted hydrogen. In this condition, mixed with carbonate of lime, it occurs in the Township of Charlotteville, (Lot 3, Con. 12,) Norfolk County, Ontario. It is found also here and there, as first pointed out by Dr. Bigsby, in the form of minute crystals, and in earthy coatings, on some of the lower thin-bedded limestones around Niagara Falls.

B.—NATIVE METALS.

2. *Native Gold:*—Golden yellow ; malleable: Regular in crystallization, but occuring chiefly in small granular or leafy particles imbedded

in quartz or other rock-matters, or in the form of small nuggets or fine grains mixed with sand and gravel. $H = 2.0 - 3.0$; sp. gr. $15.5 - 19.5$ according to purity : usually about 16 to 17.5. BB, easily fusible, but not oxydizable or otherwise affected. Insoluble in nitric acid, but soluble in aqua regia.

Native gold is always alloyed with a small amount of silver, by which its colour is rendered paler, and its specfic gravity lowered. The average amount of silver in specimens from the Eastern Townships is about 12 p. c., or from 10 to 15 p.c In the gold from the Hastings district, it appears to vary from about 2 to 10 p.c. ; whilst in much of the gold from Nova Scotia, it does not exceed 2 or 3 per cent.

As regards Ontario and Quebec, gold occurs in rock formations of three distinct ages. First, in quartz veins or bands in the Laurentian Series .* more especially in the Townships of Madoc, Marmora and Elzevir, in the County of Hastings, in Ontario. Secondly in veins—mostly of quartz intermixed with ferruginous calcspar or dolomite—in the Metamorphic Series of the Eastern Townships of the Province Quebec, south of the St. Lawrence (as well as in altered strata of the same general age in Nova Scotia) ; and thirdly, in gravel and other detrital accumulations of Post-Cainozoic age, or in part apparently of somewhat older date. These latter deposits occur chiefly at the base of the Drift-Formation (see Part V.) throughout the Eastern Townships and adjacent region generally. They usually yield, by washing, a considerable residuum of black ferruginous sand, with which the gold is intermixed—sometimes in nuggets weighing several ounces, but more commonly in very minute grains. The sands of most of the streams and rivers which traverse this district are, thus, more or less auriferous. The St. Francis, Chaudière, Famine, Metgermet, Du-Loup, Guillaume or Des-Plantes, and Gilbert or Touffedes-Pins, may be mentioned more especially in this connexion. A good deal of alluvial gold has been taken out of cracks and hollows in the slaty rocks forming the bed of these rivers, as at the Devil's Rapids on the Chaudière ; also on small streams near Ste. Marie and St. George and elsewhere. The gold-bearing veins of this district have been noticed chiefly in Vaudreuil, Aubert-Gallion, and Linière,

---

* The characters and relations of the various rock groups referred to in this Division, are ully described in Parts III and V.

in the County of Beauce ; St. Giles, in Lotbinière County ; and Leeds, in Megantic (Nutbrown's location), &c. The gold is distributed very irregularly throughout the veinstone, some samples yielding upwards of $100 per ton, and others nothing, or a mere trace (See a valuable Report by A. Michel and Dr. T. Sterry Hunt : Geological Survey of Canada, 1866).

In the older Laurentian area of Hastings and adjoining district, in Ontario, the gold occurs only in quartz or quartzo-dolomitic veins or bands in gneissoid strata. Most of these bands carry auriferous mispickel and pyrites, the so-called " free gold " being comparatively rare ; but in certain localities, as at the Richardson and some other mines in the immediate vicinity of Eldorado in Madoc, in the 2nd and 9th concessions of Marmora, and in parts of Elzevir, some rich shews have been obtained. Up to the present time, however, gold-mining in this region has met with but very partial success.

The presence of gold in Arsenical and Iron Pyrites, Blende, &c., will be referred to in the descriptions of these minerals. Auriferous varieties occur more esepcially in Hastings, and in veins on the north-west shore of Lake Superior, as well as in the Eastern Townships. Samples of Copper and Iron Pyrites mixed with much rock-matter, from the Lake Superior region, yielded the writer amounts of gold corresponding to nearly an ounce troy in the ton of 2,000 lbs. ; and some rich samples of crystalline mispickel from Marmora held nearly seven ounces per ton.

4. *Native Platinum :*—Tin-white or greyish-white. In small loose grains or scales. Sp. gr. 16 — 20. Infusible. Insoluble in nitric acid. Occurs very sparingly with native gold in the sands of the Rivière du Loup, and in some of the other iron-sands of the Eastern Townships, Province of Quebec, accompanied in places by steel-grey grains of Irid-Osmium.

5. *Native Silver :*—Metallic-white, but usually with dark surface-tarnish. Regular in crystallization, but found chiefly in small granular, leafy, or filiform masses, usually imbedded in quartz or calcspar. Malleable. H = 2.5 — 3.0 ; sp. gr. 10 — 11. BB, easily fusible, but not otherwise altered. Readily dissolved by nitric acid. A white curdy precipitate of chloride of silver, is thrown down from the solution by hydrochloric acid, or solution of any chloride, as common salt. The precipitate blackens on exposure to light, and is readily

soluble in ammonia : characters which distinguish it from chloride of lead.

Native silver occurs in a broad vein of calc spar at Prince's Mine, Spar Island, and on the adjacent main land, on the north-west shore of Lake Superior. It is associated at this spot with blende, galena, amethyst, quartz, &c., and contains, according to Dr. Sterry Hunt, a small amount of gold ; but the mine has been prematurely abandoned. East of this location, around Thunder Bay, several broad veins occur, in which native silver has been found in still larger quantities. The veinstone consists in part of amethystine and colourless quartz, and partly of crystalline calc spar, accompanied by heavy spar, fluorspar, blende, galena, and pyrites. The silver is also associated here and there with silver-glance or black sulphide of silver. It does not appear to contain gold. Silver Islet, near Thunder Cape, is one of the more remarkable of these localities, but the accessible portion of the vein at this spot appears to be now worked out. This metal occurs also in the native state, but in sparing quantities, associated with copperglance in a calcspar and quartz vein on the Island of Saint Ignace ; and with native copper on the Island of Michipicoten, further east. Native silver has likewise been seen occasionally, in small filaments, among the copper ores of the Acton Mine, in the Province of Quebec.

The occurrence of silver in galena, blende, pyrites, and other minerals, will be noticed under the descriptions of these substances.

6. *Native Copper :*—Copper-red ; malleable ; Regular in crystallization, but occurring generally in arborescent groups of minute indistinct crystals, or in masses of irregular form. $H = 2.5 - 3.0$; sp. gr. $8.8 - 8.95$. BB, easily fusible into a shining globule which becomes covered, on cooling, with a coating of black oxide. Readily soluble in nitric acid. The diluted solution is rendered intensely blue by addition of ammonia.

Native copper, although so abundant on the south shore of Lake Superior, has not been found, as yet, very abundantly in Canada. It occurs, however, in many of the amygdaloidal traps and greenstones, of the Upper Copper-bearing series of the north and east shores of the lake, associated with prehnite, epidote, chlorite, &c. Here and there it has been obtained in irregular masses of the weight of several pounds ; but it occurs most commonly scattered through the

6

trap in small grains which frequently present a rounded or semi-fused appearance. The principal localities comprise Battle Island and the Islands of St. Ignace and Michipicoten ; also Maimanse and Cape Gargantua. According to the Report for 1863 of the Geological Survey, Native Copper occurs likewise in thin plates in red shales of the Quebec series, on the Etchemin River, below St. Henri, and at Point Levis, opposite Quebec ; as well as in a kind of amygdaloidal greenstone underlying these shales at St. Flavien, in the same district. It is stated to have been found, moreover, in small dendritic and other masses, accompanying copper pyrites, apatite, and a silvery-white mica, in a quartz vein in the Township of Barford.

7. *Native Lead :*—Lead-grey ; soft and malleable. BB, fuses easily, and becomes gradually volatilized, coating the charcoal with a yellow ring of lead oxide.

Native lead is of very rare occurrence. The only specimen discovered in Canada, is in the form of a thin string in colourless quartz. It was obtained by Mr. McIntyre of Fort William, Lake Superior, from the vicinity of the Kaministiquia, Thunder Bay. As the quartz contains a few scales of specular iron ore in a perfectly normal condition, it is evident that the lead cannot have arisen from the reduction of galena by the action of heat.

8 *Native Bismuth :*—Silver-white with reddish tinge, but usually tarnished. Sectile, but not malleable. Hemi-hexagonal in crystallization, but commouly in small masses of lamellar structure. $H = 2.0 - 2.5$; sp. gr. about 9.7. BB, melts easily and volatilizes, coating the charceal with yellow oxide. Soluble in nitric acid ; the solution yields a white precipitate of bismuthic oxide on the addition of water in excess.

The only examples of Native Bismuth hitherto met with in Canada, were recognized by the writer in some rolled pieces of quartz, obtained from near Echo Lake, on the north-west shore of Lake Huron.

9. *Native Antimony ;*—Tin or greyish-white. Brittle. Chiefly in small masses of lamellar or fine granular structure. $H = 3.0 - 3.5$; sp. gr. 6.65 — 6.75. BB, melts and volatilizes, tinging the flame pale-green, and depositing a copious white crust on the charcoal. The only known occurrence of Native Antimony in Canada, is in the Eastern Township of South Ham (lot 27 of first range), where, mixed with antimony glance, &c., it forms several narrow veins in a clay slate of the Quebec Group.

APPENDIX TO GROUP I.

10. *Meteoric Iron:*—Dark steel-grey; malleable; strongly magnetic; H = 4.5 ; sp. gr. about 7.4 ; fracture, hackly.   BB, infusible.

An irregular mass of malleable iron weighing about 750 lbs. was discovered in 1854, on the surface of the ground, in the Township of Madoc.   Its examination by Dr. Sterry Hunt showed the presence of 6.35 per cent of Nickel, with other characters belonging to ordinary examples of meteoric iron.   It exhibits a dark coating of oxide, and contains a small amount of intermixed phosphide of iron (Schreibersite) and magnetic pyrites.   Nitric acid brings out on the polished surface the so-called Widmannstädt's figures, or intersecting lines and zigzag markings indicative of an irregular crystalline structure.

II.—ARSENIDES AND SULPHIDES.

[This sub-division contains the various compounds of arsenic and sulphur with metallic bases, hitherto found in Central Canada.   These may be conveniently described under five groups, as follows :—Sulphides of Silver, Lead, and Zinc; Sulphides of Copper; Arsenides and Sulphides of Nickel and Iron ; Sulphide of Molybdenum ; and Sulphides of Bismuth and Antimony.]

A.—SULPHIDES OF SILVER, LEAD, AND ZINC.

11. *Silver Glance or Argentite :*—Black, or dark lead-grey ; malleable ; Regular in crystallization, but occurring commonly in small irregular masses, or in leafy or delicate arborescent forms.   H = 2.0 — 2.5 ; sp. gr. 7.2 — 7.4.   BB, melts with bubbling, and yields a globule of metallic silver.   100 parts contain normally : Sulphur 12.90, Silver 87.10.   Hitherto, only found with native silver, &c., at Prince's Location, Lake Superior, and in the silver veins of Thunder Bay. At the " Withers Mine," at a depth of nearly sixty feet from the surface, several crystals, combinations of cube and octahedron, measuring the fourth of an inch across, were obtained by the writer ; and some others of still larger size were found in the same shaft by Mr. McIntyre of Fort William,   One of these (sp. gr. 7.31) yielded : sulphur 13.37 ; silver 86.44 ; copper, slight trace.   The adjacent mine of the Thunder Bay Company has also furnished some good specimens.

12. *Galena :*—Lead-grey ; more or less sectile, but not malleable ; Regular in crystallization, and often met with in cubes (fig. 36)

and in combination of the cube and octahedron (fig. 37), and other related forms : also in irregular masses, mostly with well-marked lamellar structure. Cleavage cubical and easily effected. H = 2.5 ; sp. gr. 7.2 — 7.7. BB, decrepitates (as a general rule) and becomes reduced to metallic lead. The

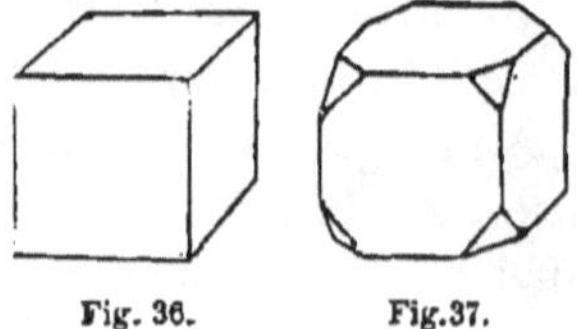

charcoal is entrusted partly with a yellow ring of lead oxide, and beyond this, with a white deposit of mixed sulphate and carbonate of lead. 100 parts of galena contain: sulphur 13.4, lead 86.6 ; but a minute portion of the sulphide of lead is almost invariably replaced by sulphide of silver. In most Canadian samples however, the amount of silver does not exceed ten or twelve dwts. in the ton, and is consequently insufficient to defray the cost of extraction. The known or reported exceptions to this statement are mentioned below.

Galena, as a mineral, is very widely distributed throughout Canada : both in veins, and in small crystalline masses, &c., scattered through rocks of various kinds, more especially in metamorphic and other limestones or dolomites. It is thus present in almost every mineral vein on the north shore of Lake Superior, in association with zinc blende, copper and iron pyrites, &c. Also, here and there, throughout the wide Laurentian area between the northern lakes, and the Ottawa ; in the limestones and dolomites of the Niagara and other formations in Ontario ; in the dark calcareous shales around Quebec ; in the metamorphic region of the Eastern Townships ; and in the limestones of Gaspé. Some of the more special localities comprise : —Prince's Location, Lake Superior ; the silver vein of Thunder Bay and Thunder Cape ; many veins holding copper pyrites, &c., north of Thunder Bay ; the region around Black Bay, where it is associated with auriferous copper-pyrites in broad veins ;* the district north of Garden River, between Lake Superior and Lake Huron, where it holds a considerable amount of silver ; in well-defined veins of much

---

* A surface-sample obtained personally from a quartzose vein in the Upper Copper bearing rocks of this district, gave me: 47.56 per cent. metallic lead, 8.10 per cent. metallic copper (another sample gave 11.62 per cent.), with an amount of gold equivalent to 16 dwts. 13 grs. per ton of 2000 lbs. of ore, and 2 oz. 12 dwts. of silver. The amount of gold in different samples varied from 14 to 19 dwts. per ton, according to the amount of pyrites. This vein is about 10 feet wide, and carries in its centre a solid lode at least 4 feet in width, of a mixture of copper pyrites and galena.

promise, with gangue of highly crystalline calc-spar heavy spar, in gneiss, in the Township of Galway, Peterborough County, and in the adjoining Township of Sommerville; in Lake, Tudor, Limerick, and Marmora, when numerous veins occur in gneissoid strata, * in the Township of Loughborough in Frontenac in broad veins, traversing gneiss and crystalline limestone; under similar conditions in Bedford in the same county; in Lansdowne, Leeds County; and Ramsay, in Lanark County. Galena occurs also in narrow, deceptive, gash veins (see under "Mineral Veins" in Part III.) in the Niagara dolomites of Mulmur (Simcoe County), Eramosa (Wellington County), and Clinton (Lincoln County).

In the Province of Quebec, this mineral occurs especially in the copper-ore veins of the Eastern Townships, as in Acton, Upton, and Ascot, and in many of the quartz veins of the Chaudière valley. Galena, apparently in workable quantities has also been noticed by the Geological Survey at Gaspé Cove and Indian Cove, near Cap^e Gaspé (Report, 1863, p. 400). Argentiferous galena (properly so-called) occurs, according to Dr. Sterry Hunt, at the St. Francis Rapids on the Chaudière, associated with Arsenical Pyrites and Blende, and at Moulton Hill, near Lennoxville. The actual amount of silver appears to vary greatly, probably from intermixed particles of native silver. Three *dressed samples* from the Chaudière yielded respectively —32 oz., 256 oz., and 37 oz., per ton of 2240 ℔s. A dressed sample from Moulton Hill yielded 65 oz. per ton. Other argentiferous varieties are reported to occur on Lake Superior (Meredith's Location, Maimanse, and elsewhere), but the silver, found in some of these may be due to intermixed scales and filaments of native silver and silver-glance.

13. *Zinc Blende or Sphalerite :*—Lustre, sub-metallic or resinous. Colour (in Canadian examples) brown, black, yellow, &c. ; streak, mostly pale-brown. Regular in crystallization, but occurring commonly in small irregular masses, or indistinct crystals, with well-marked lamellar structure. H = 3.5 — 4.0 ; sp. gr. 3.9 — 4.2. BB, infusible, or fusible on the edges only; but when strongly ignited with carb. soda on charcoal, it yields a white incrustation of zinc oxide, which assumes a green colour when moistened with nitrate of

---

* Some of these veins are apparently cut off, at a comparatively slight depth, by the walls coming together, and their working has beeen thus abandoned; but if the sinking were continued, they would probably be found to open out again.

cobalt and then subjected to ignition (see Part I, p. 35). Warmed, in powder, with hydrochloric acid, it emits an odour of sulphuretted hydrogen. Some of the yellow blendes emit a phosphorescent light when scratched or broken. 100 parts contain (normally) sulphur 33, zinc 67 ; but in the dark varieties a certain amount of iron is always present, and many specimens contain a small percentage of cadmium, manganese, &c.

This mineral occurs with galena in almost all the localities given in the description of that substance, (see under No. 12, above), and lately it has been found in large quantities north of Thunder Bay. Brown and yellow varieties are scattered through all the silver-bearing veins of Thunder Bay, and some of the latter have yielded traces of gold, not exceeding, however, 2 dwts. in the ton. Small crystalline masses and grains occur also in most of the lead veins of Peterborough, Frontenac, Hastings, &c., and some of a wax-yellow colour are occasionally seen in fossil shells, or associated with gypsum in small cracks and cavities in the limestone beds around Niagara Falls, as well as in the older limestones of Kingston, Montreal, &c. Zinc Blende is seen likewise in many of the veins of the Eastern Townships, as in the valley of the Chaudière, and elsewhere. An auriferous variety is stated by Dr. Sterry Hunt to accompany argentiferous galena, &c., in a quartz vein at the St. Francis Rapids on the Chaudière.

B.—SULPHIDES OF COPPER.

14. *Copper Glance :*—Dark lead-grey, often with blue or green tarnish ; streak, black and slightly shining. Crystallization Rhombic, but the crystals have mostly a pseudo-hexagonal aspect. Found commonly, however, in small granular or other masses. H 2.5 — 3.0 ; sp. gr. 5.5 — 5.8. BB, melts with strong bubbling or spitting, colours the edge and point of the flame green, and yields a globule of metallic copper covered by a dark scoria or crust. One hundred parts contain : Sulphur 20.2, Copper 79.8.

This ore, often termed " vitreous copper ore " (although its lustre is perfectly metallic), occurs in small quantities in many of the mineral veins of lake Superior and Lake Huron : as on Spar Island, Pigeon River, St. Ignace, Point Porphyry, Michipicoten, Point-aux-Mines, Batchewahning Bay, Echo Lake, Bruce Mines, &c. It occurs also in many of the copper-ore veins of the Eastern Townships, as in Leeds (at the Harvey Hill and other mines), Halifax, Sutton, Brome, Shefford, Stukely, Brompton, Acton, Melbourne, Cleveland, &c. Also reported from Côteau St. Geneviève, near Quebec.

15. *Purple or Variegated Pyrites* (Bornite, Erubescite, Horse-flesh Ore) :—Pale brownish-red, but always presenting a rich purple or variegated tarnish ; streak, greyish-black. Regular, but rarely crystallized ; mostly in irregular masses. Brittle. H = 3.0 ; sp. gr. 4.5 — 5.5. BB, fusible into a dark magnetic globule. Composition somewhat variable, but averaging : Sulphur 25, Copper 60, Iron 15. A sample from Lake Huron gave the author : Sulphur 24.03, Copper 63.19, Iron 11.86.

This valuable mineral (the " horse-flesh ore" of miners) occurs in large and small masses, imbedded in, or scattered through, many of the altered strata of the Eastern Townships; and also, though less abundantly, in quartz veins traversing these strata. Some of the more important localities comprise the celebrated Acton mine in Acton Township, the Halifax mine in the township of the same name, Sweet's mine in Sutton, Cold Spring mine and Balrath mine in Melbourne, the St. Francis mine in Cleveland, and the Harvey Hill mine in Leeds; but it occurs also in other parts of these townships, as well as, more or less, throughout the entire district, associated most commonly with the ordinary or yelloy pyrites, and frequently with earthy malachite, copper glance, native copper, galena, &c. The country rock is usually a dolomitic limestone, or a chloritic or micaceous slate. See further, under Copper Pyrites, below.

In other parts of Canada, this ore occurs but sparingly. It has been found at the Wellington and Bruce mines on Lake Huron ; and in veins at Point-aux-Mines, Maimanse, and elsewhere, on Lake Superior. Lake Huron specimens sometimes exhibit pseuodmorphs (tetrahedrons of the Tetragonal System after Copper Pyrites.

16. *Copper Pyrites* (Chalkopyrite) :—Brass-yellow, often with variegated tarnish ; streak, dark green, or greenish-black. Tetragonal in crystallization, but commonly found in irregular masses. Brittle. H = 3.5 — 4.0 ; sp. gr. 4.1 — 4.3. BB, melts into a dark magnetic globule ; after roasting, yields, with carb. soda, metallic copper. One hundred parts contain : Sulphur 34.9, Copper 34.6, Iron 30.5

This is the common ore of copper. It is familiarly known as " yellow copper ore." It occurs in small quantities, both in veins and in scattered masses, among the Laurentian strata of various localities :

more especially in the townships of Lake, Madoc, Elzevir, Hungerford, &c., in the County of Hastings; North Burgess in Lanark; Escott and Bastard in Leeds, and throughout the gneissoid region generally between the Ottawa and Lake Huron. The accompanying veinstone is mostly calcspar, but in some places it consists of quartz, or is of a granitic nature. Specks of galena, blende, and iron pyrites, usually accompany the copper ore. This mineral has been found also in calcspar veins traversing gneiss in Kildare, Joliette County, in the Province of Quebec.

In the Huronian strata, this ore is far more abundant. Numerous veins, with quartz gangue, occur on the north shore of Lake Huron. Many of these veins carry workable quantities of copper pyrites, accompanied in most cases by small portions of variegated pyrites, and also by copper glance, iron pyrites, &c. The best known are those of the Bruce and Wellington Mines; but others occur at Copper Bay, White Fish River, Spanish River, Garden River, Root River, Echo Lake, and elsewhere in that district. Very large quantities have lately been found in the vicinity of Sudbury.

Copper Pyrites occurs also in many localities on the east and north shores of Lake Superior, in veins traversing strata apparently of Cambrian age (see Part V). These are known as the Copper-Bearing series of Lake Superior. Among other localities may be enumerated : Bachewahning Bay, Maimanse, Point-aux-Mines, Mica Bay, Black River, Black Bay, Thunder Bay, and locations between Thunder Bay and Dog Lake on the Kaministiquia. Some of rhese veins carry but small quantities of ore, but others are exeedingly rich : those especially which occur in the vicinity of Black Bay, and in the country north of Thunder Bay. Samples from these latter districts, collected personally, and others obtained by Mr. S. J. Dawson, have yielded amounts of gold varying from a few dwts. to about an oz. troy in the ton of 2000 lbs. of ore. The gangue of these veins is either quartz, or a mixture of calcspar, heavy spar, amethystine quartz, and fluor spar ; and the copper ore is generelly accompanied by galena, zinc blende, and iron pyrites.

Finally, Copper Pyrites is widely distributed throughout many of the Eastern Townships in the Province of Quebec. In some places, the copper of this region is entirely in the form of yellow pyrites ; in others, chiefly in the state of purple or variegated ore (No 15, above). The more important localities of the yellow ore, lie in the townships of Stukely (Grand Trunk)Mine, &c.), Ely (Ely Mine, &c.), Bolton (Huntington Mine, Ives Mine, &c.), Leeds (Harvey Hill Mine, &c.), Halifax (Black Lake Mine), Inverness, Tringwick, Chester, Ham, and others. Also in the townships of Ascot (Ascot Mine, Belvidere Mine, Lower Canada Mine, Albert Mine, Capel or Eldorado Mine, Victoria Mine, Marrington Mine, Griffith's Mine, Clark Mine, &c.), Sutton, Brome, Melbourne (Coldstream M., Balrath M.), and Cleveland.* Copper Pyrites occurs also in true veins in this district. as at the Harvey Hill and Nutbrown mines in Leeds, as well as in Inverness, and elsewhere.

C.—ARSENIDES AND SULPHIDES OF NICKEL AND IRON.

17 *Arsenical Nickel Ore :*—Pale copper-red, with dull greyish tarnish. Hexagonal in crystallization, but mostly in irregular masses. Brittle. H = 5.0 — 5.5 ; sp. gr. 6.7 — 7.3· BB, emits a strong odour of garlic, and melts into a dark (sometimes magnetic) globule. One hundred parts contain: Arsenic 56, Nickel 44, but some of the nickel is commonly replaced by iron, and sometimes by cobalt.

The above characters are those of the ore in its normal state. In Canada, this ore, however, has only been found in admixture with other metallic compounds. A mixture of this kind, in small nodular masses associated with calcspar, occurs in amygdaloidal trap on Michipicoten Island, Lake Superior. The amount of nickel according to analyses by Dr. Sterry Hunt and Prof. Whitney, varies from about 17 to 37 per cent. The colour of this variety is between tinwhite and bronze-yellow: sp. gr. 7.3 — 7.4. The composition indicates a mixture of arsenides of nickel with arsenides of copper (Domeykite).

Another nickleferous compound of a steel-grey colour apparently a mixture of arsenide and sulphide of nickel with arsenical pyrites, occurs sparingly at the Wallace Mine, Lake Huron. It was first made known by Dr, Sterry Hunt. The surface is commonly covered. more or less, with minute hair-like crystals of nickel and iron sulphates, arising from the partial decomposition of the ore.

18. *Millerite or Sulphide of Nickel :*—Brass or bronze yellow.

---

* A detailed list of all the copper ore localities of the Eastern Townships will be found in the valuable Appendix of the Geological Surrvey Report for 1886.

Hemi-hexagonal, the crystals mostly acicular and very minute; also found in imbedded grains and small globular masses. H = 3.0,—3.5 (but not easily ascertained); sp. gr. 4.6 — 5.6. BB, melts into a dark globule. One hundred parts contain; Sulphur 35, Nickel 65.

Occurs very sparingly, in small specks, with calcspar and minute green crystals of chrome garnet, in the Township of Orford (Lot 6, Range 12), where it was first recognized by Dr. Sterry Hunt.

19. *Magnetic Pyrites* (Pyrrhotine) :—Bronze-yellow, with black streak. Crystal-system, Hexagonal, but crystals very rare; found commonly in granular and irregular masses. H = 3.5 — 4.5; sp. gr. 4.4 — 4.7. Slightly magn_tic, many specimens exhibiting polarity. The magnetism is best shewn by bringing a specimen of some size near a suspended needle. As a general rule, a bar or horseshoe magnet will only take up very small particles. BB, emits sulphurous fumes, and melts into a dark slag-like mass. By roasting, becomes very easily converted into red oxide. Soluble in hot hydrochloric acid, with emission of sulphuretted hydrogen odour. One hundred parts yield, on an average, Sulphur 39.5, Iron 60.5; but many varieties contain 3 or more per cent. of nickel, replacing part of the iron. A variety from Madoc, (lot 10, con. 2), yielded the writer: Sulphur 39.88, Iron 59.56, and contained no trace of cobalt, nickel, or gold.

Occurs in veins and irregular beds among the Laurentian strata north of Thunder Bay, and in other localities a short distance inland from the north shore of Lake Superior. Also, under similar conditions, near Balsam Lake, &c. ; and more or less throughout the Laurentian district between Georgian Bay and the Ottawa. Likewise in a calcspar vein in Portneuf, Province of Quebec; and still more abundantly in St. Jerome, Terrebonne. Magnetic Pyrites occurs also in the higher metamorphic district south of the St. Lawrence, generally accompanying copper ores : as in the Townships of Barford, St. Francis, and Sutton, and at the Ives and Huntington mines in Bolton.

20. *Iron Pyrites* (Cubical Pyrites, Mundic, &c.) :—Pale brass-yellow —often brown on the surface from partial conversion into brown iron oxide ; streak, greyish-black. Regular in crystallization, and frequently found in cubes (usually with striated faces, the striæ on one face running at right angles to those on the adjacent face); also in

combinations of cube and octahedron, in simple octahedrons, pentagonal dodecahedrons, &c. (Figs. 37 — 41.) Found still more frequently in granular, nodular, and other irregular masses. H = 6.0 — 6.5 ; sp. gr. 4.8 —5.2. BB, emits sul-

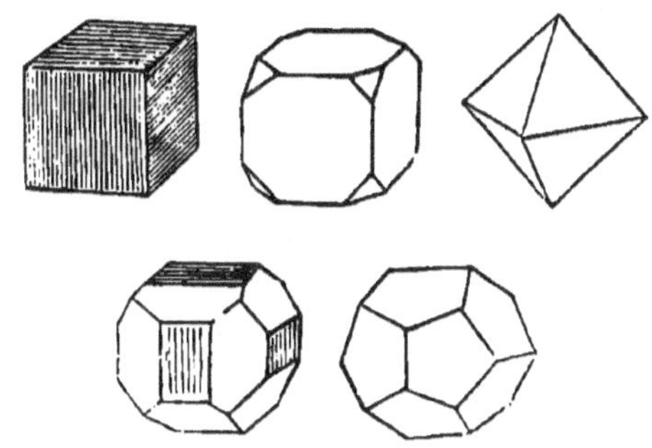

Figs. 37 to 41.

phurous fumes, and melts into a dark magnetic globule. One hundred parts contain : Sulphur 53.3, Iron 46.7, but a small portion of the iron is occasionally replaced by cobalt or nickel. Many varieties, also, contain traces of both gold and silver. (In this connection, it may be observed that a percentage of 0.01 is equivalent to 2 oz. 18 dwts. 8 grs. (troy) in the ton of 2000 lbs., or to 3 oz. 5 dwts. 8 grs. (troy) in the British ton of 2240 lbs.)

Iron Pyrites is of exceedingly common occurrence. It is present, more or less, in almost every mineral vein ; and occurs also, in crystals, grains, and irregular masses, in rocks of all ages and of various kinds. It sometimes forms the substance of organic remains, as in examples of Trilobites, &c., from the Utica Slate of Whitby and other localities. In this condition it arises most probably from the alteration of carbonate of iron.

In the Laurentian rocks of North Hastings and adjacent counties, in the copper-bearing series of Lake Superior, and in the altered strata of the Eastern Townships south of the St. Lawrence, auriferous varieties have been noticed ; but the amount of gold in these is scarcely sufficient to defray the cost of its extraction. In Elizabethtown (Lot 19, Range 2), near Brockville, and elsewhere in this vicinity, some large beds or veins of a cobaltic variety occur. Large veins occur also in Clarendon, on the Ottawa ; in Terrebonne and Lanoraie ; in Hastings, and throughout that district ; as well as on the north shores of Lakes Huron and Superior. Extensive deposits are likewise seen in some of the Eastern Townships (Garthby, Ascot, &c.)— all of which are likely to become available at no distant day, in the manufacture of sulphuric acid. Cubical crystals of large size occur in a copper-ore vein, on Lot 8, Range 1, in Melbourne Township. Small but very symmetrical octahedrons are obtained occasionally from the thick-bedded Trenton Limestone on the Bay of Quinté, near Belleville. Cubes, pentagonal dodecahedrons, and other crystals, occur in many of the veins and gneissoid rocks of Madoc, Elzevir, Tudor, &c. Occasionally also, well crystallize examples are seen in the veins, and also in the trap dykes, of Lake Huron and Lake Superior; and small brilliant crystals occur in the white compact trachyte of Montreal. Finally, it may be mentioned, without attempting however to name all the localities of this mineral in Canada, that peculiar nodular or concretionary masses occur in the shales of the Island of Orleans, and elsewhere near Quebec; and in the more modern bituminous shales of the Portage Group, at Cape Ibberwash or Kettle Point, Lake Huron.

21. *Prismatic Pyrites or Marcasite* (Radiated Pyrites, Cockscomb Pyrites, &c.) :— Light brass-yellow; Rhombic in crystallization, the prismatric crystals mostly in radiated aggregations, or united in rows, as in Fig. 42. Composition and other characters as in the common or cubical pyrites, the two

Fig 42.

minerals thus presenting an example of Dimorphism—*i.e.*, the assumption of two distinct sets of forms by the same chemical substance. The prismatic species is especially subject to decomposition, yielding iron vitriol.

The occurrence of prismatic pyrites in Canada was first made known by the author, who met with it in 1865 in a quartz vein (carying copper pyrites, galena, heavy spar, &c., together with examples of cubical pyrites), in the remote Township of Neebing, a few miles east of the Kaministiquia River, on the north west shore of Lake Superior.* Other examples have come under his notice on subsequent visits to this district, from some of the silver-bearing veins of Thunder Bay; and he has obtained recently a large and fine specimen from a vein in Laurentian rock in the Township of Hinchinbrook, in Frontenac County. Many of the spherical masses of pyrites with radiated structure and crystallized surface, it should be observed, though commonly referred to Marcasite, belong really to the cubical species.

---

* Lot 25, Con. 5. Canadian Journal, 2nd Series, Vol. X., 408.

22. *Arsenical Pyrites or Mispickel:*—Colour between silver-white and pale steel-grey, often obscured by yellowish or pale-blue tarnish; streak, greyish-black. Crystallization, Rhombic : the crystals mostly small and short rhombic prisms, terminated by two nearly flat and striated planes (Fig. 43). Occurs also in granular and irregular masses. $H = 5.0 -$

Fig. 43.

6.0; sp. gr. $6.0.—6.4.$ BB emits a strong odour of garlic, and melts into a dark magnetic globule. A garlic-like odour is also more or less perceptible when the mineral is broken by a smart blow. One hundred parts contain : Sulphur 19.6, Arsenic 46.0, Iron, 34.4 ; but a small portion of the iron is occasionally replaced by cobalt.*

This mineral is useless as an ore of iron, but it serves for the production of arsenious acid, the " arsenic " or " white arsenic " of commerce, and it frequently contains minute portions of gold. In Central Canada, it occurs in the Laurentian strata of Marmora and Tudor. Specimens from Marmora have yielded the author amounts of gold ranging from 1 oz. to over 6 ounces in the ton of 2000 lbs.† In Tudor, small crystals of mispickel ‡ accompany Bismuth Glance. The copper-ore veins of the Huronian rocks, also show here and there small crystals and granular masses of this mineral, as at the Bruce and Wellington Mines ; and it occurs in small quantities in some of the argentiferous veins of the Upper Copper-bearing Series around Thunder Bay, Lake Superior. The altered rocks of the Eastern Townships, south of the St. Lawrence, likewise contain it in places, as near the Chaudière Rapids in the County of Beauce,

---

* In this case, the roasted ore when fused with borax will import a more or less decided blue colour to the glass. For details respecting this and other blowpipe processes and reactions see Part 1.

† An amount of this kind, it will of course be understood, although rendering the ore of great commercial value, does not practically affect the normal composition of the mineral. One ounce per ton of 2000 lbs., for example, is equivalent only to a percentage of 0.00343.

‡ Although a reference to minute crystallographic details is opposed to the plan of the present work, it may be stated, here, that these Tudor crystals present the combination shown in the annexed Figure, in which the common brachydome $\frac{1}{2}\bar{\infty}$ is replaced by $\frac{1}{2}\bar{\infty}$ and $\bar{\infty}$. The form $\frac{1}{2}\bar{\infty}$, the summit angle of which equals $118° 30'$, is a comparatively rare form, but it appears to be always present in the cobaltiferous varieties of Mispickel, and in the allied species Glaucodot. The Tudor crystals, as shewn by a blowpipe examination, contain a small percentage of cobalt.

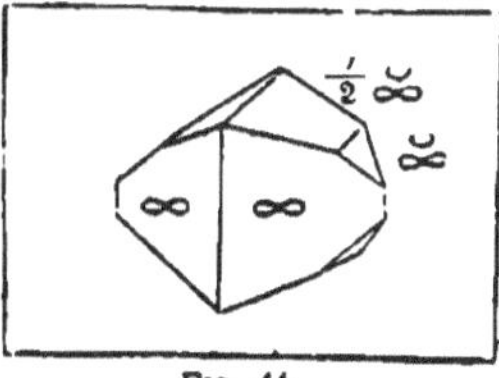

Fig. 44.

where it occurs with argentiferous galena in quartz veins; and also, according to Dr. Sterry Hunt, under similar conditions at Moulton Hill in Lennoxville. In Nova Scotia, mispickel is of exceedingly common occurrence in the gold-bearing quartz bands, and it appears invariably to be more or less auriferous.

D. SULPHIDE OF MOLYBDENUM.

23. *Molybdenite :*—Light lead-grey, with greyish-black metallic streak. Hexagonal in crystallization, but occurring commonly in the form of small scales, or in leafy or fine granular masses. Very sectile; slightly greasy or soapy to the touch, leaving a black trace on paper, and a dull greyish-green trace on smooth porcelain. H = 1.0—2.0 ; sp. gr. 4.4—4.8. BB, imparts (in the forceps) a distinct green coloration to the point of the flame, but remains infusible. In a continued blast on charcoal, however, it deposits a white coating of molybdic acid on the support. Forms with carb. soda an alkaline sulphide (see Part I, p. 44), by which, with other characters, it may be distinguished from Graphite. One hundred parts contain : Sulphur 41, Molybdenum 59.

This mineral is at present of little commercial value.* It occurs in small scales disseminated through many of the crystalline limestones of the Laurentian series, in the Counties of Frontenac, Hastings, Peterborough, Victoria, &c. According to the Reports of the Geological Survey, it has been found in much larger quantity near the mouth of the River Quetachoo, in Manicougan Bay, on the north shore of the Gulf of the St. Lawrence. It occurs also in some abundance at Sea-beach Bay, near Black River, on the north shore of Lake Superior, in several veins, accompanying copper pyrites in quartz. Samples from this locality have yielded nearly 4½ per cent. of molybdenite, or about 100 lbs. per ton of ore, (Can. Jour. Vol. X., p. 409). Terrace Cove is another locality in which molybdenite has been found on Lake Superior. This mineral occurs also in quartz veins at Harvey Hill, in the Township of Leeds, in small rounded masses of fine granular structure associated with copper pyrites and crystallized dolomite.

---

* As Molybdenite is quoted in chemical price-lists at from 50 cents to a dollar or more per lb., an idea is sometimes expressed that it would pay to work, if found in sufficient quantity. Inquiries, however, made in London, Paris, Hamburg, Berlin and other cities, have demonstrated the fact that a very few tons would completely overstock the market.

E. SULPHIDES OF BISMUTH AND ANTIMONY.

24. *Bismuth Glance* :—Light lead-grey, often with yellow or blue-ish tarnish ; streak, black. Rhombic in crystallization, but occurring commonly in lamellar and fibrous masses. $H = 2.0$; sp. gr. about 6.5. BB, melts very readily into a black globule, which gradually volatilizes, with deposition of a yellow ring of oxide (and, beyond this, a greyish-white coating of sulphate) on the charcoal*. A small residum is sometimes left ; this generally shews with borax or phosphor-salt the reactions of copper and iron (see Part I.). Dissolves, with separation of sulphur, in nitric acid. The solution dropped into excess of water forms a milky or opaline liquid. Not affected by caustic potash. One hundred parts of the pure mineral contain : sulphur 18.75, bismuth 81.25.

Bismuth glance is a comparatively rare mineral. It has not hitherto been discovered, at any locality, in sufficient quantity to form a commercial ore. In Canada, it occurs in small lamellar and sub-fibrous masses in a quartz vein, with numerous interpenetrating crystals of black tourmaline, at Hill's Mine, in the rear of Tudor, one of the northern townships of the County of Hastings. Some small samples have also been found near Cornwall, Ont.

25. *Antimony Glance* or *Grey Antimony Ore* :—Light lead-grey, often with dark, or iridescent, tarnish. Rhombic in crystallization, but occurring mostly in fibrous or granular masses. $H = 2.0$, sp. gr. 4.52—4.62. Melts *per se* in the flame of a candle. BB, melts rapidly, and becomes volatilized in dense white fumes, a white oxidized coating being deposited on the charcoal. The point of the flame, if directed on this, is tinged pale blueish-green. A hot solution of caustic potash converts the powdered ore into an orange-coloured compound of similar composition. One hundred parts contain : sulphur 28.2, antimony 71.8.

Of rare occurrence in Canada. Hitherto, found only in small quantities, with iron pyrites and mica, in a band of crystalline dolomite, in the Township of Sheffield (Lot 28, Con. 1), in Addington County ; and in small masses mixed with tremolite, under similar conditions, in Marmora. Also, in radiating fibrous masses with Native Antimony in narrow veins traversing slates of the Quebec Series, in the Eastern Township of South Ham.

---

*All compounds of Bismuth when fused on charcoal with a mixture of potassium iodide and sulphur form a vivid scarlet coating on the support, as first shewn by Merz and Von Kobell.

Note :—A plumbiferous variety of Antimony Glance, apparently a mixture of that ore with Zinkenite or Jamesonite, has been sent to me lately from Belleville, with the intimation that it was obtained in Elzevir.   It forms small fibrous or sub-fibrous masses, intimately mixed with calc-spar, and with numerous acicular crystals of Tremolite, and some massive Hornblende, in quartz.   Partially soluble in caustic potash, hydrochloric acid precipitating orange-coloured flakes from the solution.

26 *bis. Menighinite* :—Lead-grey ; H=2.5 ; sp. gr. 6. 33.   Easily fusible with antimonial fumes.   Contains sulphur, antimony and lead.   Occurs near Marble Lake, Township of Barrie, Ont., (Dr. B. J. Harrington, Trans. Roy. Society of Canada, 1883).

26. *Red Antimony Ore* (Kermesite) :—Dark cherry-red, somewhat lighter in the streak ; lustre adamantine, or approaching semi-metallic.   Monoclinic in crystallization, but occurring almost always in small radiating fibrous tufts, associated with Antimony Glance. H=1.0—1.5, sp. gr. 4.5.—4.6.   BB, melts on the first application of the flame, and becomes rapidly volatilized.   The composition is somewhat remarkable, presenting the union of a sulphur and oxygen compound.   One hundred parts contain : sulphur 19.8, oxygen 4.9, antimony 75.3.

Occurs in small feathery masses, with Native Antimony and Antimony Glance, in the Eastern Township of South Ham.

III. OXYGEN COMPOUNDS.

[This sub-division comprises the various Oxides of natural occurrence, *i. e.* combinations of oxygen with various metals ; and also the ternary oxygen compounds, or so-called oxygen salts, commonly regarded as combinations of an oxygen acid (silicic acid, carbonic acid, &c.) with an oxidized metallic base (lime, magnesia, alumina, iron oxides, &c.).   These latter compounds form the groups of Silicates, Carbonates, Sulphates, and so forth.   See the remarks on Chemical Nomenclature in Part 1., and also the observations prefixed to the various groups below.]

A. COPPER OXIDES.

27. *Red Copper Ore* (Ruby Copper, Ruberite, Cuprite) :—Red, with red streak.   Normally, in Regular crystals (chiefly the octahedron and rhombic dodecahedron) which are commonly converted on

the surface into green carbonate of copper; also massive and earthy. H=4.0 or less; sp. gr. 5.8—6.1. BB, imparts a green colour to the flame, and becomes reduced to metallic copper. One hundred parts contain : Oxygen 11.20, Copper 88.80.

In Canada, this mineral occurs in traces merely, in some of the copper ore deposits of the Eastern Townships (Halifax, Acton, &c). Spots and stains of a more or less bright red colour, are frequently the only indications of its presence. Stains of a similar appearance, are more commonly produced, however, by the weathering of iron ores.

28. *Black Copper Ore* (Melaconite) :—Black, with black streak. Mostly in dull earthy masses. BB, colours the flame green, and yields metallic copper. One hundred parts of the pure mineral contain : oxygen 20.15, copper 79.85. Occurs in traces only in some of the copper ore deposits of the Eastern Townships.

B. IRON OXIDES.

[This group comprises the mineral species which consist simply of oxygen and iron; and those, of a closely related character, in which part of the iron is replaced by titanium or chromium. These species fall into three natural groups : (1) The *Hematite* group, consisting of anhydrous sesqui-oxides (or analogous compounds), Hexagonal, or rather Hemi-Hexagonal, in crystallization; (2) the *Magnetite* group, compounds (apparently) of oxides and sesqui-oxides, Regular in crystallization; and (3), the *Limonite* group, consisting of hydrated sesqui-oxides.

(1) HEMATITE GROUP OF IRON OXIDES.

29. *Hematite (Specular Iron Ore, Red Iron Ore, Red Ochre)* :— This mineral occurs under several more or less distinct conditions, and especially : (1) In Hemi-hexagonal crystals, chiefly groups of modified rhombohedrons, and in lamellar and micaceous masses, with steel-grey colour, often iridescent on the surface, and with strongly marked metallic lustre (= *Specular and Micaceous Iron Ore*); (2) In botryoidal masses of fibrous structure, and in irregular lamellar masses, with blueish or brownish-red colour, and lustre between metallic and semi-metallic (= *Hematite* of old authors, *Red Iron Ore*), and (3), In brick-red, more or less earthy and granular masses (= *Reddle or Red Ochre*). In these varieties, the streak or powder is equally of a red colour. H = 5.5 – 6.5 in the crystals and crystal-

7

line or semi-crystalline masses, but only 1.0 – 2.0 in the earthy and ochreous varieties. Sp. gr. 4.3 – 5.3. BB, becomes magnetic, but on charcoal remains unfused, although a very thin splinter in the forceps may be rounded at the point. One hundred parts contain, normally : oxygen 30, iron 70 ; but many specimens, it should be observed, are intimately mixed with quartz, chlorite slate, or other rock matter, by which the parcentage of iron is much reduced.

This valuable ore occurs in Canada in strata of various periods of formation. One of its more important localities is in the Township of McNabb, in Renfrew, where it forms a bed of about 30 feet in thickness, associated with crystalline limestone of the Laurentian Series, and overlaid by a magnesian limestone of Lower Silurian age. It occurs also in the township of Bristol, and in Templeton and Hull, on the opposite side of the Ottawa. Other Laurentian localities comprise various spots in the counties of Addington, Hastings, Peterborough,* etc. ; and on Iron Island, on Lake Nipissing, where it also occurs in connection with crystalline limestone. In Huronian strata, it has been found near the Wallace Mine on Lake Huron, and still more abundantly on Lake Superior, as in the Bachewahnung District on the east shore of the lake ; on the north side of Michipicoten Harbour : and in widely-extended beds in the vicinity of Pic River : mostly in green, chloritic, pyroxenic, or hornblendic slates. Micaceous and other varieties occur in the metamorphic strata of the Eastern Townships : as in St. Armand, Brome, and Sutton, mostly in chloritic schists, as well as in the auriferous copper-ore veins of Leeds and Halifax. In Silurian strata, hematitic or specular iron ore has been noticed in small quantities in the Potsdam Sandstone of Bastard and Ramsay. Lastly, it may be mentioned, that an earthy impure variety is found in bands and small masses interstratified with the red ferruginous shales of the Clinton or Middle Silurian Series, near Dundas, in Flamborough West.

Note.—Small octahedrons, and related crystals, having the composition of Red Iron Ore, are occasionally found. These form the species *Martite* of some authors, but they are probably due to the alteration of Magnetic Iron Ore. See under that mineral, No. 31.

---

* Special localities in Hastings and Peterborough counties are given in a paper by the author, with analyses of the ores, in Vol. III. of the Transactions of the Royal Society of Canada, 1885

30. *Titaniferous Iron Ore* (Ilmenite, Menaccanite in part) :— Iron-black ; streak-powder, brownish-black to chocolate-brown. Hemi-Hexagonal, but commonly in lamellar and granular masses. When pure, not magnetic ; but sometimes feebly-magnetic, probably from intermixed magnetic iron ore. H = 5 – 6 ; sp. gr. 4.3 – 5.0. BB, like Hermatite ; but the glass formed with phosphor-salt, after exposure to a reducing flame, has a distinctly red colour. Composition, essentially iron, titanium, and oxygen, in variable proportions. The Titaniferous ore from Baie St. Paul, on the Lower St. Lawrence, as deduced from Dr. Sterry Hunt's analysis, contains Titanium 29.63, Iron 36.11, Oxygen 29.10, in addition to 3.60 per cent. of magnesia.

This ore occurs in Canada, in vast beds or masses interstratified with feldspathic rocks of the so-called Labrador or Upper Laurentian Series, at Baie St. Paul, below Quebec. At this locality, it exhibits a peculiar structure : an aggregation of coarse granular concretions composed of irregular lamellæ. Small grains of rutile are scattered in places through the mass. The principal bed is ninety feet in thickness and of great extent, but the ore at present is comparatively useless. This substance occurs also in grains and thin bands in a similar anorthosite or feldspathic rock (see Part III) in the neighbouring parish of Chateau Richer, and likewise under the same conditions in the Township of Rawdon, in Montcalm County. It has been detected also by the officers of the Geological Survey, amongst the iron ores of the metamorphic strata of the Eastern Townships : as in St. Francis, in Beauce County, and in Brome and Sutton.

(2) MAGNETITE GROUP OF IRON OXIDES.

31. *Magnetic Iron Ore or Magnetite* :—Iron-black, with black streak, and in general a sub-metallic lustre. Strongly magnetic, most specimens exhibiting polarity (see under *Magnetism*, Part I). Regular in crystallization, and often found in octahedrous and rhombic dodecahedrous (Figs. 45 and 46), the faces of the latter commonly striated parallel with the position of the edges of a plane of the octahedron. Occurs also still more frequently in lamellar,

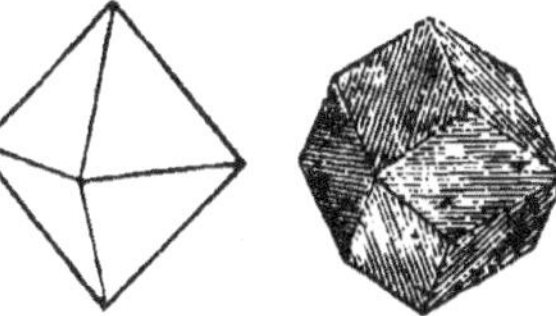

Figs. 45 and 46.

granular, and other masses, sometimes forming large beds or

" stocks." Also in the form of black sand. $H = 5.5 - 6.5$; sp. gr. 4.9 – 5.2. BB, on charcoal, infusible, but a fine splinter in the forceps may be rounded at the point. One hundred parts of the mineral contain : Oxygen 27.6, Iron 72.4 (or, oxide of iron 31.03, sesquioxide 68.07).

This ore, the most valuable of all the ores of iron, occurs in great quantity, and of good quality, in numerous localities of the Laurentian area of Canada. It is usually found in the form of large " stocks " or irregular masses, mostly associated with pyroxene, and in contact, as pointed out by Sir William Logan, with crystalline limestones of the Laurentian Series ; but it occurs also interstratified with gneissoid and schistose strata of the same group, and in grains and small masses scattered through these rocks. Sometimes, likewise, it forms true veins, traversing Laurentian strata. It occurs also in beds amongst the altered Silurian rocks of the Eastern Townships ; and, in the form of sand (usually mixed with Iserine), it belongs to comparatively recent deposits.

The principal or more interesting Laurentian localities lie in the following Townships :—Hull and Templeton in Ottawa County (several beds, one nearly 90 feet in thickness ; the ore, here and there, mixed with layers of hematite, and also with scales of graphite) ; Buckingham, in the same county (in crystalline masses in broad feldspathic veins) ; Wentworth, Grenville, and Grandison, in Argenteuil County ; Ross, in Renfrew County (in reticulating veins in cryst. limestone) ; South Crosby (bed of 200 feet in thickness), and Escott, in Leeds County ; South Sherbrooke, in Lanark County ; Bedford, in Frontenac County ; Madoc, Elzevir, Marmora, Tudor, Wollaston, Faraday, Herschel, etc., in Hastings County (many large and valuable deposits, although intermixed here and there with pyrites) ; Belmont, Cardiff, Monmouth, Glamorgan, Snowdon, Minden, etc., in Peterborough and Haliburton Counties,* forming deposits of great extent. Magnetic Iron Ore in cleavable masses, associated with Hematite, occurs also near the mouth of the Little Pic River, on the north shore of Lake Superior, and minute octahedrons are sometimes observable amongst the layers of hematite from this region.

The Eastern Townships of Sutton, Leeds, Bolton, Orford, &c., likewise possess deposits of magnetite, chiefly in masses and disseminated

---

* Special localities in the Hastings, Peterborough and Haliburton district are given, with analyses of the ores, in a paper by the author in Vol. III. of the Transactions of the Royal Society of Canada.

crystals, as well as in continuous bands, in dolomite, chlorite slate, serpentine, and other magnesian strata. Much of the ore from these localities, however, contains titanium or chromium. Lastly, in the form of black sand, alone, or mixed with Iserine, the ore occurs very commonly on the shores and islands of Lake Superior, Lake Huron, Erie, and Ontario, and on those of many of our smaller lakes. Also, here and there, on the north shore and gulf of the St. Lawrence; and mixed with the auriferous gravels of the Chaudière, St. Francis, Gilbert, and other rivers of the Eastern Townships.

Note :—Magnetite occasionally becomes altered by higher oxidation into *Hematite*, without change of form. The streak is then more or less red, and the magnetism scarcely perceptible. Some small octahedrons (with truncated edges) of this character, the *Martite* of some authors, were observed by the writer in a gneissoid boulder from Bass Lake, a few miles north of Orillia.

32. *Iserine, or Titaniferous Magnetic Ore :*—Black, with black streak and sub-metallic lustre. More or less strongly magnetic. In minute octahedrons, sand grains and pebbles. Other characters like those of Magnetic Iron Ore, but the glass obtained by fusion in a reducing flame with phosphor-salt has always a distinct red or red-brown colour. Composition, essentially, magnetic oxide of iron, with part of the iron replaced by titanium. A small amount of magnesia is also generally present. Forms a certain portion of most of the black magnetic sands of our lake, island and river shores, referred to under No. 31.*

33. *Chromic Iron Ore :*—Black or brownish-black, with, normally, a dark brown streak, and sub-metallic aspect ; but the streak is often greenish or greenish-grey, from the presence of intermixed serpentine or other silicious matter. In general, slightly magnetic: if strongly magnetic, the substance is mixed with magnetic iron ore, and the streak is more or less black. Regular in crystallization, but occurring commonly in irregular masses, mostly of granular structure. H=5.5 ; sp. gr. 4.3—4.6. BB, like magnetite, infusible or but slightly rounded on the thin edges. With borax and phosphor-salt, yields a more or less pure green glass, the green colour becoming clearer and more distinct as the glass cools. Composition, theoretically, oxide of iron and sesqui-oxide of chromium, but the latter is

---

* For the detection of Titanium in iron ores, generally, the reader is referred to the author's *Blowpipe Practice*, p. 61.

always replaced to some extent by alumina, &c., and the iron by a certain amount of magnesia. The sesqui-oxide of chromium thus varies from about 40 to about 60 per cent., in different samples. A variety from Bolton yielded Dr. Skerry Hunt 45.90 per cent., and another from Lake Memphramagog gave 49.75 per cent.

Occurs abundantly in beds and scattered grains amongst the metamorphic strata of the Easter Townships and Gaspé, mostly in connection with serpentine or other magnesian rocks, the green colour of these being partly due to the presence of oxide of chromium. The principal localities comprise : Mount Albert in the Shickshock Range of Gaspé, and the Townships of Bolton, Ham and Melbourne. Chromic Iron Ore is largely used in the preparation of chromate and bi-chromate of potash.

(3) LIMONITE GROUP OF IRON OXIDES.

34. *Brown Iron Ore* or *Limonite* (including Bog Iron Ore and Yellow Ochre) :—Brown, brownish-black, or dull-yellow ; streak, yellowish-brown or ochre-yellow. Aspect, sub-metallic in some of the dark varieties, silky and earthy in others. Occurs commonly in masses with botryoidal surface and fibrous structure, or in granular or earthy masses. $H=1.0—5.5$; sp. gr. $3.5—4.0$. Heated in the bulb tube, it gives off water, and becomes converted into red oxide. BB, turns red, and then blackens and becomes magnetic. A thin scale, in the forceps, may be rounded on the thin edges ; otherwise infusible. Composition, essentially, hydrated sesquioxide of iron ; but the amount of water varies considerably, and the more earthy varieties always contain a certain percentage of phosphoric acid, with frequently silica, alumina, oxides of manganese, and humic or other organic acids. In the sub-metallic and silky varieties, the average amount of metallic iron is equal to about 58 or 60 per cent. ; in the average bog ores it equals about 45 or sometimes 50 per cent. ; and in the ochres, it varies from about 10 to 40 per cent. The average amount of water is about 15 per cent. or from 10 to 20 per cent. Brown and Bog Iron Ores are often smelted, and the Iron Ochres are valuable as a paint material.

The varieties of this mineral hitherto found in Canada, comprise the more earthy varieties, Bog Iron Ore and Yellow Ochre. These belong to comparatively modern deposits, and, in places, indeed, they are now under process of formation. The iron is taken up by water percolating through ferruginous strata, and is held in solution for a time as bicarbonate or in combination with organic acids ; and afterwards, by absorption of oxygen, it becom's converted into insoluble sesquioxide, and is thus deposited in a hydrated condition, mixed more or less with earthy and other impurities.

In the Province of Ontario, Bog Iron Ore occurs in small quantities in almost every Township, but some of the more important deposits lie in the townships of Charlotteville, Middletown and Windham, in Norfolk County, on Lake Erie ; also in Camden Township in Kent, West Gwillimbury in Simcoe, Bastard in Leeds, March and Fitzroy, and also Vaudreuil, on the Ottawa, and elsewhere. Ochres occur also at the la'ter locality, associated with the bog ore ; and extensive beds have been discovered in various places in the County of Middlesex ; in Walsingham Township, Norfolk Co. ; at Limehouse in Halton Co. ; as well as near Owen Sound in the township of Sydenham in Grey County, and in Nottawasaga Township in Simcoe.   Also in Elzevir, Leeds and other Townships.

Bog Iron Ore, in still more valuable deposits, occurs abundantly in the Province of Quebec.   The most important localities lie perhaps in the Three Rivers District, or between the Rivers St. Maurice, Batiscan and St. Anne.   The old St. Maurice forges, so celebrated for their castings, were fed by the ore of this neighbourhood ; and the more recently established Radnor forges, at Batiscan, draw their supply from the same district.   Other deposits of bog ore occur in Lachenaie in l'Assomption County, Kildare in Joliette County, and elsewhere in that section ; also in Templeton, Hull and Eardley, on the left bank of the Ottawa. South of the St. Lawrence, the ore occurs more or less abundantly in the Eastern Townships of Stanbridge, Farnham, Simpson, Ascot, Stanstead, Ireland, &c., and in St. Lambert, St. Vallier, Villeray, Caconna and elsewhere. Valuable deposits of ochre occur especially near the mouth of the St. Anne, in Montmorenci, below Quebec ; and at Cap de la Madeleine and Point du Lac, near the St. Maurice, in the Three Rivers District.   Also in the township of Mansfield, on the Upper Ottawa. A bed of ochre occurs likewise in Durham, and elsewhere, in the Eastern Townships.   These ochres are frequently, in places, of a dark-brown or greenish-black colour, from intermixture with earthy manganese ore.

C. MANGANESE OXIDES.

35. *Manganite :*—Steel-grey, with brownish streak, and metallic

or sub-metallic lustre. Rhombic in crystallization, but occurring chiefly in fibrous masses. H=3.5—4.0; sp. gr. 4.3—4.4. B.B infusible. Yields water by ignition in the bulb-tube, and forms a "turquoise enamel" with carb.-soda (see Part I., p. 39). Composition, if pure : sesquioxide of manganese 89.8, water 10.2.

Said to occur in a broad vein, with quartz, calc spar and fluor spar, traversing trap rocks, on the south shore of Bachewahnung Bay, Lake Superior, but no examples have yet come under the author's observation.

36.' *Earthy Manganese Ore* (Wad, Bog-Manganese, Manganese Ochre) :—Black or blackish-brown, in dull, earthy and often nodular masses. Very soft. BB, infusible. Yields water in the bulb-tube, and forms with carb.-soda a "turquoise enamel," green whilst hot, greenish-blue and opaque when cold. Composition, essentially, hydrated oxide of manganese, but always mixed with earthy matters, and often with iron ochre. Some varieties contain baryta, others oxide of cobalt, copper, &c. The manganese is usually present both as protoxide and sesquioxide.

This substance occurs principally in recent deposits throughout the district south of the St. Lawrence, as, more especially, in Cleveland, Bolton, Stanstead, Tring, Aubert Gallion, Ste. Marie (Beauce), St. Sylvester, Lauzun, &c. Deposits of this ochre have also been found on the north shore, as in Seigniories of Ste. Anne and Cacouna, and in the immediate vicinity of Quebec. In Ontario, it has only been observed, as yet, in the Township of Madoc ; and, in admixture with iron ochre, on the north-east shore of Thunder Bay, Lake Superior. A sample from the latter locality, yielded the author :

Sesquioxide of Iron........	33.68	
Sesquioxide manganese.....	16.54	
Protoxide manganese.......	5.08	Carbonate manganese.. 8.23
Lime	0.81	Carbonate of lime...... 1.44
Carbonic Acid......... ...	3.78	
Sulphuric acid.........trace only		
Phosphoric acid..very slight trace		
Water....................	3.82	
Silicious rock matter.......	36.12	
	99.83*	

* The small amount of water held by this ochre is somewhat remarkable. As regards the Dominion of Canada, workable amounts of Pyrolusite or Black Ore of Manganese appear to occur only in the Lower Carboniferous strata of Albert and King's Counties, New Brunswick.

D. URANIUM OXIDES.

37. *Uran Ochre :*—Yellow, in earthy crusts. BB, blackens, but does not fuse. Composition, probably, sesquioxide of uranium and water. In Canada, observed only as a coating on magnetic iron ore with intermixed actynolite, at the Seymour Mine in Madoc.

38. *Black Uranium Ore* or *Pitch-blende* (Coracite, &c.) :—Black, greyish-black, greenish-black, with greyish or brownish streak. Aspect between sub-metallic and vitreo-resinous. Mostly in nodular or other uncleavable masses. $H = 5.5$ when pure, but frequently less from intermixed earthy matters; sp. gr. $6.6 — 7.0$ when pure, but sometimes as high as 8.0, and often only 4.0 or 4.5, from impurities. BB, infusible, or rounded only on the thinnest edges. Composition, normally, protoxide of uranium 32.10, sesquioxide 67.90; but, in many instances mixed with carbonate or silicate of lime, lead, bismuth, copper and other compounds.

The only known locality in which this substance occurs in Canada, is at Maimanse, on the east shore of Lake Superior. The variety found at this spot was first described by Dr. Le Conte under the name of Coracite. It is mixed with carbonate of lime and other impurities, by which its sp. gr. is reduced to between 4.3 and 4.4 (4.378 Le Conte), and its hardness to about 3.5 or 4.0. It yields also, according to the analyses of Whitney and Genth, about 5 or 6 per cent. of water (Dana's Mineralogy : 5th ed. p. 155).

E. TUNGSTENUM COMPOUNDS.

39. *Wolfram :*—Brownish-black, with strong, sub-metallic lustre, and blackish-brown or red-brown streak. Rhombic in crystallization, but occurring frequently in irregular masses of lamellar or columnar structure. $H=5.0-5.5$; sp. gr. 7.1—7.6. BB, melts into a dull iron-grey globule with striated or crystalline surface. Consists of Tungstic acid combined with oxides of iron and manganese.

The only known examples of Canadian wolfram, were found by the writer, some years ago, in a large boulder of gneiss, on the north shore of Chief's Island, Lake Couchiching. (See description in *Canadian Journal*, 2nd series, Vol. 1, p. 308. Also, for analysis by Dr. Sterry Hunt, Vol. V., p. 303).

F. TITANIUM OXIDES.

[See also *Ilmenite* and *Iserine* under the Iron Ores.]

40. *Rutile :*—Dark-red, with peculiar adamantine lustre : streak,

pale-brown or greyish.   Tetrag. in crystallization, the crystals often in geniculated twin-combinations. Commonly, also, in columnar and fibrous masses, and sometimes in small grains or scales (imperfect or flattened crystals).   H=6.0 –6.5 ; sp. gr. 4.15—4.3.   BB, infusible.   With borax in a reducing flame, it forms a dark, amethystine glass, which is transformed into a light-blue opaque enamel by exposure to an intermittent flame (see Part I.).   Composition : oxygen 39, titanium 61.

Small grains or indistinct crystals of Rutile occur in the beds of Ilmenite at Baie St. Paul, below Quebec ; and at other localities, in Laurentian strata, associated with this ore.   Tolerably distinct crystals, half-an-inch in length, have been found in crystalline limestone on Green Island, Moira Lake, in Madoc.*   Acicular crystals occur sparingly in quartz cavities at the Wallace Mine, Lake Huron. Small crystalline grains and flattened crystals also, in the chloritic schists of some of the Eastern Townships, more especially in Sutton. Minute grains of Rutile occur likewise in many of the black ferruginous sands described uuder Nos. 31 and 32, above.

G. ALUMINA AND ALUMINATES.

[This group includes but two minerals of Canadian occurrence : *Corundum* and *Spinel*.   The first, by crystallization and atomic constitution, is related to *Hematite*, amongst the Iron Ores, and the second to *Magnetite*].

41. *Corundum :*—Blue, blueish-white, red, brownish, greenish, dark-grey ; streak, white or greyish ; aspect vitreous or stony.

Hexagonal in crystallization, but occurring frequently in grains and small granular masses.   H = 9.0 ; sp. gr. 3.9 – 4.2.   BB, infusible.   Not dissolved by carb. soda. Consists, normally, of alumina.   Transparent blue

Fig. 47.   Fig. 48.   varieties from the *Sapphire* of commerce, and red varieties, the *Ruby*.   Coarse, dull-coloured varieties are known as *Common Corundum* or *Adamantine Spar ;* and opaque, dark-grey, granular varieties (often mixed with magnetic iron ore) constitute *Emery*, a substance largely used as a polishing material.   Some of the finer varieties of corundum exhibit, especially when cut, a peculiar opalescence, frequently in the shape of a six-rayed star. These are known as *asteria sapphires, rubies,* &c.

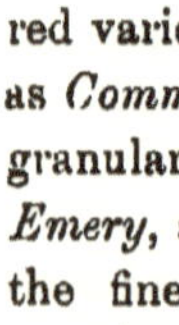

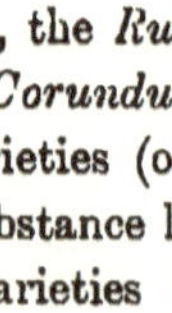

<hr>

* This locality was first pointed out by the late T. C. Wallbridge, of Belleville.

In Canada, this mineral has hitherto been noticed only in the form of blueish and pale-red grains in the crystalline Laurentian limestones of the Township of Burgess, Lanark County, Ontario. At one locality (Lot 2, Con. 9), it is associated with quartz, orthoclase, pearly-white mica and sphene.

42. *Spinel :* – Red, blueish, dark-green, black ; streak, white or grey ; aspect, vitreous or stony.   Regular in crystallization, and commonly occurring in octahedrons, either simple, or united in twin-forms (Figs. 49 and 50).   H = 8.0 ; sp. gr. 3.5 – 4.5.   BB, infusible.   The red and transparent varieties consist essentially of alumina and magnesia : normally of alumina 72, magnesia 28, *per cent.* ; in the black varieties (Pleonaste, Ceylanite), the magnesia is largely replaced, however, by oxide of iron ; and in the dark green or greenish-black varieties (Gahnite, Automolite) it is almost entirely replaced by oxide of zinc.

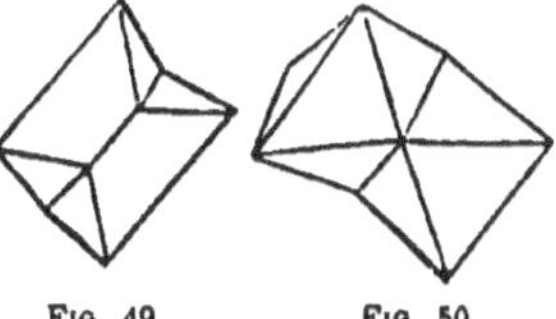

FIG. 49.   FIG. 50.

Small octahedrons and grains of a pale-blue colour (much resembling the spinel which occurs under similar conditions at Aker, in Sweden) are found in a crystalline limestone in the Seigniory of Daillebout, Jolliette County, in the Province of Quebec.   Large and often very symetrical black crystals occur in crystalline limestone in Burgess, Lanark Co. ; and less perfect examples of a similar colour, accompanying fluor spar, apatite, and white orthoclase crystals, are found in a vein of flesh-red calcite, in the Township of Ross, in Renfrew County, on the Ottawa.

H. SILICA AND SILICATES.*

[This division comprises the different varieties of Quartz and Opal, or silica in the free state ; together with the natural compounds of silica with various bases, such as alumina, the iron oxides, magnesia, lime, soda, potash, and the like.   Some of these silicates yield water when ignited ; others are anhydrous in their normal condition, but frequently yield traces of water as the result of incipient decomposition.   It is not possible to arrange the silicates strictly in accordance with their bases, without separating, in many instances, substances which in general characters are closely allied ; and in some cases, an

---

* For details respecting the crystallization characters of the silicates generally, the reader may consult the Notes attached to the Mineral Tables in the author's Blowpipe Practice.

arrangement of this kind would lead to a separation of varieties
of one and the same mineral. In the garnets, for example, certain varieties contain magnesia, and others lime or oxide of iron,
&c., in place of magnesia, these bases being capable of mutual
substitution without the general or essential character of the substance being altered by the change—a peculiarity known as isomor-
phism. The silicates possess representatives of all the crystal systems. In their hardness they vary from 1.0 (in tale) to 8.0 (in
topaz). Their aspect is most commonly vitreous, resino-vitreous,
stony, or pearly, but the micas and some few other silicates
(bronzite, &c.) exhibit a pseudo-metallic lustre (see Part I). The
colour frequently varies greatly in examples of the same species,
as it is due chiefly to minute and accidental proportions of
foreign matters, or to variations in the isomorphous bodies which
form the base. Thus, where protoxide of iron or ferro-ferric oxide is
largely present, the mineral will generally possess a dark-green or
black colour, but where these bases are replaced by lime or magnesia
in greater or less proportion, the same mineral may be quite pale or
light in colour, or even colourless. The different garnets, pyroxenes,
amphiboles, tourmalines, &c., are familiar examples of this fact. The
streak, however, is always white (or nearly so) under normal conditions, but it may exhibit a slight or indefinite tinge of grey, green,
or brown, in a very dark or ferruginous variety, especially if the substance be slightly altered or decomposed. Many silicates unless previously ignited or fused with potash or alkaline carbonates, resist
altogether the action of acids. Others become partially attacked or
decomposed (some by boiling hydrochloric acid, and others by sulphuric acid), the silica separating in a granalar, slimy or gelatinous
condition (See under "Action of Acids," in Part I). Some silicates,
which do not gelatinize in their ordinary state, exhibit this peculiarity if previously fused or strongly ignited. Certain silicates are
quite infusible in the blow-pipe flame. Others, if held, in the form
of a thin or pointed splinter, in the platinum forceps (Part I), become rounded and vitrified at the point or edges ; and others, again,
melt into a perfect globule. In some cases, the substance exfoliates,
or swells up and forms an intumescent branching mass, on the first
application of the flame; and in many instances the fusion of a silicate is accompanied by continued bubbling. Silicates which contain
a large proportion of silica form a clear transparent glass with carb.

soda, if the latter be added litt'e by little until the proper quantity be obtained ; but phosphor-salt is a far more characteristic reagent for these bodies. When a silicate is exposed in a bead of phosphor-salt to the action of the blowpipe, the bases (lime, magnesia, alumina, &c.) become gradually taken up, whilst the silica remains wholly or in chief part undissolved. A small portion may be taken up by the hot flux, but as this cools, the silica is precipitated, rendering the glass opaline or milky. The undissolved silica, if a small fragment or scale-like particle of the mineral be subjected to the test, forms a thin, translucid, flocculent mass, technically known as a "silica skeleton," in the centre of the bead. A silicate may thus be readily distinguished from a phosphate, carbonate, sulphate, &c., as these latter bodies are rapidly and entirely dissolved (the carbonates with effervesence) by phosphor-salt under the action of the blowpipe.

The silicous minerals, hitherto discovered in Canada, are described in this work under twelve groups or sub-divisions ; but some of these, it should be observed, are rather groups of convenience than strictly natural collocations. Their distinct characters are given below. The groups, themselves, comprise : (1) Quartz group ; (2) Basic Silicates ; (3) Pyroxenic Silicates ; (4) Chrysolitic Silicates ; (5) Feldspathic Silicates ; (6) Calcareo-Feldspathic Silicates ; (7) Nephelitic Silicates ; (8) Zeolitic Silicates ; (9) Micaceous and Chloritic Silicates ; (10) Talcose Silicates ; (11) Kaolinic Silicates ; (12) Copper and Nickel Silicates.

(1) QUARTZ GROUP.

[This group includes the different conditions of Silica in its free or uncombined state. These conditions are principally three : the crystalline anhydrous condition—yielding the different varieties of Quartz ; the Calcedonic condition ; and the uncrystalline, generally hydrated condition, giving rise to the various Opals.

43. *Quartz* :—Colourless or variously coloured, and vitreous or stony in aspect. Streak (normally) white. Frequently found in Hexagonal crystals, consisting almost invariably of a six-sided prism, transversely striated, and terminated by the planes of a six-sided pyramid. In many examples, however, the pyramidal planes, more especially, are of very unequal size, some of the faces being often abnormally developed so as to produce the partial or complete obliteration of the rest. The point of the pyramid is thus often ex-

tended into an edge, as in some of the accompanying figures. Quartz occurs also, and more frequently, in masses of irregular shape, as

well as in nodular and stalactitic forms, and in small grains. Cleavage, scarcely observable: fracture conchoidal and uneven. H = 7.0; sp. gr. 2.5—2.8, mostly about 2.65.

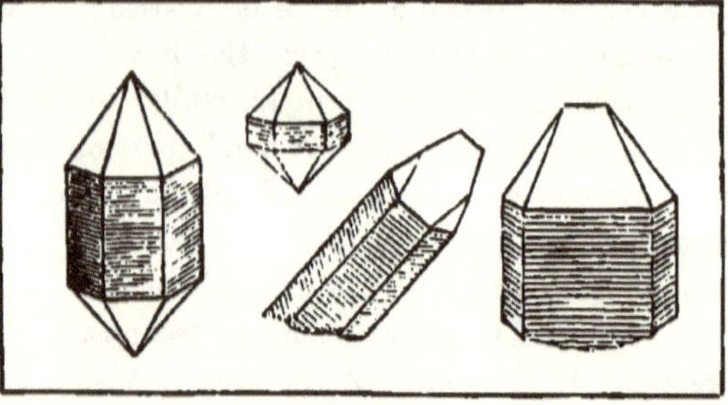

FIGS. 51 TO 53.

BB, *per se*, quite infusible ; with carb. soda, melts with effervescence (due to the expulsion of the carbonic acid of the flux) into a transparent glass.   Insoluble in the ordinary mineral acids.   Consists, normally, of pure silica, the tints of the coloured varieties being due to accidental amounts of iron and manganese oxides, bituminous matter and other inessential ingredients.   The principal varieties of Quartz, hitherto met with in Canada, are as follows :—

(*a*) *Common Quartz, Rock Crystal :*—Vitreous or stony ; mostly colourless, but sometimes pale reddish, yellowish, greenish or grey. Forms an essential component of granite, mica-schist, gneiss, quartz-rock and various other crystalline rocks, and is thus present throughout the wide area occupied by our Laurentian strata, as well as in many localities where Huronian rocks prevail, and amongst the crystalline strata of the Eastern Townships (see Part V.). Very common also in mineral veins : as in those of Thunder Bay, Lake Superior ; the Bruce Mines, Lake Huron ; Harvey's Hill Mine, in Leeds ; and elsewhere.   Occasionally present likewise, in fissures and cavities in limestone rocks, as in the vicinity of Quebec, where the crystals are known as Quebec diamonds.

(*b*) *Smoky Quartz :*—In brownish crystals : Thunder Bay, Lake Superior ; also near Quebec ; and elsewhere.

(*c*) *Amethyst :*—In violet-coloured crystals, sometimes of large size.   Fine specimens, associated with fluor spar, calcspar, pyrites, native silver, &c., occur in veins on Thunder Bay and throughout that district ; also on Spar Island, farther west, on Lake Superior. Many of these crystals present a deep reddish-brown colour on the outer surface, arising from a deposition of numerous minute spots of jasper or sesquioxide of iron. The colouring matter appears to consist in certain cases of a minute trace of some silver compound.

(d) *Chalcedony* :—In nodular semi-translucent masses of a yellowish, grey, or reddish colour. Occasionally present in the amygdaloidal traps of Lake Superior. Also in thin bands or veins, with Jasper, on the River Ouelle in Kamouraska. The nodules pass more or less into semi-opal.

(e) *Agate* :—In nodular masses of various clouded or banded colours, either feebly translucent or opaque. Very abundant in the amygdaloidal traps of St. Ignace, Agate Island, Michipicoten, etc., on the north shore of Lake Superior, and in the shingle beaches of these islands. Also in the conglomerates of Gaspé; and in the pebbly beaches along the shores of Gaspé Bay, arising from the destruction of these conglomerates.

(f) *Jasper* :—In opaque rounded masses, and in beds, of a brown, red, green and other colour; sometimes striped or banded; and always more or less dull or earthly-looking on the fractured surface. Some remarkable quartz-rocks, evidently altered conglomerates, containing pebbles of red Jasper, occur on the north-west shore of Lake Huron. Many of the dark green and striped slates of Lake Huron, also, may be regarded as closely akin to Jasper. At Bachewahnung, on the east shore of Lake Superior, bands of red Jasper are associated with hematitic iron ore; and layers and imbedded nodules occur in the copper-bearing series of the north shore, around Thunder Bay, &c. Many of the so-called agates of this region are properly Jaspers. Beds and layers of red Jasper, in places very ferruginous, are found in the crystalline strata of the Eastern Townships, as in Sherbrooke, Shipton, Broughton, &c., and on the River Ouelle. Jasper pebbles are associated also with agates in the conglomerates and Shingle beaches of Gaspé.

(g) *Chert or Hornstone* :—Yellowish, brownish, reddish-white, grey, black, &c. Mostly in nodular and irregularly-shaped masses, and occasionally in beds and veins which often present a cellular or brecciated structure. Translucent to nearly opaque. Closely allied to Chalcedony and Flint. Occurs in the form of veins traversing syenite in the township of Grenville, at first pointed out by Sir William Logan. Also in layers, &c., in the upper copper-bearing series of Thunder Bay, Lake Superior, and abundantly in imbedded nodular masses, and in thin layers in the Corniferous Formation of the Devonian series of Western Canada, on the shore of Lake Erie, &c.; as well as occasionally under similar conditions in limestones of the Niagara and Trenton groups. Hornstone, or related silicious matter, forms the fossilizing substance of most of the corals and brachipods of our Western Devonian beds, as well as that of many of the organic remains found in Silurian strata, as at Pauquette's Rapids on the Ottawa, and elsewhere.

(*h*) *Sandstones; Sands; Gravel* :—Sandstones consist essentially of quartz grains, cemented together, or consolidated by pressure (see Part III); whilst sands and gravel consist of the same substance in loose grains and pebbles. These rock matters, although occasionally colourless, usually exhibit various shades of yellow, brown, or red, from the presence of sesquioxide of iron. Sandstones are also occasionally of a green or greyish-green colour, in which case part of the iron is in the condition of protoxide. Some of our purest sandstones and quartz sands are found at the following localities : Pittsburg township (near Kingston); Charleston Lake, in Escott; Vaudreuil, on the Lower Ottawa; Beauharnois; the Grès Rapids, on the St. Maurice; Township of Batiscan; and also near Brockville, Perth, Owen Sound, Dundas, &c. (see Part V.)

(2) GROUP OF BASIC SILICATES.

[This group includes a small number of silicates in which the percentage of silica varies from 30 to 40. The specific gravity is comparatively high (=3.0 to 4.75); and the hardness sufficient in all cases to scratch glass strongly (=5.5 to 7.5, but mostly over 6.0)].

44. *Zircon* :—Brown, red, reddish-yellow, with resino-vitreous aspect. In Tegragonal crystals, mostly square prisms, terminated at each extremity by a four-planed pyramid (Figs. 54, 55); occasionally also in small granular masses. H = 7.5 . sp. gr. 4.0 — 4.75. BB, quite infusible. Not attacked by acids. Consists of : silica 33.2, zirconia 66.8. Occurs with plumbago, wollastonite, pyroxene,

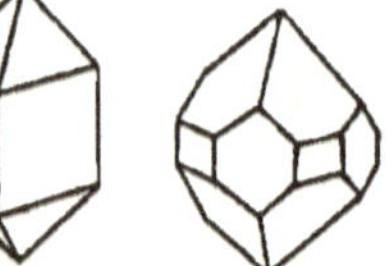

FIG. 54.   FIG. 55.

&c., in the crystalline limestone of the Township of Grenville, in Argenteuil County, and more or less throughout the phosphate deposits of Buckingham, Templeton, &c., often in fine crystals.

Also in granitic veins, with tourmaline, on the North River, in St. Jerôme, Terrebonne County ; and, according to the Reports of the Geological Survey, in a syenitic rock, composed of red feldspar and black hornblende, on Pie Island, Lake Superior. Transparent varieties of this mineral are employed in jewellery, under the name of *Jargon* or *Hyacinth*.

45. *Chiastolite :*—Grey or pale-red. Occurs in rectangular and rhombic prisms, mostly of narrow diameter, and in compound groupings, which present the appearance of a simple prism with dark cross on the transverse section (Fig. 56), the cross consisting of slate or other rock matter, in which the prisms are imbedded. Found also in granular masses. H = 5.5 — 7.7 ; sp. gr. 3.1 — 3.2. BB,

Fio. 56.

quite infusible. The powder by ignition with nitrate of cobalt (p. 34) assumes a fine blue colour. General composition : silica, 37, alumina 63. Occurs in somewhat indistinct crystals imbedded in argillo-micaceous slates, in the immediate vicinity of intrusive masses of granite, on Lake St. Francis, in Megantic County.

46. *Tourmaline :*—Of various colours—green, blue, black, brown, yellow, red, and sometimes colourless; but Canadian varieties are either black, brown or brownish-yellow. The black variety is commonly known as *Schorl*, and is quite opaque. Hexagonal (or rather Hemi-Hexagonal) in crystallization, the crystals being almost invariably three-sided prisms (or these, with bevelled edges, producing a prism of nine sides). The cross fracture is thus as a rule more or less distinctly triangular. The prisms are often longitudinally striated, and are frequently much broken, especially when imbedded in quartz (Fig. 58). Tourmaline occurs also very generally in columnar, acicular and fibrous masses. H=

Fig. 57.

Fig. 58.

6.5—7.0 ; sp. gr. 3.0—3.3. BB, the black and most of the brown varieties melt very easily, the other varieties being for the greater part quite infusible. Nearly all exhibit electrical properties when heated. Composition somewhat variable, but the essential components consist of : silica (averaging about 38 per cent.), boracic acid

8

(4—9 per cent.), alumina (30—44 per cent.), with more or less sesquioxide of iron, magnesia, protoxide of iron, protoxide of manganese, lime (under 2 per cent.), soda, potash and sometimes lithia. A small amount of fluorine is also generally present.

Tourmaline is of comparatively common occurrence in the Laurentian strata of Canada. It is met with both in the crystalline limestones and in many of the gneissoid or quartz beds of that formation, as well as in some of the granitic veins by which these beds are traversed. In the Ottawa district, it occurs especially in crystalline limestones, as at Calumet Falls (yellowish-brown and black, with Idocrase, &c.); in Clarendon Township, County of Pontiac; in North Burgess and Elmsley, Lanark County; in Grenville, and at Lachute, in Argenteuil County and elsewhere. In Ross, Elmsley, Bathurst, Blythfield (near the High Falls of the Madawaska), St. Jérôme (Terrebonne County), Galway (Peterborough County), and on Yeo's Island, Stoney Lake, Charleston Lake, &c., it is found in granitic and syenitic veins. Also in quartz veins and beds associated with gneissoid strata, in various parts of Madoc, Tudor, Elzevir, and more or less generally throughout the back country between the Ottawa and Georgian Bay.

47. *Garnet* :—Variously coloured—most commonly red, brown, black, green or yellow: rarely colourless. Regular in crystallization : the crystals almost invariably either rhombic dodecahedrons or trapezohedrons (Figs. 59 and 60). The mineral occurs also, very commonly, in granular and lamellar masses. H=6.5—7.0 ; sp. gr. 3.5 —4.2. BB, most varieties melt more or less readily, the dark-red yielding a magnetic globule ; but the bright-green chrome garnet and some light-coloured varieties are infusible. After fusion or strong ignition, most varieties gelatinize in boiling hydrochloric acid (see under "Action of Acids," Part I.). Composition exceedingly variable, but essentially silica (33 to 43 per cent.), alumina or sesqui-oxide of iron (or both), with either lime, magnesia, protoxide of iron, or protoxide of manganese, or several of these bases combined. In the bright-green garnet *(Ouvarovite)*, the sesquioxide of iron is chiefly replaced by sesquioxide of chromium, and the monoxidized portion of

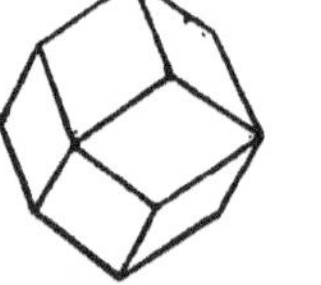
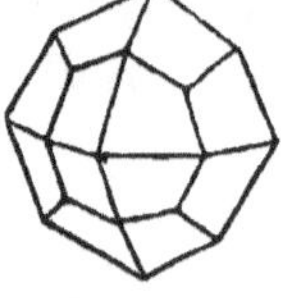

Fig. 59.   Fig. 60.

the base is essentially lime. All the deeply-coloured garnets are strongly ferruginous ; whilst in the light-coloured varieties, iron is chiefly replaced by alumina, lime or magnesia.

Garnets occur in Canada in many crystalline strata of the Laurentian series ; also in the less ancient crystalline beds of the Eastern Townships (see Part V.) ; and in some of the trappean rocks of Lake Superior. In Laurentian strata, they affect principally the beds of hornblende-rock, and gneiss, which lie in contact with, or adjacent to, the interstratified bands of crystalline limestone, but they occur also apart from these limestones. The best-known Laurentian localities comprise : the banks of the River Rouge and adjacent country near the " Three Mountains," in the township of Clyde, Ottawa County (pink and red ferro-magnesian varieties in gneiss and quartz rock) ; Seignory of St. Jérôme, on the Ottawa (red and very abundant in gneiss); Rawdon Township, in Montcalm County (in quartrock) ; Townships of Chatham, Chatham Gore and Grenville in Argenteuil County (red and yellowish-red varieties) ; Hunterstown, in Maskinonge County ; Bay St. Paul (red, in quartz-rock); Murray Bay (large crystals and rounded masses in gneiss) ; Madoc Township (Lot 11, Con. 11, in hornblende rock with iron pyrites, &c.) ; Townships of Elzevir, Barrie, &c. (dark-red, in hornblende rock) ; Marmora (in quartz rock, &c.). It occurs thus in the Laurentian area generally between the Ottawa and Georgian Bay. In the altered strata of the Eastern Townships, yellowish-red or pale-brown garnets occur in pyroxene rock on Brompton Lake, and minute grains and crystals of bright-green chrome garnet are thickly disseminated through a calc-spar vein at the same locality. Red garnets are found also in crystalline magnesian limestone, with talc, magnetic and chrome iron ores, &c., in the Townships of Broughton and Sutton ; and with black hornblende in the serpentines of Mount Albert in Gaspé. In Orford, Dr. Sterry Hunt has discovered a peculiar variety of a white or light-coloured calcareo-aluminous garnet, in rounded masses of somewhat waxy aspect, mixed with serpentine ; and he has described the occurrence of a similar · variety in more or less compact beds, holding specks of native gold, in St. Francis (*Rep.* 63 : p. 496). Finally, it may be observed, garnets of a pale red-brown colour occur sparingly, with epidote, &c., in amygdaloidal traps, at Maimanse, on the east shore of Lake Superior.

48. *Vesuvian* or *Idocrase* :—Yellow, brown, yellowish-red, &c.
Tetragonal in crystallization : otherwise, both in composition and
general characters, identical with garnet.   H = 6.5 ; sp.
gr. 3.3—3.45.   BB, more or less readily fusible.   Occurs
in some of the crystalline limestones of the Ottawa District : principally in brown crystals, with tourmaline, at
Calumet Falls, and in the township of Clarendon ; and
also in small reddish-yellow crystals, with zircon, pyroxene, graphite, &c., in the township of Grenville.

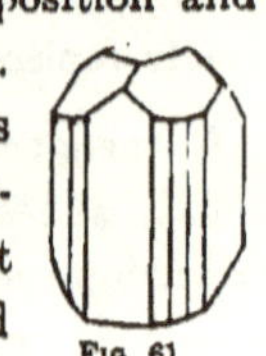
Fig. 61.

49. *Epidote* : — Green, yellowish-green, blackish-green, grey, &c.
Monoclinic in crystallization, but well defined crystals are of rare
occurrence in Canada ; mostly in acicular crystals, and in columnar,
reniform and more or less compact masses, or in imbedded grains.
H = 6.0—7.0 ; sd. gr. 3.25—3.35.   BB, swells up and forms a dull
slag-like mass, with rounded edges.   This is generally magnetic,
but unlike the beads formed by hornblende, pyroxene, vesuvian, &c.,
it resists further fusion.   After strong ignition, epidote gelatinises
in boiling hydrochloric acid.   (See under " Action of Acids," in
Part I.)   General composition : silica 34—40, alumina 18—28, sesquioxide of iron 7—17, lime 20—25.   This mineral is comparatively
rare in the Laurentian rocks of Canada, but it occurs, although with
more or less indistinct characters, in the gneissoid strata associated
with the iron-ore of Belmont, Seyour, and Marmora ; and in irregular
layers in a reddish gneiss at Carleton Place, in the township of Beckwith, Lanark County, the rock, when polished, forming a handsome
ornamental stone.   Epidote is far more abundant in the metamorphic strata of the Eastern Townships, and it occurs thus in St.
Armand, Potton, Shipton, Melbourne, and elsewhere throughout
that region.   Some of the best defined examples are found in spheroidal masses of a peculiar slate-rock in the seignory of St. Joseph, the
epidote in these masses being associated with calcite, serpentine,
chlorite, and quartz.   Epidote occurs also in some of the amygdaloidal traps and greenstones of Lake Superior, as at Maimanse (with
mesolite, chlorite, brown garnet, &c.), and on the Island of Michipicoten.

50. *Allanite* (Orthite) :—Black, brown, yellowish-brown.   Monoclinic, and like Epidote in crystallization, but occurring generally in
granular and amorphous masses, with strong resino-vitreous lustre.

Fracture, conchoidal. $H = 5.5 - 6.0$; sp. gr. $3.1 - 4.2$. BB, intumesces strongly, and melts into a black and usually magnetic globule. The powder is readily decomposed by hot hydrochloric acid, silica separating in a gelatinous state. General composition; silica (30 to 38 per cent.), alumina (8 to 17 per cent.), iron oxides (8 to 20 per cent.), oxide of cerium (usually about 15 or 16 per cent., but in some examples nearly 30 per cent.; generally replaced, however, in part, by oxides of lanthanum, yttrium, &c.), lime (6 to 12 per cent.), with a small amount of magnesia, and a little water, the latter indicating incipient decomposition. Allanite is of comparatively rare occurrence in Canada. Hitherto, only recognized in pitch-black masses or grains in Laurentian strata, as in the upper Laurentian feldspathic rocks around Bay St. Paul and Lake St. John (as first made known by Dr. Sterry Hunt); and in the form of a narrow vein in gneissoid strata at Hollow Lake, the head waters of the South Muskoka. (See a notice, by the writer, in *Canadian Journal*, Vol. IX., p. 103).

51. *Sphene* or *Titanite* :—Brown, black, yellow, greenish. Monoclinic in crystallization (the crystals most commonly, as in Fig. 62); but occurring also in small granular masses, and in veins or strings of more or less compact structure. $H = 5.6$; sp. gr. $3.4 - 3.6$. BB melts with bubbling into a dark glass or enamel, but sometimes on the edges only. In powder, decomposed by hot sulphuric acid. Consists of: silica, about 32 per cent., titanic acid, 40, lime 28, but part of the latter is usually replaced by a little oxide of iron and magnanese. Occurs in small dark-brown opaque crystals in the Laurentian gneissoid rocks of Tudor, Madoc, Lutterworth, Muskoka, &c. Also in crystalline limestone in Grenville, Burgess, North Elmsley; and at Lachine and Calumet Falls, in the Ottawa country. Sphene is also found in small amber-coloured grains and crystals in the granitic trachytes of the Eastern Townships (Brome, Shefford, Yamaska), and in thin veins or strings with micaceous or slaty iron ore in the altered rocks of Sutton.

FIG. 62.

(3.) GROUP OF PYROXENIC SILICATES.

This group consists essentially of non-aluminous silicates of lime and magnesia, these bases being partly replaced, however, in dark varieties, by protoxide of iron. Alumina is only exceptionally present, and rarely exceeds 4 or 5 per cent. Crystallization, essentially clinorhombic. Sp. gr 2.9—3.3. Scarcely, if at all, attacked by acids.]

52. *Amphibole* (including *Tremolite, Actinolite, Hornblende, &c.*):— Green of various shades, greenish-white or almost colourless, brown, black. Clinorhombic in crystallization, the crystals mostly oblique rhombic or six-sided prisms, with the obtuse prism-angle (V on V in the accompanying figures) = 124° 30'; but occurring commonly in acicular forms, and in fibrous, lamellar and granular masses. H = 5.5 — 6.0; sp. gr. 2.9 — 3.4 (mostly 3.0—3.2) BB, melts more or less easily, the dark varieties yielding a magnetic bead.

Fig. 63. Fig. 64.

Scarcely or not at all attacked by acids. The greenish-white and colourless or pale-grey varieties of this mineral are usually known as *Tremolite;* the bright-green, or dark-green, acicular and fibrous varieties, as *Actinolite;* and the green massive varieties, as well as those in green, brown, or black, thick crystals, are commonly termed *Hornblende,* a name applied by many authors to the species generally. A soft, silky variety, in fibrous masses, belonging, however, partly to Pyroxene (No. 53), is also known as *Asbestus* or *Amianthus,* but this variety does not appear to occur in Canada, our so-called asbestus being a fibrous serpentine, containing about 12 or 14 per cent. of water. (See under No. 83, below.) Average composition : silica (40 — 60 per cent.) magnesia (15 — 25 per cent.), lime (12 — 15 per cent.), with, in most varieties, a small amount of protoxide of iron, &c. Alumina, when present, varies in amount from less than one to above 15 or 16 per cent., but the latter is only found in a few dark-coloured hornblendes of exceptional occurrence.

Amphibole is an essential constituent of many eruptive and metamorphic rocks, such as syenite, diorite or greenstone proper, syenitic gneiss, hornblende slate, &c.; and it is present accidentally in many crystalline limestones and other rocks. It occurs in various localities throughout the large area occupied by the Laurentian series of Canadian strata (see Part V), and also in the more modern metamorphic district of the Eastern Townships. Examples of Tremolite occur more especially in the crystalline (Laurentian) limestones of the Ottawa region, as at Calumet Falls, and in the townships of Algona, Blythfield, and Dalhousie. Dark-green Amphibole, in good crystals, occurs with diopside at the High Falls of the Madawaska, and elsewhere on that river. A fibrous and acicular pale-grey or greenish variety (Raphilite) is found near Perth, in Lanark County. Actinolite occurs here and there amongst the magnetic iron ores of Madoc and Belmont. Beds of hornblende rock range through Frontenac, North Hastings, &c., in the Laurentian area lying between the Ottawa and Georgian Bay; and syenitic or hornblendic gneiss occurs abundantly throughout the Laurentian area, generally. See Parts III. and V. Black and dark-green hornblende is seen in distinct crystalline masses and grains in many syenites and diorites; notably in the large development of syenite in the townships of Grenville, Chatham, and Wentworth, on the east side of the Ottawa: (See parts III. and V.) South of the St. Lawrence, green hornblende occurs in well-defined examples in the township of Potton : and actinolite is found, with talc, chlorite, fibrous or asbestiform serpentine, &c., in the townships of Brome and Sutton, as well as in beds of fibrous structure in St. Francis, Beauce County. Black hornblende, with garnets, is associated with the Serpentines of Mount Albert, in Gaspé; and small grains and crystalline masses occur in the diorite and granitic trachytes (Part III.) of Mount Johnson, Yamaska, Brome and Shefford.

53. *Pyroxene* (including *Diopside, Sahlite, Augite,* &c.) :—Green of various shades, greenish-white or almost colorless, brown, black. Clinorhombicic in crystallization, the crystals mostly eight-sided prisms with sloping terminal planes, as in the annexed figures. The prism-faces v, v meet (over v̄) at an angle of 87° 5′; v inclines to v̄ at an angle of 133° 33′; v̄ and v′ form a right angle.

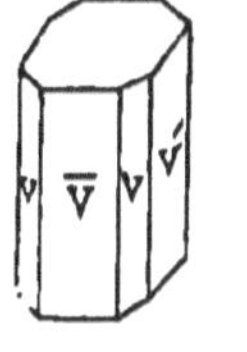
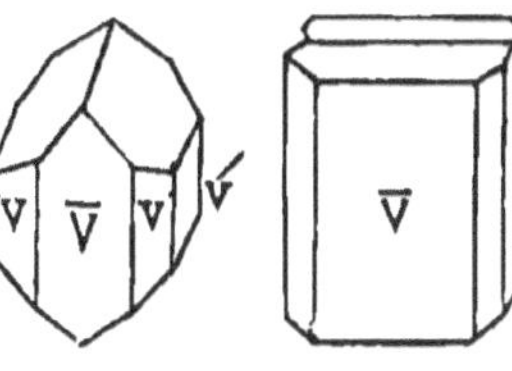

FIG. 65.   FIG. 66.   FIG. 67.

Fig 65 is the combination usually presented by the light or dark coloured varieties of our Laurentian crystalline limestones. Fig. 66 represents the ordinary *augite* crystals of basaltic rocks : good examples occur in the trap of the Montreal Mountain. Fig. 67 represents a twin or compound crystal from Orford, presented to the writer by

Dr. Sterry Hunt. It consists of two crystals of diopside, like Fig. 65, united by a front vertical face, and much extended or flattened in this direction. Pyroxene occurs also very commonly in acicular and fibrous groups, and in cleavable and also granular masses. Cleavage planes meet at angles of 87° 5' and 92° 55'. H (except in altered or abnormal varieties) = 5.5 — 6.0 ; sp. gr. = 3.2 — 3.5. BB, melts in general without difficulty, the dark varieties yielding in most cases a magnetic bead. Scarcely or not at all attacked by acids. In composition, essentially a bisilicate of magnesia and lime, with part of these bases replaced by protoxide of iron, &c. A small amount of alumina is likewise occasionally present, as in Amphibole, the composition of these two minerals being practically identical. Pyroxene and Amphibole are also closely allied by crystallization and physical characters, but their crystals have a more or less distinct aspect, and the cleavage angles are not alike. Pyroxene exhibits also, as a general rule, a somewhat higher density, its sp. gr. varying usually from 3.25 to 3.35, whilst that of Amphibole lies most commonly between 2.9 and 3.2. Light-coloured varieties of pyroxene are usually known as *Diopside* (also as Sahlite, Malacolite, Traverselite, Alalite, &c.), whilst the term *Augite* is generally applied to the dark varieties. (Jeffersonite, Hudsonite, Hedenbergite, Coccolite, &c., are other synonyms of this species.)

Pyroxene is of common occurrence in eruptive and metamorphic rocks (see Part III). It is especially characteristic of modern volcanic products, but occurs also in many of the more ancient trappean formations. In Central Canada, it is found in white and pale-green crystals in many of the Laurentian crystalline limestones, as at Calumet Falls and elsewhere in the Ottawa district. Also in large crystals, with amphibole, at the High Falls of the Madawaska ; with mica, apatite, &c., in Bathurst; in well-defined crystals with pyrites in a quartz vein near Belmont Lake ; in the anorthosites or Upper Laurentian strata of Château Richer ; and elsewhere in these older metamorphic strata. In the higher series south of the St. Lawrence, it occurs with garnets, &c., in Orford Township. Finally, well-defined black crystals are imbedded in the trap of the Montreal Mountain, and also in the erupted traps or dolerites of Rougemont and Montarville. *Acmite,* a related silicate in thin, black crystals, occurs according to Dr. B. J. Harrington, in the nepheline syenites of Montreal and Beloil.

54. *Hypersthene* :—Brown, brownish-black, brownish-green ; with pale-grey streak, and a more or less pearly-metallic lustre. Mostly in laminar or foliated masses. H = 5.0 — 6.0 ; sp. gr. 3.3 — 3.42. BB, gives a black magnetic bead. Essentially a silicate of magnesia and iron oxide. A specimen from Château Richer yielded Dr. Sterry Hunt : silica 51.35, alumina 3.70, protoxide of iron 20.56, lime 1.68, magnesia 22.59, volatile matter 0.10. Occurs in association with anorthosites or feldspar rocks, more especially in Château Richer and St. Urbain, near Baie St. Paul, below Quebec.

(4) GROUP OF CHRYSOLITIC SILICATES.

[The minerals of this group are essentially silicates of magnesia, the latter base being more or less replaced, however, by protoxide of iron. Sp. gr. 3.1 — 3.5. Infusible. Gelatinizing in heated hydrochloric acid.]

55. *Chrysolite* or *Olivine* :—Yellow, green, brownish-green. Rhombic in crystallization, but rarely occurring otherwise than in small grains and granular masses imbedded mostly in eruptive rocks. H = 6 — 7 ; sp. gr. 3.3 — 3.5. BB, loses its colour but remains unfused, except in the case of certain highly ferruginous varieties not yet found in Canada. The powder becomes decomposed, with separation of gelatinous or flocculent silica, in both hydrochloric and sulphuric acid. This species occurs in the trap mountains of Montreal, Rougemont, and Montarville, usually in the form of small green and yellow grains, but occasionally in indistinct crystal-masses. An analysis by Dr. Sterry Hunt yielded : silica 37.17, magnesia 39.68, protoxide of iron 22.54.

56. *Chondrodite* :—Yellow, brownish-yellow. In small granular masses, mostly imbedded in crystalline limestone. H = 6.0 — 6.5 ; sp. gr. 3.1 — 3.25. BB, infusible. Gelatinizes in acids. Consists essentially of silica, magnesia, and fluoride of magnesium ; or is an oxygen compound with part of the oxygen replaced by fluorine. The latter element appears to vary in amount from about $2\frac{1}{2}$ to nearly 10 per cent. Chondrodite occurs in many of our crystalline Laurentian limestones, frequently accompanied by scales of graphite. Newboro', in the township of North Crosby, in Leeds County ; Grenville, in Argenteuil County ; and St. Jerôme, in Terrebonne County, have yielded examples.

(5) Group of Feldspathic Silicates.

[This group is composed essentially of alkaline and non-magnesian silicates, containing a high per-centage of silica (62 to 69), and about 20 per cent. of alumina. H=6 to nearly 7 ; sp. gr. 2.4 — 2.7. Crystallization, Clinorhombic or Triclinic. Fusible in thin splinters only. Insoluble in acids.]

57. *Orthoclase* or *Potash Feldspar* :—White, red, flesh-red, apple-green, grey, &c. Lustre more or less pearly on cleavage planes, otherwise vitreous or stony. Crystallization Clinorhombic, but with two well-marked cleavage directions (parallel with base and side vertical) meeting at right angles. Crystals often compound, as in the more common twin combination shewn in Fig 70. Found usually, however, in lamellar and granular masses. H = 6.0 ; sp. gr. 2.5 — 2.6. BB, fusible with difficulty, unless in the form of a thin pointed splinter, in which case the edge and point become quickly rounded. Practically, unattacked by acids. Average composition : silica 64.8, alumina 18.4, potash 16.8 ; but many varieties contain a small percentage of soda, replacing a portion of the potash.

Orthoclase is one of the component minerals of many crystalline rocks, granite, syenite, gneiss, &c. ; it occurs also in many trappean rocks, and forms the essential com-ponent of trachytes and ordinary lavas.

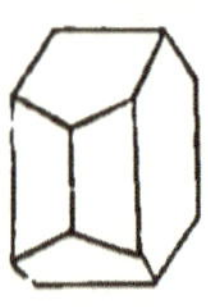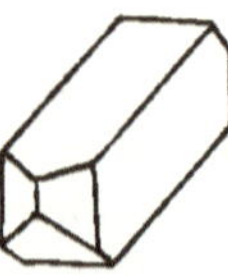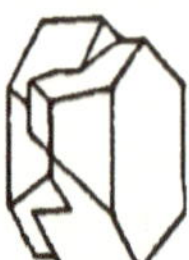

Fig. 68.   Fig. 69.   Fig. 70.

In the Laurentian strata so widely developed throughout the more northern portions of Canada (see Part V), this mineral is consequently largely present ; and well defined cleavable masses, mostly of a flesh-red or greyish-white colour, may be obtained in almost every district in which gneissoid rocks occur, more especially from the coarser granitic veins by which these rocks are so commonly traversed. Some of the more remarkable Laurentian localities comprise : the townships of North Burgess, Elmsley, Grenville, Chatham, &c. : also the township of Ross, and other places in the neighbourhood of Calumet Falls ; and several spots on the north shore of Lake Huron. In Burgess (Lot 3, Con. 6), among other varieties, a striped red and brownish ortho-clase occurs. This presents iridescent reflections, and is the variety known as *Perthite.* It contains soda as well as potash. In Ross (Renfrew County) large white crystals occur with apatite and spinel in calcite veins. Pale-red and other varieties are found in the Ottawa region, especially in the phosphate deposits. Orthoclase occurs also in the metamorphic strata south of the St. Lawrence, as in veins cutting altered slates in the townships of Inverness, Leeds, and Sutton ; and it is likewise present in many of the eruptive rocks of this district, notably in the porphyritic trachyte of Chambly, and in the trachytes of Montreal and Brome. These varieties, according to Dr. Sterry Hunt's analysess (Report : 1863, p. 476) contain nearly equal amounts of potash and soda. Orthoclase, in common with other feldspathic silicates, yields by atmospheric decomposition, a white earthy clay, largely used, under the term of *Kaolin*, in the manufacture of porcelain.

58. *Albite* or *Soda Feldspar :*—White, red, greyish-white, &c., sometimes with pale-blueish or pearly opalescence. Triclinic in crystallization, but occurring commonly in lamellar masses, readily cleavable in two directions under angles of 93° 36' and 86° 24'. One of the cleavage planes usually exhibits a delicate striation. In other respects, Albite closely resembles Orthoclase. $H = 6.0$; sp. gr. 2.55 — 2.65. Average composition : silica 68, alumina 20, soda (with trace of potash, &c.) 12. Albite is a constituent of many trappean rocks, and it occurs also in certain granites and syenites, and in various metamorphic strata. An opalescent variety, known as *Peristerite*, occurs in the township of Bathurst (Lot 19, Con. 9), and also on the north shore of Stony Lake, in Burleigh. A white variety in cleavable masses of considerable size forms the feldspathic portion of certain granitic and gneissoid rocks of the more northern districts of the county of Ottawa. Fine crystals are also said to occur in a vein on Lake Massawippi (Stanstead), in the metamorphic region of the Eastern Townships.

59. *Oligoclase :*—White, greenish, pale-grey. Triclinic in crystallization, and closely allied in all its characters to Albite, but containing a somewhat smaller percentage of silica. The cleavage planes meet at angles of 95° 50' and 86° 10'. Occurs in Canada, according to Dr. Sterry Hunt, associated with black amphibole in the eruptive mass of Mount Johnson, in the district of Iberville, near the east shore of the River Richelieu.

(6) Group of Calcareo-Feldspathic Silicates.

[The minerals of this group are very closely related to those of the preceding division, but they are essentially lime-holding, and contain a lower per centage of silica. They are more readily fusible, moreover ; and are decomposed, or at least strongly attacked, by hydrochloric acid.]

60. *Labradorite* or *Lime Feldspar:*—Grey, greyish-white, greenish white, greyish-blue, with frequently reflections of a beautiful blue, green, orange or other colour. Triclinic in crystallization, but rarely occurring otherwise than in cleavable lamellar masses, the cleavage planes, which usually present a delicate striation, meeting at angles of 93° 40′ and 86° 20′. H=6.0 ; sp. gr. 2.66—2.76. BB, in thin splinters, readily fusible. Decomposed, or strongly attacked, in powder, by hydrochloric acid. Average composition : silica 53, alumina 30, lime 12.5, soda 4.5. This species enters into the composition of various trappean rocks, and it also forms, both alone and in admixture with other triclinic feldspars, large masses of crystalline structure associated with gneiss and other metamorphic strata. In this latter condition, it predominates amongst the Upper Laurentian or so-called Labrador series of Canada (see Parts III. and V.). Fine examples occur in St. Jérôme, Morin, Abercrombie and Mille Isles, in the County of Terrebonne, north-west of Montreal ; and in boulders (probably from the above sources) scattered over Grenville Township, on the Lower Ottawa. The Labradorite of these localities is frequently opaque-white on the surface from semi-decomposition or weathering. A pale-blue and greyish variety, without opalescence, occurs in Château Richer (Montgomery County), below Quebec. Pale greenish-blue and other opalescent examples have been obtained from boulders in the Townships of Drummond and Lanark, west of the Ottawa ; and a range of feldspathic rocks, presenting fine examples of colour-reflecting Labradorite, occurs on the north shore of Lake Huron, east and south-east of French River. The occurrence of Labradorite at the latter locality was first made known by Dr. Bigsby. A granular variety, in which the opalescent tints only shew under the magnifying glass, occurs with scattered pyrites in a vein traversing Laurentian gneiss in the vicinity of Haliburton, Ontario.

61. *Andesite :*—This is a somewhat doubtful species apparently intermediate in character between Albite or Oligoclase and Labradorite. A reddish, feldspathic mineral in cleavable and striated masses, from the Labrador rocks of Château Richer, below Quebec,

is referred to it by Dr. Sterry Hunt. Report for 1863, p. 478. Average composition : silica $59\frac{1}{2}$, alumina $25\frac{3}{4}$, lime $7\frac{3}{4}$, soda 5, potash 1. Sp. gr. 2.66—2.67.

62. *Anorthite :*—This species is also very closely related to both Labradorite and Albite. It occurs in Triclinic crystals and cleavable masses of a greenish-white, reddish, pale-grey and other colour, with H = 6 — 6.5, and sp. gr. 2.66 =2.79. Fusible, and more or less readily decomposed by hydrochloric acid. Average composition : silica 44 — 47, alumina 30 — 35, lime 14 — 18; with small percentage of soda, potash, &c. A variety found in boulders in the vicinity of Ottawa city was originally described under the name of Bytownite. Some of the feldspar of Château Richer, according to Dr. Sterry Hunt, belongs probably to this species. The feldspar which enters into the composition of the diorite of the Yamaska Mountain is also referred to it by the same observer; and fine crystals of Anorthite, according to Mr. Thomas Macfarlane, occur in a large dyke of dioritic porphyry, of which several rocky islets in the vicinity of Thunder Cape, Lake Superior, are mainly composed.

.*. The two following species are placed for convenience in this group, as they are essentially lime-containing silicates, fusible, and decomposable in hydrochloric acid.

63. *Wernerite* or *Scapolite :*— White, grey, red, greenish, &c. Occurring in crystals of the Tetragonal System (mostly combinations of a square-based prism and pyramid), and in lamellar, columnar and sub-fibrous masses. H = 5.5 — 6.0 (under normal conditions, but often somewhat lower from incipient decomposition of the specimen) ; sp. gr. 2.6 — 2.8. BB, easily fusible, mostly with strong bubbling. Partially decomposed by hydrochloric acid. Average composition : silica 48, alumina 28, lime 18, soda 5, the latter sometimes largely replaced by potash. Carbonate of lime and a small percentage of water are very constantly present in altered or weathered specimens. Scapolite occurs in the Laurentian limestones of Calumet Island, and in Grenville Township, on the Ottawa, as well as in most of the phosphate deposits of the Ottawa region. Also in large crystals and cleavable masses, with sphene and augite, in the Laurentian strata of Hunterstown, Maskinonge County, Quebec; and at Golden Lake, in Algona Township, County of Renfrew. An altered or semi-decomposed variety in violet-red or

greyish-red cleavable masses, from the vicinity of Perth, in Lanark County, has been described under the name of *Wilsonite*.

64. *Wollastonite :*—White, pale-greenish, brown, grey, &c. Clino-rhombic in crystallization, and of the pyroxene type, but occurring sp. gr. 2.7 — 2.9. BB, more or less readily fusible. Decomposed, commonly in thin tabular masses of fibrous structure. H = 4.5— 5.0 ; with separation of gelatinous silica, by hydrochloric acid. Essential composition : silica 51,7, line 48.3. In Canada, fibrous Wollastonite occurs in many of the crystalline limestones of the Laurentian series, mixed more or less intimately with pyroxene, mica, quartz, and other minerals. Grenville Township in Argenteuil County, St. Jerome and Morin in Terrebonne, North Burgess in Lanark, and Bastard in Leeds County, are the best known localities.

(7) GROUP OF NEPHELETIC SILICATES.

[This group includes a small number of essentially anhydrous silicates of alumina and soda, in some of which chloride of sodium is also present, whilst others contain traces of chlorine or sulphuric acid. They fuse more or less readily, and gelatinize in acids. Canadian examples are comparatively unimportant.]

65. *Nepheline*, including *Elæolite :*—White, brownish, greenish, blueish-grey, yellowish, dull-red. Hexagonal in crystallization, but occurring commonly in cleavable masses of a more or less greasy or vitreo-resinous lustre, forming the variety known as Elæolite. H = 5.5 — 6.0 ; sp. gr. 2.5 — 2.65. BB, easily fusible. Decomposed readily by acids, with separation of gelatinous or slimy silica. Average composition : silica 44, alumina 44, soda 17, potash 5. This species is said to occur in small orange-red granular masses in boulders with orthoclase and black amphibole, on Pie Island, Lake Superior. Also, it is stated by Dr. Sterry Hunt, in white crystals in the granitic trachyte of Brome. *Cancrinite*, a closely allied silicate, occurs, according to Dr. B. J. Harrington, in the nepheline-syenites of Belœil and Montreal.

66. *Sodalite :*—Blue, greyish, colourless, &c. Regular in crystallization, but occurring mostly in small granular masses. H = 5.5 — 6.0 ; sp. gr. 2 15 — 2.35. BB, melts with bubbling. Gelatinizes in acids. Recognized in the form of small grains of a fine blue colour, by Dr. Sterry Hunt, in the granitic trachyte of Brome. In

composition, essentially a silicate of alumina and soda, combined with 6 or 7 per cent. of chloride of sodium.

(8) GROUP OF ZEOLITIC SILICATES.

[The silicates of this group are essentially hydrous species, especially characteristic of trappean or basaltic rocks. All fuse more or less readily, the fusion in many cases being preceded by intumescence, or accompanied by bubbling, whence the old name of the group from ζέω and λίθος. Most of these minerals also gelatinize in acids, or become readily decomposed with separation of granular or slimy silica. Those which occur in Central Canada may be arranged in two sub-groups, comprising (a) *Calcareous Zeolites;* and (b) *Alkaline Zeolites.*]

SUB-GROUP A. CALCAREOUS ZEOLITES.

67. *Prehnite :*—Green of various shades, greenish-white. Rhombic in crystallization, but occurring mostly in botryoidal masses with

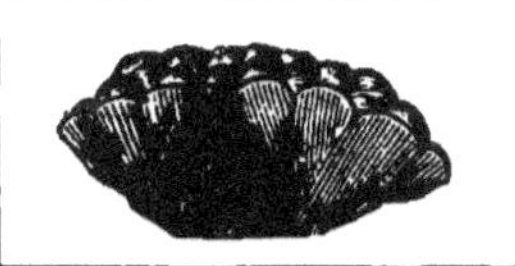

crystalline surface and radiating fibrous structure (Fig. 71). H = 6.0 — 6.5 ; sp. gr. 2.8 — 2.95. BB, easily fusible with great bubbling. Attacked by boiling acids with separation of granular silica,

FIG. 71.

but complete decomposition is not readily effected Essential composition : silica 43.5, alumina 25, lime 27, water 4.5. Occurs chiefly in the trap rocks of Lake Superior, sometimes forming distinct veins, as on Slate River, an affluent of the Kaministiquia, and with imbedded nodules of native copper on an island near St Ignace. A specimen, obtained by the author from Slate River, shewed a sp. gr. of 2.88, and yielded : silica 43.41, alumina 23.80, sesquioxide iron 1.26, sesquioxide manganese 0.53, lime 26.62, water 4.14. The " Chlorastrolite " from Isle Royale is probably a variety of Prehnite. It occurs in small nodular masses in amygdaloidal trap ; or in the form of small pebbles, left by the disintegration of the trap, on the beaches of the island. Its colour is mostly dark and light green in alternate patches, and it exhibits a radiating fibrous structure and somewhat silky aspect. Sp. gr. 2.90 to 3.20, the heavier specimens invariably containing minute particles of magnetic iron ore, forming nuclei from which the fibres radiate. The amount of water varies from a little over 4.0, to about 5.5 per cent.

68. *Datolite* :—White or pale-green.   Clinorhombic in crystallization, but occurring chiefly in botryoidal masses of fibrous structure.
H = 5.0 — 5.5 ; sp. gr. 2.95 — 3.0.   BB tinges the flame pale-green,
and melts with great bubbling.   In the bulk-tube yields about 5 per
cent. water.   Gelatinizes in hydrochloric acid.   Average composition :
silica 38, boracic acid 21.5, lime 35.5, water 5.0.   Of doubtful
presence in Central Canada, but bel ved to occur sparingly in some
of the trap rocks of Lake Superior.   Abundant on the south shore
of the lake, and present also on Isle Royale.

69. *Laumontite* :—White, greyish, pale-red, &c. Clinorhombic but
mostly in fibrous groups.   H = 3.5 — 4.0, but often less from
incipient decomposition of the mineral ; sp. gr. 2.2 — 2.35.   BB,
exfoliates, and melts to a white enamel.   Gives off a large amount
of water in the bulb-tube.   Gelatinizes in hydrochloric acid.   Average composition : silica 52, alumina 21, lime 12, soda 4.5, water
15.   Occurs in the amygdaloidal traps of Lake Superior, but
mostly in a weathered condition.

70. *Thomsonite* :—White, red, &c.   Rhombic, but most commonly found in indistinct acicular crystals and fibrous groups.   H = 5.0
—5.0 ; sp. gr. 2.3—2.4.   BB, intumesces, and melts into a blebby
glass.   Gives off water in the bulb-tube.   Gelatinizes in hydrochloric
acid.   Average composition : silica 38, alumina 30, lime 13, soda
4.5, water 13.5.   Of somewhat doubtful occurrence, but some of
the zeolites from the amygdaloidal traps of Lake Superior belong
apparently to this species.   They are mostly in a weathered and
semi-decomposed condition.

71. *Heulandite.*   72. *Stilbite.*   73. *Chabazite.*   These minerals
are essentially hydrous silicates of alumina and lime.   They are said
to occur in some of our trappean rocks, but nowhere in distinct or
well-characterized examples.

SUB-GROUP B : ALKALINE ZEOLITES.

Natrolite :—White, yellowish, &c.   Rhombic in crystallization,
but occurring most commonly in radiating fibrous groups.   H = 5.0
— 5.5 ; sp. gr. 2.15 — 2.35.   BB, tinges the flame strongly yellow,
and melts very easily without intumescence. In the bulb-tube, yields
a large amount of water.   Gelatinizes in acids.   Average composition : silica 47.5, alumina 27, soda 12, water 9.5.   Occurs, but
mostly in a weathered condition and in part altered into carbonate of

lime, in some of the amygdaloidal traps of Lake Superior. Also in trap rocks around Montreal.

75. *Analcime :* — White, greyish, pale-red, &c. Regular in crystallization : the crystals mostly trapezohedrons. Occurring also in small granular masses. $H = 5.0 — 5.5$; sp. gr. $2.2 — 2.3$. BB, imparts a yellow tinge to the flame, and melts without bubbling or intumescence into a more or less clear glass. In the bulb-tube, gives off water in considerable quantity. Gelatinizes in acids. Average composition: silica 54, alumina 23, soda 14, water 9. Occurs in the amygdaloidal traps of Lake Superior, and more especially, with native copper, on the Island of Michipicoten. It is said also to be present, here and there, in the trap rocks of the district around Montreal.

76. *Apophyllite :*—White, pale-red, &c. Tetragonal in crystallization, but occurring commonly in lamellar masses of a somewhat pearly aspect. (The crystals exhibit a pearly opalescence on the basal plane, and a vitreous lustre on the other faces.)* $H = 4.5$; sp. gr. $2.3 — 2.4$. BB, easily fusible, and yielding a large amount of water in the bulb-tube. Decomposed, with separation of floccnlent silica, by hydrochloric acid. Apophyllite differs from other zeolitic minerals in being non-aluminous. Average composition : silica 52, lime 25, water 16, potassium fluoride $6\frac{1}{2}$. Occurs in pale reddish and colourless foliated masses, mixed with calcspar, in the silver-bearing vein of Prince's Location, on Spar Island, Lake Superior.

(9) GROUP OF MICACEOUS AND CHLORITIC SILICATES.

[The silicates of this group possess an eminently fissile or foliaceous structure, in consequence of which they admit of separation into plates or scales of extreme tenuity. Although differing more or less in composition, they form a connected series, commencing with species which consist essentially of silicates of alumina and potash, anhydrous or nearly so, and terminating in hydrated silicates of alumina and magnesia, but with intermediate species in which both potash and magnesia are present, and in which the amount of water gradually increases. Many of these intermediate links, however, have not been met with, as yet, in Canada. Through the hydrated magnesian species, there is a transition into the talcose minerals of the next group.]

---

* See Types of Crystallization, in the author's Mineral Tables, page 276.

9

77. *Muscovite or Potash Mica* :—Silvery-white, grey, brown, green, black, with pseudo-metallic pearly lustre.   Rhombic in crystallization : the crystals usually six-sided tables or prisms with strongly-pronounced basal cleavage, but distinct crystals are comparatively rare. Most commonly in foliated or scaly masses, tough and flexible. H = 1.5 — 2.0 or cleavage surface, somewhat higher on edges of folia.   Sp. gr. 2.7 — 3.1.   BB, whitens, and melts on the thin edges. In the bulb-tube, usually gives off a small amount of water.   Not attacked by acids.   Average composition : silica 46, alumina 30, sesquioxide iron 4, potash 10, water (and traces of fluorine) 2 to 4. In some bright green varieties, 3 or 4 per cent. of oxide of chromium is present.

Muscovite is an essential component of ordinary granite, gneiss, mica slate, and other crystalline rocks.   It occurs, thus, more or less abundantly throughout the Laurentian area of Canada (Part V), and also amongst the metamorphic series of the Eastern Townships. Most commonly it forms small scaly masses, but,. as stated by Sir William Logan, large crystals and plates occur in a vein of graphic granite (see Part III) on Allumette Lake, north of Pembroke in Renfrew County, and with black tourmaline on Yeo's Island, in the Upper St. Maurice.   Large crystals of mica (apparently Muscovite) are also said by Dr. Bigsby to occur in granite at Cape Tourmente below Quebec.   A green chromiferous variety in the form of small scales in magnesite and dolomite has been recognized by Dr. Sterry Hunt in the Eastern Townships of Sutton and Bolton.

78. *Phlogopite* (Magnesia Mica) :—Yellowish-brown, brownish-red, olive-green, yellowish-green, blueish-grey, &c., with pearly-metallic lustre.   Rhombic :—but occurring mostly in six-sided plates and broad foliated masses, or in scaly particles, tough and elastic.   Cleavage strongly pronounced in one direction.   H = 2.0 – 2.5 ; sp. gr. 2.72 – 2.85.   BB, whitens, and generally melts at the point and edges.   In the bulb-tube, most varieties yield traces of moisture. Attacked, in powder, by hot sulphuric acid, the silica separating in fine scales.   Average composition : silica 41, alumina 13 to 18, magnesia (with some oxide of iron (&c.) 30, potash (with soda) 8 to 10, water and fluorine 1 to 4.

This species occurs most generally in association with the crystalline limestones and pyroxenic beds of the Laurentian series, and it is abundant in all the phosphate deposits of the Ottawa region.   Large plates of economic value for stove-fronts, lanterns, &c., are obtained in the townships of Grenville, Buckingham, Templeton, &c., and in North and South Burgess.   On the north shore of Rideau Lake in Burgess, large six-sided plates and prisms, associated with apatite and calcite, occur in great profusion.   Translucent greenish-yellow prisms with calcite and diopside occur also at Calumet Falls.

Note :—Lepidolite or *Lithia Mica* has not yet been recognized in Canada, although abundant in Maine and in Connecticut.   It is mostly in granular scaly masses of a pink, red, or greyish colour, melting easily and with much intumescence before the blow-pipe, and colouring the flame carmine-red.   *Biotite* is another magnesian mica, closely related to Phlogopite, described above, but Hexagonal in crystallization, and mostly black or green in colour.   It is of doubtful occurrence in Canada, but a dark-green mica from Moor's Slide on the Ottawa has been referred to this species.

79. *Seybertite* (Clintonite) :—Brown, brownish-red.   Mostly in small scaly or foliated masses with pearly-metallic aspect.   H = 4.0 or less ; sp. gr. 3.0 – 3.1.   BB, whitens but does not fuse, or melts only on the thinest edges.   Gives off water in the bulb-tube.   Decomposed, in powder, by sulphuric acid.   Contains a comparatively small amount of silica.   Average composition : silica 20, alumina 40, iron oxide 4, magnesia 20, lime 13, water 3.   Occurs sparingly in crystalline limestone with blue spinel in the seignory of Daillebout, Joliette County, Province of Quebec.

80. *Chlorite* (Pennine) :—Dark-green, greenish-grey.   Hexagonal or Hemi-Hexagonal in crystallization, but occurring principally in scaly or foliated masses, and frequently in a more or less earthy condition, or in granular slaty masses.   H = 2.5 or less ; sp. gr. 2.6 – 2.8.   Sectile.   Flexible in thin pieces, but not elastic.   BB, generally melts upon the edges.   In the bulb-tube yields water.   Decomposed, in powder, by hot sulphuric acid.   Average composition : silica 33, alumina 13, iron and chromium oxides 6, magnesia 35, water 13.   Occurs chiefly in crystalline strata south of the St. Lawrence, forming beds of chloritic slates which often carry copper ores, as in Cleveland, Bolton, Shefford, Melbourne, Ascot, and other Eastern Townships.   In Sutton, St. Armand, Brome, and elsewhere in the same district, it occurs in admixture with specular iron ore,

forming schistose beds or chloritic iron slates. Also in sub-foliate
or more or less compact and sectile beds, of economic value as po[
stones in Bolton and Broughton. In the form of chloritic rock
masses this mineral is especially characteristic of Huronian strata
and is thus largely present throughout the wide oxtent of countr
north of Lake Huron and Lake Superior.

81. *Chloritoid* :—Greyish-green, greenish-black, dark grey, I
thin lamellar masses, and also. occasionally in imbedded nodules c
foliaceous texture. H=5.5—6.0 ; sp. gr. 3.5 — 3.6. BB, in th
outer flame, becomes red ; in the inner flame, dark and magnetic
but resists fusion, or vitrifies only on the thinnest edges. Yield
water in the bulb-tube. Decomposed by sulphuric acid. A dark
greenish-grey variety from Leeds, analysed by Dr. Sterry Hun
yielded : silica 26.30, alumina 37.10, protoxide of iron 25.92, pr(
toxide of manganese 0.93, magnesia 3.66, water 6.10. Occurs i
many of the schistose strata of the Eastern Townships, more e
pecially in Brome and Leeds.

Loganite (Altered Hornblende) :—This substance is derived apparently fro
the alteration and partial decomposition of hornblende (No. 52, above). I
occurs in small dull brown crystals, resembling those of hornblende, in th
crystalline limestone of Calumet Falls on the Ottawa, associated with serper
tine, phlogopite and apatite. H about 3.0 ; sp. gr., according to Hunt, 2.6
—2.64. Infusible. Partially attacked by acids. Dr. Hunt's analysis shewe(
silica 33.28, alumina 13.30, magnesia 35.50, iron peroxide 1.92, volatile matte
(water, &c.) 16.00.

Hydrous Diallage :—This is also a product of alteration, derived apparentl
from augite (No. 53, above). It occurs in cleavable masses of a greenish-gre
or pale green and somewhat waxy lustre, associated with apatite, calcite an
sphene, in crystalline limestone in North Elmsley, and also in association wit
phlogopite in North Burgess. H=1.5—3.0 ; sp. gr.=2.3—3.55. Infusibl(
or nearly so. Composition somewhat variable, but essentially, after Hunt
analysis : silica 36.50—39.70, alumina 10.80—14.80, magnesia 25.62—28 2(
iron protoxide, 4.32—9.54, water 14.0—17.66.

Another variety, in greenish and somewhat pearly masses with H=5.0 an
sp. gr. about 3.0, from the Eastern Township of Orford, yielded Dr. Sterr
Hunt : silica 47.10, alumina 3.50, magnesia 24.58, lime 11.34, iron protoxid
8.55, water 5.5. It is related apparently to the hydrous bronzite describe
below.

Hydrous Bronzite :—Also an alteration product, derived apparently fro
augite. Occurs in bronze coloured cleavable and scaly masses in schistos
strata in the township of Ham. Dr. Hunt's analysis gives : silica 50.0
magnesia 27.17, iron protoxide 13.90, lime 3.80, water 6.30.

(10) Group of Talcose Silicates.

[The minerals of this group are essentially non-aluminous magnesian silicates, foliated or compact in texture, very sectile, and more or less greasy or soapy to the touch.]

82. *Talc*, including *Steatite* or *Soapstone* :—Silvery or greenish-white, pale-green, greyish ; often mottled in grey and greyish tints. Hexagonal in crystallization, but occurring mostly in foliated or scaly masses, the folia flexible but not elastic (*Talc*), or in beds of a sub-granular, slaty or compact texture (*Steatite* or *Soapstone*). Very sectile, and more or less soapy to the touch. H=1.0—2.0 ; sp. gr. 2.56—2.8. BB. sometimes exfoliates, but melts only on the thinnest edges. In the bulb-tube, yields, generally, a little water. Scarcely attacked by acids. Average composition ; silica 63, magnesia 33, water 4 ; but some specimens contain merely a trace of water, whilst others yield 6 or 7 per cent. Talc appears to be of rare occurrence in the more ancient metamorphic rocks of Canada, but a bed of grey and dark greyish-green steatite, mixed with magnesian carbonate of lime occurs near the village of Bridgewater in Elzevir, County of Hastings. It is also said to have been found in Galway or Somerville. In the higher metamorphic strata south of the St. Lawrence, talcose slates, on the other hand, are not uncommon, and beds of steatite are comparatively abundant. These lie principally in the townships of Bolton, Sutton, Potton, Stanstead, Leeds and Vaudreuil. As shown by Dr. Sterry Hunt, they frequently contain traces of oxide of nickel. A bed of mottled and pale-green steatite of excellent quality has been found recently by Mr. Peter McKellar near Thunder Bay, on Lake Superior. A specimen analysed by the writer, yielded : silica 62.67, magnesia 33.40, oxide of iron 0.86, water 1.88. The more compact kinds of steatite are capable of economic employment in the manufacture of fire-bricks, stoves, baths, gas burners, culinary vessels, table ornaments, &c., and other varieties are used as a paint material. Powdered talc or steatite appears also to be occasionally added to ordinary lead paint with a view to produce increased lustre ; and it has been employed to lessen friction in machinery.

Pyrallolite or *Renselaerite* :—This substauce agrees essentially in composition and general characters with steatite. but presents the cleavage and occasionally the crystalline form of augite. It is evidently a product of alteration. Mostly greenish-white, pale-green, grey, or brownish, with somewhat waxy lustre. Very sectile. H = 2.5 – 3.0 ; sp. gr. 2.6 – 2.8. Infusible, or fuses only on the thinnest edges, but differs (although perhaps only in some examples) from steatite, proper, in being partially attacked by hot sulphuric acid. Occurs in beds in the crystalline limestone of Grenville on the Ottawa. Also in the townshiphs of Ramsay, Rawdon, and Lansdowne. The Grenville variety yielded Dr. Sterry Hunt: silica 61-60, magnesia 31.06, iron protoxide 1.53, water 5.60.

83. *Serpentine* (including *Retinalite, Chrysotile, &c.*) :—Greyish green, oil-green, greyish-yellow, brown, reddish, sometimes nearly white, and often veined or mottled. Mostly in compact, granular, slaty masses, forming rock-beds. Occasionally fibrous, and then commonly known as *Serpentine-asbestos* or *Chrysotile*. Also at times in pseudomorphous crystals derived from the alteration of chrysolite, pyroxene, amphibole, spinel, and other species. H = 2.5 —4.0, but in general about 3.0. Very sectile. Sp. gr. 2.2 (fibrous varieties) to 2.66 ; usually about 2.5. BB, fusible on thin edges only. In the bulb-tube yields water. Decomposed by heated acids, especially by sulphuric acid. Average composition : silica 42, magnesia 42, iron protoxide 3, water 13. In Canada, serpentine is met with abundantly in the older crystalline or Laurentian rocks, and still more extensively in the crystalline strata of the Eastern Townships and Gaspé. In Laurentian rocks it occurs principally in connection with the crystalline or Eozoon limestones, as, more especially in the township of Grenville on the Ottawa, on Calumet Island, and in the township of Burgess in Lanark County. These Laurentian serpentines are mostly pale-green or yellowish, often with spots and streaks of a reddish-brown colour. Serpentine occurs also, in smaller quantities, in some of the iron-ore beds of Belmont and Marmora. In the metamorphic strata south of the St. Lawrence, large beds, mostly intermixed with carbonate of lime or dolomite, and thus forming serpentine marbles of more or less beauty, occur in the townships of Melbourne, Orford, Broughton, Bolton, Ham, and Garthby, and abundantly around Mount Albert in Gaspé. In many of these localities, the serpentine is closely associated with chromic iron ore. In St. Francis in Beauce County, serpentine occurs also in connection with magnetic and titaniferous iron ore ; and in Roxton, Brompton, Orford, and other townships of that district, it often carries copper pyrites and other copper ores. Finally, a fibrous asbestiform variety

is found in Melbourne, Bolton, Ham, Thetford, Coleraine, Broughton
and other eastern townships, where within the last three or four years
it has been extensively mined, and it now forms an important article
of commerce.

Aphrodite :—White, yellowish-white.   In soft, earthy, or waxy-looking
masses, strongly adherent to the tongue.   Essentially a hydrated silicate of
magnesia, allied to serpentine and also to meerschaum.   Occurs in small quantities, in a bed of steatite or pyrallolite (see under No. 62 above) in the township of Grenville on the Ottawa.

(11) Group of Kaolinic Silicates.

[The minerals of this group much resemble in aspect and general
characters the talcs and steatites of the preceding group.   Foliated
examples present a pearly lustre and talcose appearance, and compact
and granular varieties are more or less soapy to the touch and
adherent to the tongue.   Th se Kaolinic silicates, however, differ
from the talcose species in being essentially non-magnesian.   They
are hydrated silicates of alumina, or of alumina and potash, and are
evidently products of alteration, derived from the decomposition of
feldspathic and other aluminous silicates.   As in the case of all
substances of this kind, composition and physical characters are
necessarily somewhat variable, numerous so-called species might be
made out of these products if slight points of difference were taken
into consideration; but Canadian examples may be referred to two or
three types, as given below.]

84. *Kaolinite* or *Pholerite :*—Pearly-white, pale-green, greenish-
grey, and sometimes red from admixture with scaly red iron ore.
Occurs in soft unctuous scaly masses, and also in a more or less
compact and granular condition.   Very sectile, and soapy to the
touch. $H = 1.0 - 2.0$: sp. gr. $2.33 - 2.63$.   BB, sometimes
exfoliates or expands in bulk, but remains unfused.   In the bulb-
tube, yields a large amount of water.   The light-coloured varieties
assume a fine blue colour after ignition with nitrate of cobalt (See
Operation 3, under the Application of the Blowpipe, in Part I.).
Scarcely attacked by acids.   Average composition : silica 46, alumina
40, water 14.   Occurs in fissures of a sandstone of the Quebec group
near Chaudière Falls (Dr. Sterry Hunt); also, according to Dr.
Hunt, in films in the joints of some of the quartzose sandstones of
the Huronian series.   A red ferruginous variety in strongly soiling

particles which become lustrous when rubbed, occurs in Madoc and
elsewhere in the counties of Hastings and Peterborough, and probably in other parts of that region. Finally, it may be observed that
many of the metamorphic slates of the Eastern Townships appear to
owe their nacreous talcose aspect the presence of kaolinite, or to that
of the related non-magnesian silicates described under Pinite, below.

85. *Pinite* (including *Aluminous-Agalmatolite* and *Parophite*, &c.):
—Greenish or greyish-white, dull-yellow, grey, green, brown, &c. In
compact, granular, and sometimes slaty masses : also occasionally in
pseudomorphous crystals. Very sectile, and more or less unctuous to
the touch. H = 2.5 to 3.5 ; sp. gr. 2.65 — 2.8. BB, infusible, or
fusible with difficulty on the edges only. The light-coloured varieties
assume a blue tint after ignition with nitrate of cobalt, In the bulb-
tube, yields water. More or less attacked by acids. Average composition : silica 45 to 55, alumina 25 to 35, iron oxides 1 to 4, potash
6 to 10, with small amounts of magnesia, soda, &c., and from 5 to 8
per cent. of water.

The term Pinite (from the Pini mine near Schneeberg in Saxony)
was originally restricted to certain brown pseudomorphous crystals
apparently derived from the decomposition of Iolite, but it is now
applied by Dana so as to include a number of related substances of
various colours and modes of occurrence. These substances are essentially hydrated silicates of alumina and potash, much resembling the
magnesian steatites and serpentines in their physical characters. One
of the best known is the Chinese Agalmatolite or Figure-stone, but
many of the so-called agalmatolites are magnesian in composition, and
identical with steatite. Dr. Hunt refers the Wilsonite (see No. 63,
above) to this group, on account of its composition ; but its physical
characters are quite distinct from those of the typical pinites and agalmatolites. It wants the sectility and soapy feel, for example, so characteristic of these latter, whilst it possesses, on the other hand, a distinctly spathoid structure.

The agalmatolite variety occurs in beds and layers amongst the
strata of the Eastern Townships of Canada, especially in St. Nicholas
(Lévis), where it forms green and greenish-white layers in an indura-
ted clay-slate of the Quebec group (see Part V.) ; also near St. Francis
(Beauce), in yellow, waxy-looking, semi-translucent layers ; and on
Lake Memphramagog in Stanstead, where it occurs in yellowish beds,

one of which presents a sub-fibrous silky aspect, in chloritic slate. Analysis of these varieties by Dr. Sterry Hunt, will be found in the elaborate Report of the Geological Survey for 1863.

Glauconite (Green Sand) :—This substance occurs only in the form of small grains and specks of a green colour, distributed through sandstone and other rocks. These grains appear to consist essentially of a hydrated silicate of alumina, potash, and iron oxide. They occur in a sandstone of the Quebec group near Point Lévis, and on the Island of Orleans. Certain bright-green markings in the siliceous Black River limestones of Lake St. John, in Rama, have also been referred to Glauconite.

(12) Group of Copper and Nickel Silicates.

[The minerals of this group, as regards Canadian examples, are comparatively unimportant. They are essentially hydrated silicates of an amorphous or earthy structure : products of decomposition of copper and nickel ores.]

86. *Chrysocolla* :—Green, greenish-blue, occasionally passing into brown and black. In amorphous masses, and in earthy crusts on copper ores, frequently mixed with malachite. $H = 1.5 — 4.0$; sp, gr. 2.1 — 2.3. BB, blackens, and imparts a green colour to the flame-border, but does not fuse. In the bulb-tube yields a large amount of water. Attacked and decomposed by heated acids. Average composition : silica 34, oxide of copper 45, water 21. The brown and black varieties are intermixed with iron and manganese oxides, or with black oxide of copper. In Canada, found sparingly amongst some of the copper ores of Lake Superior.

87. *Genthite* (Nickel-Gymnite) :—Pale-green, greenish-yellow. Occurs in earthy crusts, and in amorphous masses sometimes with botryoidal surface. $H = 1.5 — 4.0$; sp. gr. 2.2 — 2.5. BB, blackens, but remains infusible, In the bulb-tube gives off a large amount of water. A soft earthy variety from Michipicoten yielded Dr. Sterry Hunt : silica 35.80, oxide of nickel 32.40, water 12.20; but in another specimen (less thoroughly dried before analysis) the amount of water was found equal to 17.10 per cent. Hitherto only recognized in Canada in a vein on the Island of Michipicoten, Lake Superior. The vein traverses amygdaloidal trap, and carries small grains and rounded masses of native copper and native silver.

I. CARBONATES.

[This subdivision comprises the natural compounds of Carbonic Acid (now commonly called carbon dioxide) with various bases, such as lime, magnesia, and the like. In acids these compounds become decomposed with strong effervescence, the latter effect being due to the liberation of their carbonic acid, but in many cases the application of heat is required to develop the phenomenon. The substance, in the form of a small particle or two, or in powder, may be conveniently examined, with some diluted hydrochloric acid, in a test-tube or deep watch-glass supported over a common spirit-lamp. (See under "Action of Acids," in Part I.) The carbonates, also, when fused with borax before the blowpipe, dissolve with marked effervescence, their carbonic acid being driven off. Up to the present time, only eight carbonates have been recognized amongst Canadian minerals, and five of these are altogether unimportant. We arrange the whole, therefore, simply under two groups: Anhydrous and Hydrous Carbonates, respectively.]

(1) GROUP OF ANHYDROUS CARBONATES.

[The anhydrous carbonates belong properly to several distinct groups : more especially to a *Rhombohedral Group*, typified by calcite or ordinary calc spar, and including dolomite, magnesite, siderite, &c.; and a *Prismatic Group* of Rhombic and Monoclinic species, typified by Arragonite, and including carbonates of lead, baryta, strontia, &c. But in Canada, the latter group is only represented, and that obscurely, by arragonite or prismatic carbonate of lime.]

88. *Calcite* or *Calc Spar* (Rhombohedral Carbonate of Lime) :— White, grey, reddish white, greenish-white, yellowish-white, red, black, &c., but mostly colourless or lightly tinted. Hexagonal or Hemi-Hexagonal in crystallization, with strongly pronounced rhombohedral cleavage. The crystals are chiefly obtuse and acute rhombohedrous (Figs 72 and 74); combinations of a rhombohedron and hexagonal prism, the so-called "nail-headed" crystals (Fig. 73); and more or less acute scalenohedrons (Fig. 75), the mineral in the latter form being often popularly known as

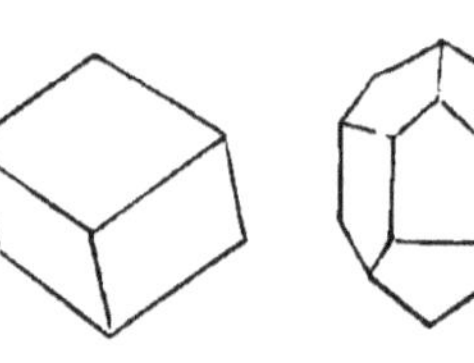

FIG. 72.　　FIG. 73.　　FIG. 74.　　FIG. 75.

"dog-tooth spar." Calcite occurs also abundantly in lamellar,

columnar, fibrous, granular, and earthy masses. The crystals and crystalline masses break readily into rhombohedrons which measure 105° 5′ over a polar edge, and 74° 55′ over other edges. In some of its conditions, this species presents a more or less pearly or silky lustre ; and all transparent specimens exhibit in certain directions a strongly-marked double refraction, as in the so-called " Iceland Spar." This is best shown by placing a rhombohedron, as obtained by cleavage, with its broader faces over a ruled line or other thin object, and turning the crystal so as to make it revolve around this. In the direction of a line joining the obtuse plane angles of the rhombic face, the two images coalesce ; but in the opposite direction they are more or less widely separated, according to the thickness of the crystal. H = 3.0 in crystals and cleavable masses, but less in earthy varieties. Sp. gr. = 2.5 — 2.75, mostly about 2.7. BB, infusible, but glows strongly and becomes caustic, the carbonic acid being expelled. Readily soluble with strong effervescence in diluted acids, without the aid of heat. Normal composition : carbonic acid 44, lime 56, but a small portion of the lime is very generally replaced by magnesia, protoxide of iron, protoxide of manganese, &c.

The varieties presented by this mineral are comparatively numerous. Those which occur in Canada may be arranged under three divisions, comprising : (*a*) Crystals, and crystalline cleavable varieties ; (*b*) Concretionary and stalactitic varieties ; (*c*) Rock varieties.

(*a*) *Crystallized and cleavable varieties of Calcite* :—Rhombohedrons and scalenohedrons of calcite occur in many of the mineral veins on the north shore of Lake Superior ; at the Bruce and Wellington Mines, Lake Huron ; in the galena-bearing lodes of Galway, Ramsey, Loughborough, &c. ; and in some of the copper lodes of the Eastern Townships. In a " pocket " or " vug " in the Shuniah vein north of Thunder Bay, the writer observed a large bunch of scalenohedral crystals, many of which measured upwards of 18 inches in length. Some large scalenohedrous have also been observed at the Wellington Mines on Lake Huron. Fine cleavable and transparent masses of calcite occur at Harrison's Location on the Island of St. Ignace, Lake Superior ; and others, perfectly fit for optical purposes, were found in abundance in the upper part of the main shaft at the Galway lead mine in North Peterborough.

Crystallized examples occur likewise in hollows and in fissures of many of our Silurian and Devonian strata, as more especially, in the Trenton limestone near Lachine, and in the same formation in the township of Huntingdon in Hastings County; in dolomitic beds of the Quebec group near Point Levis, opposite Quebec; and in the Niagara formation in the vicinity of the Great Falls, Hamilton, Dundas and elsewhere. Many of the amygdaloidal trap rocks of Lake Superior and Lake Huron, also, enclose nodular cleavable masses of calcite, and occasionally the more open amygdaloidal cavities are lined with crystals. These are almost always scalenohedrons, or combinations in which one or more scalenohedrons predominate.

(b) Concretionary and Stalactitic varieties of Calcite :— These varieties are being constantly formed by deposition of carbonate of lime from springs and streams in limestone districts, and from water percolating through limestone rocks. Carbonate of lime, consisting of equal atoms or combining weights of lime and carbonic acid, is comparatively insoluble in water, but the bicarbonate, containing two parts of carbonic acid to one of lime, dissolves to a certain extent. Water contains very generally a small amount of free carcarbonic acid, derived from the atmosphere, decaying organic matter, &c., and thus it is enabled to take up a certain quantity of carbonate of lime, this becoming converted into bicarbonate. The latter compound, however, is extremely unstable. It parts with carbonic acid very readily, even by simple exposure to the air. The insoluble carbonate thus again results, and is necessarily precipitated from the water, the precipitation often taking place upon moss, roots and other organic bodies, converting these into so-called "petrifactions."
Water issuing from limestone strata often deposits concretionary masses of carbonate of lime, in this manner, as at Hamilton, Rockwood, the Falls of Noisy River, the Banks of Beaver River in Euphrasia and Artemisia, and other places along

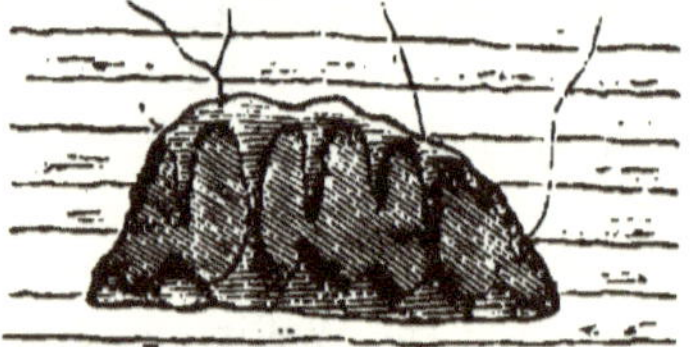

Fig. 76.

the escarpment of the Niagara Formation (see Part V.). Deposits of this kind are commonly known as *Calcareous Tufa.* Specimens from Hamilton, more especially, are hard and solid, and admit of a good polish. They are mostly of a brownish-yellow color. Caverns and hollows of greater or less extent often occur in limestone recks. Water percolating into these through minute fissures in the roof, very generally deposits on the latter a thin coating of carbonate of lime, and then dropping on the floor, deposits there a further portion of calcareous matter. In this manner, the process constantly going on, stalactites and stalagmites originate, the two occasionally meeting in the form of a pillar (as shown in Figure 76). These stalactic deposits usually exhibit a radiated fibrous structure, with frequently a botryoidal surface. Some large stalactites have been obtained from a cavern at the lower falls of the Nottawa River in Mono (Geological Report, 1863, p. 334); and others, of smaller size and less symmetrical form, have been found in adjoining townships.

(c) Rock Varieties :—These come properly under review in Parts III. and V. of this work. They comprise the various kinds of limestone, including : Crystalline Limestone, the finer varieties of which are commonly known as Marble ; Ordinary Limestone ; Lithographic Limestone ; Oolitic Limestone, composed of minute spherical concretions ; Earthy Limestone or Chalk, and so forth. In Canada, valuable beds of marble occur in the Laurentian strata of Renfrew (Arnprior), McNabb, Grenville, Wentworth, Bastard, Marmora, Elzevir, &c. ; and in the metamorphic region south of the St. Lawrence, as in St. Armand, St. Joseph, Melbourne, Orford, Dudswell and elsewhere, many of the marbles from these localities being mixed with green and other coloured serpentine. In some of the unaltered Lower Silurian strata, also, red, grey, black and brown marbles occur ; as at St. Lin, Caughnawaga, St. Dominique, Montreal, Cornwall, Point Clare and Pakenham. See further, under Part V.

89. *Arragonite* (Prismatic Carbonate of Lime) :—Colourless, and of various colours—yellow, blueish, brownish-red, &c. Rhombic in crystallization, and often in compound crystals which sometimes present a pseudo-hexagonal aspect. Also in fibrous and stalactitic masses. H = 3.5 — 4.0 ; sp. gr. 2.9 — 2.95. BB, infusible, but becomes opaqu and falls into powder. Soluble in acids with strong effervescence. Composition identical with that of calcite : carbonate of lime being thus a dimorphous body—*i.e.*,, a substance capable of assuming two distinct sets of physical characters. Fibrous arragonite

appears to occur sparingly amongst the Lake Superior traps; and occasionally in stalactitic coatings on the sides of cracks in some of our limestone rocks, as in the township of Tring, and elsewhere, but no very distinct or crystallized examples have as yet been found.

90. *Dolomite* (Pearl Spar, Bitter Spar) :—White, grey, brownish, &c. Crystallization Hemi-Hexagonal, the crystals being, mostly, rhombohedrons, the faces of which are often more or less curved. Occurs also in lamellar cleavable masses, with cleavage angles of 106° 15′ and 73° 45′, and in granular and rock masses. $H = 3.5$—4.0; sp. gr. 2.8—2.95. BB, infusible, but becomes caustic. Slowly soluble in cold acids, but rapidly dissolved with strong effervescence if the acid be gently heated. Essential composition : carbonic acid, lime, and magnesia, forming carbonate of lime 54.35, carbonate of magnesia 45.65, but small portions of the lime and magnesia are very generally replaced by protoxide of iron and protoxide of manganese, by which the cleavage angle is slightly altered. The various rhombo-hedral carbonates, Calcite, Dolomite, Magnesite, Siderite, Rhodochrosite, &c., merge, in fact, into each other by intermediate transitional forms, to some of which distinct names have been given. The ferru-ginous and manganesian dolomites become brown by weathering.

Crystals and crystalline varieties of dolomite occur in many of the metalliferous veins of Lake Superior and Lake Huron, and occasion-ally in those of the Eastern Townships and other parts of Canada. Groups of small rhombohedrons of more or less pearly aspect, have been obtained, more especially from the Wellington Mines on Lake Huron. Small rhombohedral crystals occur also in cavities and on the sides of cracks, &c., in many limestone strata : as in the dolo-mitic limestones of the Calciferous formation near Prescott on the St. Lawrence, and Rigaud on the Ottawa; and also in the dolomitic beds of the Niagara Formation in the vicinity of the Falls and else-where.

In the form or rock-masses, dolomite is of very common occur-rence in many parts of Canada. A white fine-granular crystalline variety, or dolomite marble, occurs in Laurention strata at Lake Mazinaw, in the Township of Barrie, Frontenac County; and many of the marbles from the altered strata of the Eastern Townships are more or less magnesian or dolomitic. In the unaltered Silurian series, beds of dolomite, of a more or less sub-crystalline texture, make up the strata of the Guelph Formation, as seen in the Townships of Elora, Guelph, Dumfries, Waterloo, Bentick, &c.; and dolomitic limestones, or mixtures of limestone and dolomite, belong to the various other formations of this series, more especially to Calciferous, Chazy, Niagara, and Onondaga strata, as described under these divisions in Part V.

91. *Magnesite :*—White, brownish, &c. Hemi-Hexagonal in crystallization, the crystals mostly obtuse rhombohedrons; but occurring commonly in cleavable masses (with cleavage angles = 107° 29' and 72° 31')' and in granular and rock varieties. H (in pure varieties) = 3.5 — 4.5; sp. gr. 2.8 — 3.0, or slightly higher in the brown ferruginous varieties. BB, infusible. Soluble in heated acids with effervescence. Normal composition : carbonic acid 52.4, magnesia 47.6, but part of the magnesia usually replaced by protoxides of iron and manganese. In Canada, this mineral occurs only in rock masses, forming beds in the altered strata of the Eastern Townships of Sutton and Bolton, south of the St. Lawrence, where it is associated chiefly with serpentine and steatite.

92. *Rhodochrosite, or Carbonate of Manganese :*—This species has not yet been found in Canada in distinct examples, but it occurs in admixture with many of the manganese ochres (No. 96), and is also present, in traces, in some of the altered strata of the Eastern Townships. Colour, rose-red or pale-red, weathering brown.

93. *Siderite* or *Spathic Iron Ore* (*Spherosiderite, Clay Iron Ore,* &c.) :—Yellowish, greyish, light and dark brown, green, &c. Occurs under several conditions, and more especially : (1), in rhombohedrons, scalenohedrons, and lamellar masses, with cleavage angles of 107° and 73° (Spathic Iron, proper); (2), in spherical or concretionary masses with radiating fibrous structure in trappean rocks (Spherosiderite); and (3), in nodular masses and occasionally in layers, mostly of a brown colour and earthy or dull stone-like aspect (Clay Iron Ore). Crystalline varieties of this mineral have not yet been recognized with certainty in Canada; but nodules and thin layers of clay ironstone or clay iron ore occur in the Devonian strata of Gaspé, associated with a small seam of impure coal, and with fossilized plant-remains. This variety is a mixture of carbonate of iron (more or less converted into brown iron ore) with argillaceous mat-

ter.   Although rarely yielding more than 25 or 30 per cent. of iron,
clay ironstone, as occurring in the Carboniferous strata of Europe
and the United States, supplies a large number of furnaces, and
yields metal of good quality.   The nodules have usually a strongly-
marked slaty structure ; and, when broken, they almost invariably
exhibit the impression of a fern frond, fish skeleton, or other organic
body.   Small fragments after ignition before the blow-pipe, or in a
glass tube held over a common spirit lamp, assume at first a red
colour, and then become black and magnetic.

93 *bis*. *Strontianite* :—This carbonate is stated by Dr. B. J. Harrington to occur in the form of white fibrous tufts in cracks in some
concretionary limestone masses in the Utica slate of St. Helen's
Island, Montreal.   It imparts a crimson colour to the blowpipe
flame.

(2) GROUP OF HYDROUS CARBONATES.

[This group is only represented in Canada by the somewhat problematical Dawsonite, and by the two cupreous carbonates Malachite
and Azurite ; and these latter species do not occur in well characterized examples, but merely as incrustations on Copper Ores, or in the
form of stains and small earthy masses in copper-holding rocks.]

94. *Dawsonite* :—In white, thin-bladed aggregations or coatings on
compact trachyte, Montreal, H = 3 ; sp. gr. 2.4, contains, according
to Dr. Harrington, alumina, soda, lime, carbonic acid, and water.

95. *Malachite* or *Green Carbonate of Copper* :—Green of various
shades, with pale-green streak.   Monoclinic in crystallization, but
crystals exceedingly rare.   Mostly in botryoidal masses of concentric
lamellar, and fibrous structure ; in earthy coatings on copper ores ;
and in the form of streaks and markings in copper-holding rocks.
H = 3.5 — 4.0 (in the solid state); sp. gr. 3.7 — 4.0.   BB, tinges
the flame green, and becomes rapidly reduced to metallic copper.
Soluble in acids with effervescence.   Essential composition : carbonic
acid 20, oxide of copper 72, water 8.   Occurs in small quantities with
copper glance, native silver, &c., in a calc-spar vein on Spar Island,
Lake Superior, and in small earthy incrustations and markings amongst
many of the copper ores and associated veinstones of Lake Superior
and Lake Huron, generally.   Also under similar conditions in Madoc,
Marmora, and various other localities in which copper pyrites occur
in larger or smaller quantities ; and especially in the chlorite and

other altered rocks of the metamorphic country south of the St. Lawrence, as in the townships of Leeds, Halifax, Inverness, Ham, Shipton, Cleveland, Stukely, Bolton, Brome, Sutton, &c.

96. *Azurite* or *Blue Carbonate of Copper* :—This species has hitherto been recognized only in small incrustations and stains of a blue colour, associated with malachite, at most of the localities named under No. 95, above. The blue carbonate contains : carbonic acid 25.6, oxide of copper 69.2, water 5.2.

K. SULPHATES.

[The mineral substances placed under this division may be regarded, according to the commonly received view, as compounds of sulphuric acid with one or more oxidized bases, such as baryta, lime, oxide of lead, alumina, and the like. As regards physical characters, these bodies exhibit a non-metallic aspect, and either a colourless or a very faintly-coloured streak, the colour in the latter case being green or blue, or occasionally yellow. They afford representatives of all the systems of crystallization : Rhombic and Clino-Rhombic types being especially abundant. $H = 1.0 — 4.0$. The sulphates may be easily distinguished from carbonates, phosphates, silicates, &c., by fusion in a reducing flame on charcoal with carb-soda ; or better with a mixture of carb-soda and a little borax, as the latter reagent facilitates the decomposition of earthy sulphates, and prevents the absorption of the fused mass. An alkaline sulphide is formed by this treatment. When moistened, and placed on a piece of silver or on lead test-paper (a bright coin or glazed visiting card may be used as a substitute), the fused mass produces a black or brown stain of sulphide of silver or sulphide of lead. The stain may be easily removed from the silver by friction with moist bone-ash.

Amongst the sulphates generally, several natural groups stand out with great prominence. The Rhombic group of anhydrous species, for example, containing the Sulphates of Baryta, Strontia, Lime, Lead, &c. ; the Gypsum group ; the Monometric group of Alums ; the Prismatic group of Vitriols ; and others of subordinate importance. The sulphates hitherto found in Canada, are too few, however, to admit of distribution into special groups of this kind. In the descriptions which follow, the anhydrous species Barytine and Celes-

tine are placed first. To these, succeeds the hydrous sulphate, Gypsum ; and a few sapid types of obscure or comparatively rare occurrence, close the list.]

97. *Barytine* or *Heavy Spar* :—White, yellow, reddish, pale-blue,
grey, &c. Crystallization Rhombic (Figs. 77–78, and other combinations). Occurs very commonly in lamellar masses and aggregations
of large flat crystals with cleavage angles of 101° 40', 78° 20', and 90°,
yielding a right rhombic prism. Also in masses of a granular or more
or less compact structure. H = 3.0 — 3.5 ; sp. gr. 4.3 — 4.7, mostly
about 4.4 — 4.5. BB, generally decrepitates, tinges the flame
pale-green, and melts with great difficulty, often at the point only,
into a white enamel. Dissolves entirely in carb-soda before the
blowpipe. Not attacked by acids.
Normal composition : sulphuric acid
34.33, baryta 65.67. This mineral
occurs abundantly in many parts of
Canada. In the Laurentian strata,

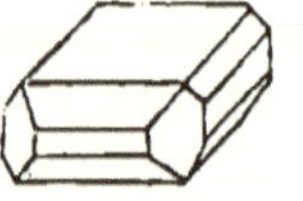

Fig. 77.

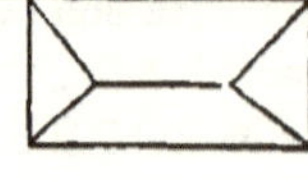

Fig. 78.

it occurs in veins *per se*, and as a gangue or veinstone with galena—
more especially in the townships of Lansdowne in Leeds County ;
Bathurst and North Burgess in Lanark County ; McNab, Renfrew
County ; Dummer and Galway, in Peterborough County ; and Sommerville in Victoria County. A broad vein of white crystalline
heavy spar is exposed along the side of the road in lot 7 of the 10th
concession of Hull, near the Gatineau River. Red crystals were
discovered by Mr. Murray on Iron Island, Lake Nipissing ; and other
examples have been met with in the copper-ore veins of Lake Huron.
Isolated pale reddish-yellow crystals (Fig. 78) were found by the
writer (*Canadian Journal*, November, 1885,) in veins in Neebing
Township near Fort William, Thunder Bay, Lake Superior, and subsequently in other mineral veins in that region. Massive and sub-
crystalline varieties from also large veins on Jarvis Island, near
Pigeon River west of Fort William, and also on Pie Island ; and
other veins of a similar character are said to occur east of Thunder
Cape, as at Edward Island in Black Bay, and elsewhere. Heavy Spar
has also been noticed in some of the serpentines and other altered
strata of the Eastern metamorphic region south of the St. Lawrence,
as on the Bras River, where a white variety occurs in small veins.
Nodular masses of a red or reddish-yellow colour occur with fibrous

and granular gypsum in the Hudson River strata of Cape Rich on Georgian Bay ; and small crystals and crystalline masses are occasionally found in cavities of the dolomitic limestones of the Calciferous and Niagara groups, as near Brockville, and in the vicinity of Niagara Falls. Heavy Spar is employed in the manufacture of paints, and is too frequently used in this connection as a fraudulent substitute for white lead. It is also the chief source of the baryta salts of the laboratory.

98. *Celestine :*—White, blue, grey, pale-red, &c. Rhombic in crystallization, the crystals frequently bearing a close resemblance to those of heavy spar. Occurring also in lamellar and crystalline masses, with cleavage angles of about 104°, 76°, and 90°, yielding a right rhombic prizm ; and in masses of fibrous or granular structure. H = 3.0 - 3.5 ; sp. gr. 3.95 — 3.97. BB, imparts a crimson colour to the point and border of the flame, and melts into a white alkaline enamel. Dissolves entirely, by fusion, in carb. soda. Not attacked by acids. Normal composition : sulphuric acid 43.6, strontia 56.4. This mineral occurs chiefly in sedimentary rock-formations : very rarely in mineral veins or in crystalline rocks. In Central Canada, it is found somewhat abundantly in the interior of small cavities in the Black River or Trenton limestone of Kingston ; and also, with crystals of dolomite, gypsum, fluor spar, blende, and other minerals, in cavities in the Niagara limestone, as in the vicinity of the Falls ; around Owen Sound ; on Drummond Island ; and on the Grand Manitoulin, Lake Huron. A red variety has been recently found by Mr. Roche in freestone of the Medina formation at the forks of the Credit. Celestine is the principal source of strontia salts, used in pyrotechny to impart a red colour to rockets and signal lights, and for laboratory purposes.

99. *Gypsum* (Hydrous Sulphate of Lime, Selenite, &c,) :—White grey, yellowish, pale-red, &c. Monoclinic in crystallization, the crystals very commonly as in Fig. 79 *a*, or in arrow-headed twins as in Fig. *b*, also in lamellar and foliated crystalline masses with strongly pronounced cleavage in one direction, and in fibrous and granular masses, the latter often forming rock deposits. The cleavage planes present a more or less pearly aspect, the other

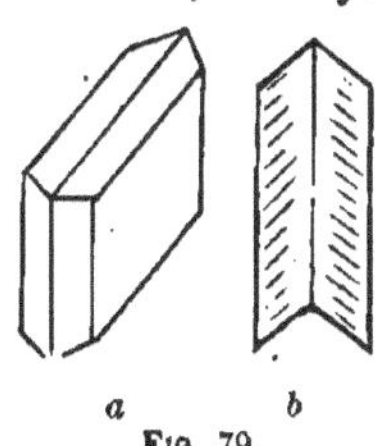

Fig. 79.

crystal-faces exhibiting a vitreous or pearly-vitreous lustre. Granular and rock varieties have mostly a dull earthy aspect. H = 1.0, 2.0 ; sp. gr. 2.25 — 2.35. Sectile, and, in thin lamellæ, somewhat flexible. Becomes opaque when held at the edge of a lamp or candle-flame. BB, exfoliates, and melts into a white caustic enamel. In the bulb-tube yields a large amount of water. Soluble in hydrochloric acid. Dissolves also, if in fine powder, in a large amount of water, and more readily in a solution of rock salt. Normal composition : sulphuric acid 46.51, lime 32.56, water 20.93. The transparent crystals and cleavable varieties are commonly termed *selenite ;* and the fibrous and fine granular varieties form the *alabaster* and *satin spar* of lapidaries, but these names are also bestowed on similar varieties of carbonate of lime. When deprived of its water by exposure to a low red heat, gypsum is converted into *plaster* or *Plaster of Paris.*

Crystalline and fibrous masses, and occasionally distinct crystals of gypsum, associated with crystals of quartz, dolomite, &c., occur in cavities of many of the Silurian strata in Canada, and thin bands are interstratified in places with the shales and limestones of some of these formations. Gypsum occurs under these conditions in the Calciferous formation of Beauharnois, the Hudson River formation of Point Rich on Georgian Bay, the Medina formation of St. Vincent, and in the Clinton and Niagara strata in the vicinity of the Falls, Hamilton, Dundas, and elsewhere.

Rock masses of granular and compact gypsum, more or less mixed with carbonate of lime, characterize the Onondaga Formation of Western Canada, and occur largely in the valley of the Grand River : more especially in the townships of Dumfries, Brantford, Oneida, Seneca, and Cayuga ; as well as throughout the tract of country, generally, between the eastern extremity of Lake Erie and the mouth of the Saugeen. (See under the Onondaga Formation, in Part V.) The greater part of the gypsum from these localities is ground for agricultural use.

100. *Epsomite* (Epsom Salt) :—White or greyish. Soluble : taste, strongly bitter. Rhombic in crystallization, but occurring chiefly in fibrous tufts and earthy or botryoidal incrusting masses. H = 2.0, or less. BB, runs at first into liquid fusion, and then forms an alkaline infusible crust which assumes a flesh-red colour if moistened.

with a drop of nitrate of cobalt and again ignited. Yields a large
amount of water in the bulb-tube. Normal composition : sulphuric
acid 32.52, magnesia 16.26, water 51.22. Occurs in Canada, as an
efflorescence or incrustation, on exposed surfaces, and on the edges of
the planes of bedding, of shales and other strata, where it is formed
apparently by the action of percolating water containing soluble
matters derived from the decomposition of pyrites. It occurs thus
in some of the slaty talcose layers associated with the iron ores of
Marmora, and also on the weathered shales of the Utica series, near
Montreal, Quebec, and Collingwood ; and still more abundantly on
some of the dolomitic beds of the Clinton and Niagara Formation, as
near Dundas and elsewhere. Sulphate of magnesia occurs also in
solution in the Tuscarora water, and in some other mineral springs.

101. *Iron Vitriol* (Green Vitriol, Copperas, Melanterite, &c.) :—
Pale-green, greenish-white ; brownish-yellow by partial decomposition.
Monoclinic in crystallization, but occurring mostly in efflorescent
crusts and minute hair-like indistinct crystals. Soluble : taste, inky
and metallic. $H = 2.0$, or less. BB, blackens and becomes magnetic. In the bulb-tube yields a large amount of water, and gives
off sulphurous acid. The aqueous solution gives a deep-blue precipitate with "red prussiate of potash ;" and in general also with
the yellow prussiate, from the presence of more or less sesquioxide
of iron. Normal composition : sulphuric acid 28.8, protoxide of
iron 25.9, water 45.3. Occurs on decomposing pyrites and marcasite,
and on the exposed surfaces of rocks in which these minerals are
present. It is thus found, in small quantities, on many of the ores
from the mineral veins of Lake Superior, Lake Huron, the Hastings
region, and other parts of Canada. A specimen of iron pyrites from
the Galway Lead Mine in the northern part of the county of Peterborough, became covered in the course of a few weeks with delicate
tufts of minute acicular crystals of this mineral.

102. *Nickel Vitriol* (Morenosite):—Pale-green, greenish-white. In
efflorescent tufts of minute crystals on nickel ores. Soluble : taste,
strongly metallic. BB, evolves sulphurous acid, swells up, and forms
a dark grey mass. With borax, gives reactions of nickel oxide (see
Part I.). In the bulb-tube yields a large amount of water. If free
from iron, the aqueous solution does not yield a blue precipitate with
red or yellow "prussiate of potash." Normal composition : sulphuric

acid 28.5, oxide of nickel 26.7, water 44.8.   Detected by Dr. Sterry
Hunt, as an efflorescence on an aresenical nickel ore from the Wallace
Mine, Lake Huron.   (See No. 17, above.)

103. *Alum :*—Normally, white, but sometimes stained of a yellowish or brownish colour by sesquioxide of iron and other impurities.
Monometric in crystallization, but occurring commonly in earthy
efflorescent crusts.   Soluble : taste, sharp and more or less bitter.
BB, froths up and forms a white earthy mass which assumes a fine
blue colour if moistened with a drop of nitrate of cobalt, and again
ignited.   Normal composition : sulphuric acid 33.75, alumina 10.82,
potash 9.95, water 45.48.   Occurs in considerable abundance on the
exposed face of some high bluffs of argillaceous shale (belonging to
the Animikie series) on Slate River, a tributary of the Kaministiquia,
about twelve miles west of Fort William, Lake Superior.

L. PHOSPHATES AND ARSENIATES.

[These compounds are composed of phosphoric acid or arsenic acid
with various bases.   They present a vitreous or other non-metallic
aspect.   Phosphates when moistened with a drop of sulphuric acid
(and many without this addition), impart a green colour to the point
of the blowpipe flame.   When fused, in powder, with carb. soda in
a platinum spoon, an alkaline phosphate is formed, soluble in water.
The clear solution decanted from the insoluble residuum, and acidified
by a few drops of nitric acid, yields a canary yellow precipitate with
a drop or two (or small fragment) of ammonium molybdate.
Arseniates, when mixed in powder with some carb. soda, and ignited
on charcoal in a reducing flame, emit a very distinct odour of garlic.
Canadian examples, of this group, amount to only three in number,
as given below ; but, one of these, the lime fluor-phosphate. Apatite,
occurs in comparative abundance, and is a substance of great commercial value.]

104. *Apatite* (Phosphate of Lime) :—Green, blueish-green, violet-
red, rose-red, brownish, greenish-white, &c.—shades of green and
dull-red being often present in the same specimen.   Lustre, vitreous
and vitreo-resinous, with frequently a slight opalescence on one of
the cleavage planes: Crystallization, Hexagonal : the crystals consisting most commonly of six-sided prisms, often of large size, and

frequently with rounded edges. Occurs also in
lamellar cleavable masses, and occasionally in
globular and other shapes with fibrous structure.
H = 4.5 – 5.5, normally 5.0. Sp. gr. 2.9 – 3.3,
most commonly about about 3.18 to 3.2. BB,
in most cases quite infusible, but some varieties

Fig. 79.*

vitrify slightly at the point of the assay-fragment after exposure to
a long-sustained blast. The powder moistened with sulphuric acid
tinges the flame-point distinctly green. Melts and dissolves readily
in borax and phosphor-salt, forming a glass which becomes opaque on
cooling or when flamed. (See under Blowpipe-reactions in Part I.)
Easily soluble in nitric or hydrochloric acid. The diluted solution,
saturated with ammonia, yields a copious white precipitate of phosphate of lime. This precipitate assumes a canary-yellow tint if
treated with a solution of nitrate of silver, or if a crystal of that
substance be laid in it whilst still moist. The presence of phosphoric
acid may also be rendered evident in the diluted nitric acid solution
by the formation of a clear-yellow precipitate with molybdate of
ammonium.* Apatite consists essentially of phosphate of lime (or
calcium, phosphorus and oxygen) combined with in general about 8
or 10 per cent. of fluoride of calcium or chloride of calcium, or with
a mixture of both, the fluoride usually preponderating. Canadian
examples appear to be essentially fluor apatites. The normal composition of an apatite of this kind is equivalent to : phosphoric acid
42.26, lime 55.60, fluorine 3.37 ; or tribasic phosphate of lime 92.26,
fluoride of calcium 7.74 ; but samples even when dressed for shipping
usually contain a good deal of intermixed calcite and mica scales, and
rarely run higher in tribasic phosphate than about 80 per cent.

Extensive deposits of this mineral, chiefly in the form of veins,
occur in the Laurentian strata of North Burgess and North Elmsley
in the County of Lanark. These veins cut the enclosing strata
transversely, and vary in width from an inch or two to several feet.
The apatite, in crystals, and in cleavable and granular masses, is
associated with phlogopite, pyroxene, and other silicates. Where

---

* The test-solution is prepared by dissolving some of the crystallilzed molybdate in a very
small quantity of water, nitric acid being added to the solution until the cloudiness or thick
recipitate, which forms at first, becomes redissolved. When this is added to the solution
of the mineral, the whole must be gently warmed. A yellow coloration, succeeded by a
yellow precipitate, then quickly ensues.

the veins occur in contact with crystalline limestones, these latter contain in many places detached crystals and grains of apatite, with occasional masses of that substance. The most important phosphate region, however, lies on the left bank of the Ottawa, in Buckingham and adjacent townships, where the apatite occurs in the form of large lenticular masses and crystals in broad veins, with pyroxene, magnesian mica (phlogopite), calcite, and other minerals.* Apatite occurs also in connexion with crystalline limestone, associated with fluor spar and octahedrons of black spinel, in the township of Ross in Renfrew county on the Ottawa; and with quartz and and calcite, at Calumet Falls. Small shews also are seen in many of the limestone bands throughout the Laurentian country between the Ottawa and Georgian Bay. Transparent pink and purple crystals are also reported by Dr. Sterry Hunt to occur in association with crystals of augite in a mass of erupted dolerite (see Part III) at St. Roch on the River Achigan. Apatite has likewise been found, in a quartz vein carrying copper pyrites and native copper, with large plates of white mica, in the township of Burford, in the metamorphic district south of the St. Lawrence.

Finally, it may be observed, small nodular masses consisting in great part of phosphate of lime, mixed with carbonates of lime and magnesia, sand, and other matters, are scattered through a conglomerate of the (Lower Silurian) Chazy formation at the Allumette Rapids; and similar nodules occur in limestone strata of the same formation in the townships of Hawkesbury and Lochiel, west of the Ottawa; as well as in strata of the Quebec group at Point Lévis, and on the River Ouelle. These phosphatic nodules present a chocolate or blackish-brown colour, and contain in some cases fragments of the shells of lingulæ (see Part IV) and other organic bodies. They are supposed to be coprolites or fossilized excrementous matters. When heated, they emit an odour of burnt animal matter, and evolve ammonia. Phosphate of lime, when converted into superphosphate by treatment with sulphuric acid, constitutes an agricultural fertilizer of the highest value.

---

* Crystals of apatite consist most commonly of a simple hexagonal prism with large basal plane, but our Canadian crystals, when unbroken, are terminated by the planes of an obtuse hexagonal pyramid, the basal plane being thus entirely suppressed. This combination has hitherto been only seen in the so-called *spargelstein* of German mineralogists, from the mountains near Jumilla in the south east of Spain, and in the variety known as *moroxite* from Arendal in Norway.

104. *Vivianite* (Hydrated Phosphate of Iron) :—Blue, bluish-green (normally, colourless, but becomes blue on exposure); streak pale-blue or blueish-white. Monoclinic in crystallization, with very perfect cleavage in one direction, but found more commonly in bladed and fibrous varieties, and in earthy masses, often forming, when in the latter condition, beds or layers of a certain extent. $H = 1.0 - 2.0$; sp. gr. $2.55 - 2.7$. BB, tinges the flame-point pale-green (from presence of phosphoric acid), and yields a dark magnetic globule. In the bulb-tube gives off a large amount of water. Normal composition : phosphoric acid 28.30, iron protoxide 43.00, water 28.70, but the iron in the coloured varieties is always partly in the state of sesquioxide, and the earthy varieties moreover are usually mixed with a certain amount of clay, sand, iron ochres, manganese ochre, or other foreign matters. In Canada, this mineral has only been found in an earthy condition, underlying a bed of bog iron ore, in Vaudreuil, on the Lower Ottawa.

105. *Cobalt Bloom* (Erythrine, Arseniate of Cobalt) :—Occurs only (as regards Canada) in the form of a slight efflorescence or in-crustation, of a peach-blossom red colour, on the silver-holding calc spar of Prince's Location, on the north-west shore of Lake Superior; and also, but in traces only, in the more recently discovered silver bearing vein near Thunder Cape. Normal composition : arsenic acid 38.25, oxide of cobalt 37.85, water 23.90; but sometimes mixed with arsenious acid.

IV. FLUORIDES AND CHLORIDES.

[This subdivision comprises the compounds of Fluorine and Chlo-rine, respectively, with metallic bases, such as sodium, calcium, alu-minum, lead, silver, and the like. These compounds present a non-metallic aspect; and they exhibit a general resemblance, also, in other characters, to many so-called oxygen salts, more especially to certain phosphates, borates, carbonates, and sulphates. Amongst Canadian minerals, however, as at present discovered, we have but a single representative of each group.]

A. FLUORIDES.

[The only Fluoride as yet discovered in Canada, is the fluoride of calcium, long known under its popular name of Fluor Spar. In a strictly natural classification, this mineral should occupy a place in

the immediate vicinity of the Apatite and Calc Spar groups. The
fluorides generally, when treated, in powder, with hot sulphuric acid,.
evolve fumes of hydrofluoric acid which exert a strongly corrosive
action on glass. The powdered substance may be warmed with some
sulphuric acid in a platinum or lead crucible covered with a glass
plate, when the under surface of the latter will be quickly corroded.
In making the experiment, great care must be taken not to inhale the
evolved fumes, as these are highly injurious. See also under "Blowpipe Reactions," in Part I.)

106. *Fluor Spar :*—Occasionally colourless, but more commonly
violet or amethyst-blue, dark blueish-green, pale-green, pale blueish-
grey, yellow, brownish, or rose-red, the edges and angles of many
crystals being more deeply tinted than the other parts, or sometimes
presenting a distinctly different tint or shade of colour. Streak,
white. Crystallization, regular; the crystals mostly cubes, or cubes
with bevelled edges (Figs. 80 and 81.) The corners of these cubes
break off very readily, in consequence of the strongly-pronounced
octahedral cleavage possessed by the mineral. H = 4.0; sp. gr. 3.1
—3.2. Emits a blueish or other coloured phosphorescent light, when
moderately heated in the form of powder. BB, generally decrepitates violently (see Part I), and fuses into an opaque white bead,
which becomes caustic after strong ignition. Decomposed, with evolution of corrosive fumes, by hot sulphuric acid. The evolved fumes
consist of hydrofluoric acid, which strongly corrodes the surface of
glass. Average composition: fluorine 48·72, calcium (the metallic
base of lime) 51.28.

Fluor Spar occurs very
generally in association with
metallic ores in veins. It
also forms *per se*, or in connection with calcite, the
substance of many narrow

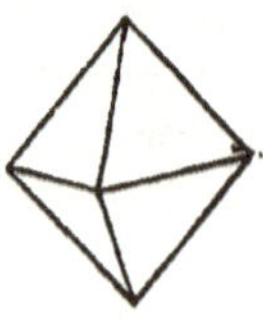

Fig. 80.   Fig. 81.   Fig. 82.

veins; and it occurs likewise in cavities and small fissures in limestone and other rocks, and is occasionally disseminated through beds
of crystalline limestone. The finest examples hitherto discovered in
Canada, have been obtained from a large vug or cavity in a vein of
amethyst-quartz on the north-east shore of Thunder Bay, Lake
Superior. The fluor spar from this spot forms large cubes of two or

three inches in diameter, which rest on equally large pyramids of amethyst-quartz, and are coated with iron pyrites in minute cubes, the whole being surmounted, here and there, by scalenohedrons of calcite. The fluor spar is partly of a pale greenish tint, but mostly of a violet or amethystine colour. Pale green and purple cubes occur also in most of the metalliferous veins of Thunder Bay and the surrounding region, mostly with quartz, calcite, blende, galena, and copper and iron pyrites, as at Prince's Mine, the Shuniah Mine, in several veins in the township of Neebing, and in others near Black Bay and Terrace Bay, on Fluor Island in Neepigon Bay, and elsewhere. Also in amygdaloidal greenstone, near Cape Gargantua. Fluor spar occurs likewise, according to Mr. Murray, in association with specular iron ore, in crystalline limestone on Lake Nipissing. It occurs also, with apatite, in crystalline limestone in the township of Ross, in Renfrew county on the Ottawa, and also, with heavy spar, in Hull, and elsewhere in that district. Also in veins, with galena and calcite, in Trenton limestone in contact with gneiss at Baie St. Paul ; and in narrow veins In the Trenton limestone of the vicinity of Montreal, and the Utica slates of Quebec. Small crystals have likewise been obtained, from fissures and cavities of the Niagara strata, in the neighbourhood of the Falls, and on the escarpment at Hamilton.

B. CHLORIDES.

[This group is represented in Canada by a single type, the highly important Chloride of Sodium, or Rock Salt. The presence of chlorine in mineral bodies is easily ascertained by the blowpipe. Some phosphor-salt, with a few particles of black oxide of copper, is fused in a loop of platinum wire, so as to produce a deeply-coloured glass. To this, a small portion of the test-substance, in powder, is added, and the glass during fusion is held just within the point or edge of the flame. The latter, if chlorine be present, will assume a rich azure-blue colour from the volatilization of chloride of copper. Many chlorides are soluble in water. None possess a metallic lustre, nor is the degree of hardness in any species sufficient to scratch ordinary glass.]

108. *Rock Salt* :—Colourless, and also variously coloured by accidental impurities, as sesquioxide of iron, organic matters, &c., the

imparted tints being mostly red, brownish, violet-blue, yellowish, or
pale-green.   Streak, white.   Crystallization, Regular :
the crystals usually cubes, often with hopper-shaped
depressions on each face—the larger crystals being
composed of numerous minute cubes so arranged as to
produce this peculiarity.   Occurs also in lamellar and
and granular masses.   Cleavage, cubical.   H = 2.0
—2.5 ; sp. gr. 2.1 — 2.25.   Taste, strongly saline.

Fig. 83.
"Hopper-shaped"
cube of salt.

BB, decrepitates strongly (unless very dry), and melts into an
opaque bead, which colours the outer flame intensely yellow.   Normal composition : chlorine 60.66, sodium 39.34 ; but usually, small
portions of chlorides of magnesium and calcium, and sulphates of
lime, magnesia, and soda, are also present.   Most samples contain
likewise a certain admixture of clay or other impurities.

A deep boring on the bank of the River Maitland near the town
of Goderich, commenced at the close of 1865 in quest of rock oil, has
yielded an abundant supply of strong brine of remarkable purity—
thus indicating the presence of a very extensive deposit of salt below
this section of the country.   The boring has been carried down from
the surface gravel and underlying Corniferous Formation into and
apparently through the Gypsiferous or Onondaga strata (see Part
V.), the total depth from the surface being a little over one thousand
feet.   According to Mr. Platt, who conducted the boring, salt in
solid layers was reached at 964 feet from the surface, and the total
thickness of these layers, exclusive of some thin partings of salt-
bearing clay, averages about thirty feet.   An analysis of the brine by
Dr. Sterry Hunt, shews it to be a saturated solution, containing over
26 per cent. of saline matter : 25.90 (equal to 99.018 per cent.) of this,
being pure salt or chloride of sodium.   (See Geol. Reports 1868 and
1869, for various comparative analyses, and much valuable information
on the Goderich and other brines, by Dr. Sterry Hunt.)   Salt has also
been subsequently reached by other borings at Kincardine, Clinton,
and Seaforth, in the same district.   At Seaforth, from information
received from Dr. N. Coleman, a more or less solid bed was struck
at a depth of about 1040 feet.   Chloride of sodium occurs also in
solution in many of our mineral springs, but only in small quantity,
and always accompanied by much chloride of calcium or chloride of
magnesium, sulphate of lime, and other saline compounds, which

interfere with its separation for economic purposes. The Hallowell Spring contains from $3\frac{1}{2}$ to nearly 4 per cent.; the St. Catharines' water about 3 per cent.; and other springs still lower amounts. Quite recently, an announcement of the discovery of rock salt in the township of Combermere, in Renfrew County, has been made in the newspapers, but this requires verification. In all probability, the mass said to have been found was placed there by some of the earlier settlers to prevent cattle from straying in the woods, or by hunters to attract deer. Large masses of rock salt are brought by Quebec ships from Liverpool, in ballast; and blocks of this salt, taken into the woods, have often given rise to pretended discoveries. A block of 80 or 100 lbs. weight will remain undissolved for many years.

V. BODIES OF ASSUMED ORGANIC ORIGIN.

This division includes many salts, resins, coals, and other carbonaceous matters, to which an organic origin is generally attributed; but the supposed derivation of all matters of this kind from organic bodies is by no means free from doubt. The group is represented in Central Canada by a single salt, an oxalate, and by two or three bituminous and carbonaceous substances, one of which, the fluid petroleum or rock-oil, is of great economic value.

109. *Humboldtine* (Oxalate of Iron):—Only known, in Canada, as a yellow incrustation on the bituminous (Devonian) shales of Kettle Point or Cape Ipperwash in the township of Bosanquet on Lake Huron. BB, becomes black and magnetic when gently heated, and is finally converted into red oxide of iron. In the bulb-tube, blackens, and yields a large amount of water. Normal composition: oxalic acid 42.40, protoxide of iron 41.13, water 16-47.

110. *Petroleum* or *Naphtha* (Rock Oil):—Fluid, passing into a semi-fluid and viscous condition. Colour, yellowish-brown or brownish-black in petroleum: pale-yellow, occasionally with a blueish tinge, in naphtha. Highly inflammable. Essential composition: carbon 83 — 88 per cent., hydrogen 12 — 17 per cent. Occurs in rocks of various kinds and of different periods of formation, and is usually thought to have originated from the slow decomposition of imbedded vegetable and animal matters. This view, however, is exceedingly problematical as applied to petroleum generally: regard being had

to the enormous quantities of this substance occurring in so many different parts of the earth; to the unceasing flow of vast numbers of petroleum springs in many localities, age after age, from the earliest periods of history; to the fact that petroleum occurs in many rock-formations—even in ancient gneissoid strata—which lie far below the great Carboniferous and Devonian series (the first, apparently, in which land vegetation has been detected); and to the absence in petroleum-bearing rocks of any special organic remains or peculiar characters suggestive of naphtha-forming capabilities, as compared with strata in which petroleum has not been found. Regard being had to these and other related facts, it is scarcely possible to refer the enormous quantities stored up in subterranean reservoirs, or poured out in flowing springs from age to age, simply to the decomposition of sea-weeds or the soft parts ordinary mollusca, radiata, or lower types, entombed in rock deposits: evidences of these organic bodies being wanting, moreover, in many petroleum-holding rocks, and being far less abundant in others than in various strata in which no traces of petroleum are met with. It might be pretended, with almost an equal show of probability, that all the water on the earth had come from organised bodies, simply because these bodies contain or yield water. A suggestion of this kind would probably have been attempted if water were a substance of comparatively limited occurrence.

In the province of Ontario, petroleum occurs abundantly in springs or wells, arising apparently from reservoirs in the Corniferous (Devonian) Formation, in many parts of the region lying between the more southern point of Lake Huron and the north-west shore of Lake Erie: more especially in the township of Enniskillen; and, less abundantly, in Oxford, Mosa, and Dereham. Small quantities have also been obtained from a well in the Utica (Lower Silurian) Formation of the Great Manitoulin Island in Georgian Bay—the shales of this formation, both there and elsewhere, being more or less saturated with bituminous matter, and thus yielding petroleum on distillation. Many of the calcareous strata of the Niagara, Trenton, and other Silurian Formations, are also more or less bituminous; and liquid and viscous petroleum is occasionally found in the cavities of fossil shells, enclosed in these beds, as well as fosilized corals, &c., of Devonian rocks. Petroleum springs occur likewise in the Devonian strata of Gaspé in Eastern Quebec, as near Douglastown on the St. John River, and on a branch of a small stream known as Silver Brook in the adjacent country, as first made known by the officers of the Geological Survey. Viscous petroleum is cited also, in the Geological Report for 1863, as occurring in cavities, many of which are lined with chalcedony, &c., in a greenstone dyke at "Tar Point" in Gaspé Basin. Indications of petroleum or asphalt have also been noticed in other eruptive dykes of that region.

111. *Asphalt*:—Black, blackish-brown. In solid and also in spongy or semi-viscous masses. H, in the solid varieties, $= 1.0—2.0$; sp. gr. $1.0 — 2.0$. Very inflammable—melting easily, and burning with a yellow flame and emission of bituminous odour. Consists essentially of Carbon, Hydrogen, Oxygen, and Nitrogen, in somewhat variable proportions. In many, if not in all cases, asphalt is derived from petroleum, the two substances passing into each other by insensible transitions. Petroleum thickens and assumes a darker colour under certain conditions of exposure, and finally becomes solid and partially oxidized. The so-called "gum beds" or "mineral-tar deposits" of Enniskillen may be referred to this variety. These beds, which have evidently resulted from the drying up of ancient overflows of petroleum, occupy, in the southern part of the township, two detached areas of about an acre, each, in extent; and they present a thickness varying from a couple of inches to two feet. A small deposit, covered by ten or twelve feet of drift clay, and resting on gravel, occurs in the northern part of the township. This deposit is partly of a leafy texture, somewhat resembling the so-called "paper coal" from the lignite deposits of the Rhine, &c., and its shaly layers exhibit the impressions of leaves and insects in various places. Being mixed moreover with much earthy matter, or "ash," the deposit has all the characters of a small coal-seam.

112. *Anthraxolite*:— Black, lustrous, resembling anthracite in general characters, but very brittle. $H = 2.25 —2.5$; sp. gr. $1.35 — 1.55$. Generally decrepitates when heated. BB, a small fragment loses its lustre, but exhibits no further change. Composition, essentially, carbon, with from 3 to 25 per cent. of volatile matter, including a small amount of moisture. The ash, as at present observed, varies from 0 to 10 or 11 per cent. When present, it exhibits under the microscope no trace of organic structure. This substance, in all probability a product of alteration from petroleum

or asphalt, occurs in narrow veins in rocks of various kinds, and in small masses and thin layers or coatings in strata of the Utica and other formations. Occasionally also, it is found in the interior of orthoceratites and other fossil shells. As it differs essentially by these conditions of occurrence from anthracite proper, the name anthraxolite has been given to it, but simply as a convenient term for present use. It occurs in narrow veins, associated with quartz, amongst the altered strata of Lotbinière, in the Eastern Townships; and also, in regularly banded veins with quartz and iron pyrites, on Thunder Bay, Lake Superior. A variety from the latter district, shewed a sp. gr. of 1.43, and gave the writer: moisture 2.08, additional loss in closed vessel 3.56, ash 0.00, fixed carbon (by difference) 94.36 (Canadian Journal, vol. x. 411). The substance occurs likewise in narrow broken veins, or filling small cracks, *per se*, at Acton and other localities in the Eastern Townships, as well as on the Island of Orleans, at Beauport and Point Levis near Quebec, and elsewhere in the neighbourhood of the latter city. The variable percentage of volatile matter (exclusive of moisture) is evidently due to the greater or less amount of alteration to which the original bituminous matter has been subjected.

113. *Coal :*—Black (often with iridescent tarnish) in anthracite and bituminous coal; brown, in brown coal or lignite. H = 1.0 (or less) — 2.5; sp. gr. 1.0 —1.7. BB, anthracite is scarcely altered; bituminous coals take fire, and many exhibit a kind of fusion. True coal, in its different varieties, occurs in regular beds or layers, mostly associated with bituminous shale, nodules of iron-stone or impure carbonate of iron, and numerous fossilized plants. Anthracite consists almost wholly of carbon (exclusive of a small amount of mineral matter or "ash"). Anthracitic coals contain, in addition, a small percentage of hydrogen, oxygen, and nitrogen; and in bituminous coals, these components are more largely present. Many coals also contain sulphur, derived in chief part, or perhaps wholly, from intermixed pyrites. A thin seam of bituminous coal occurs in the Devonian sandstones of Gaspé, the only known locality within the old limits of Canada in which true coal has been found.*

---

* The great, workable, coal beds of the Dominion of Canada occur at two widely separated geological horizons, namely in the true Carboniferous Formation of Nova Scotia, and in the Cretaceous and Cainozoic deposits of the North-West Territories and British Columbia. Much of this latter coal closely resembles ordinary bituminous coal in general character, and in some places anthracitic varieties occur.

114. *Peat*.—This substance is simply vegetable matter—consisting chiefly of semi-aquatic mosses—in a peculiar state of decomposition. It presents in its more typical form, a brown or blackish-brown colour, with an earthy, or, in places, a sub-slaty or sub-fibrous, texture. Sp. gr. 0.33 — 1.0 [illegible] burning with a pleasant odour and yellow flame. Composition, essentially carbon, hydrogen, oxygen, and a large amount of water (in dried peat, normally from 15 to 25 per cent.) with from 2 or 3 to 10 or 12 per cent. of mineral matter or ash. This valuable fuel, occurs in large beds of more or less modern origin, in various parts of Ontario and Quebec, mostly overlying deposits of shell-marl. The principal localities lie within the townships of Humberstone and Wainfleet on Lake Erie ; Sheffield in Addington County; Beckwith, Huntly, Goulbourne, Westmeath, Nepean, Gloucester, Cumberland, Clarence, Plantagenet, Roxborough, Osnabruck, and Finch, between the west bank of the Ottawa and the St. Lawrence ; Grenville, Harrington, Mille-Isles, and adjacent localities on the east side of the Ottawa ; the Seigniories of Assumption, St. Sulpice, Lavaltrie, and Lanoraie, on the north shore of the St. Lawrence, above Lake St. Peter ; St. Etienne, Champlain, and other places between the St. Maurice and Quebec ; Sherringtoñ, Hemmingford, Longueuil, Ste. Marie de Monnoir, Ste. Roselie, and other localities on the south shore of the St. Lawrence ; the Seigniories of Rivière Ouelle and Rivière du Loup, farther east : near the Métis, Rimouski, and Madaswaka Rivers, in Gaspé ; and largely in the Island of Anticosti. Peat in a properly dried and compressed condition, has been shown of late years to form a good fuel for the use of locomotives, and also for many metallurgical operations.

11

PART III.

ROCKS AND ROCK-PRODUCING AGENCIES.

I. GENERAL CLASSIFICATION OF ROCK MASSES.

The term "rock" in its geological acceptation, includes all the stony and earthy masses — whether consolidated, as granite, limestones, &c., or composed of loosely coherent particles, as sands and gravels—which make up the outer or visible portion of the earth. The mean radius of the earth-mass, or distance from centre to surface is equal to 3956 miles. The elevations and depressions which occur upon the earth's surface, forming mountain-chains and table-lands, valleys and the beds of seas and lakes, are thus, as compared with this radius, of but slight significance. It is necessary to bear this in mind, in order that we may not exaggerate the intensity of the forces by which these inequalities have been produced. In a section or profile in which the same scale is employed for longitudinal and vertical dimensions, the greatest inequalities become scarcely apparent. In order to render evident the differences of level existing between separate points, it is necessary in engineering drawings, and in ordinary geological diagrams, to use a greatly exaggerated scale for heights or depths as compared with horizontal distances ; and the eye unconsciously follows a somewhat similar process in taking in the contour-lines or general aspect of a mountainous region.

Our knowledge of the internal condition of the earth is necessary to a great extent conjectural ; but the weight of evidence, collected in reference to this subject, leads to the conclusion that the earth-mass, from surface to surface, is not throughout a perfectly solid body. In the opinion of some investigators, the central portion is solid, and between this and the consolidated surface-layers a zone of fluid or incandescent matter exists. According to others, the earth-mass is more or less consolidated throughout, but with enormous cavities here and there, filled with molten or fluid matter. The more commonly received opinion, again, infers the surface rocks—technically known as the earth's crust—to extend downward to a more limited depth, whilst the whole of the internal portion is in a condition of igneous fluidity. These views are practically identical, in so far as they assume the presence of molten or incandescent matter, and the existence of a high temperature, at a certain depth beneath the surface rocks; and they are sustained, more especially, by the following data :

(1.) Careful observations made in various parts of the world, shew that a constant temperature is maintained throughout all periods of the year, at a certain depth beneath the earth's surface. The depth varies in different localities, and especially where different kinds of rock occur, but it averages in temperate climates about 100 feet. At lower levels the temperature is found invariably to increase with increase of depth. The ratio of increase is not uniform, being greater or more rapid in some places than in others : but an actual and marked rise of the thermometer from point to point, below the zone of constant temperature, is always observable. The mean ratio of increase, at the limited depths to which researches have been carried, may be assumed to equal 1° Fahrenheit for each descent of 60 feet. At this ratio even, and we may reasonably infer that it would be much accelerated at lower levels, a temperature sufficiently high to maintain most mineral substances in a state of fusion, or in part even in a vaporous condition, would soon be reached.

(2.) Water brought to the surface from great depths by narrow bore-holes commonly known as Artesian wells, always exhibit a higher temperature than the mean temperature of the locality ; and if the boring be increased in depth, the temperature of the water becomes also increased.

(3.) Active volcanoes, which may be regarded as channels of communication between the surface and the internal parts of the earth, are more or less constantly pouring forth, from unknown depths, vast streams of molten rock or lava, accompanied by other products of igneous action. About two hundred and seventy volcanoes are now known to be from time to time in eruption, and many others are apparently in a permanently quiescent state. Eruptions also frequently take place on the bed of the sea.

(4.) Certain rock-masses, in districts now remote from centres of volcanic action, have evidently been forced upwards, from deeply-seated sources, in a molten or more or less incandescent state, amongst previously consolidated rocks. The latter exhibit at the points of contact, and for some distance beyond, changes of colour, and other effects, that can only have resulted from the direct or indirect action of heat. These effects are not seen in all cases of rock-intrusion, but in the great majority of instances they have undoubtedly occurred.

In different localities, as a general rule, the rocks which form the surface of the ground, or which become visible to us on the sides of cliffs and river-banks, in quarries, railway cuttings and the like, are more or less distinct in composition and other characters. This must be familiar to the most casual observer. Thus, around the Falls of Niagara, and extending far and wide across that section of the Province, we find vast beds of dolomitic or magnesian limestone presenting several varieties of texture. About Hamilton and Dundas, with other rocks, ferruginous shales and beds of red marl and grey sandstone are seen. At Toronto, our rock-masses consist of layers of gravel and clay, overlying grey and greenish sandstone-shales. Near Collingwood, and again at Whitby, we observe dark-brown, highly-bituminous shales, containing the impressions of trilobites and lingulæ (see Part IV.), often in great numbers. At Kingston, we meet with limestone rocks differing from those of the Niagara district, and giving place, as we proceed north and east of the city, to beds of crystalline rock of granitic aspect, geologically known as Gneiss. Some of the "Thousand Islands" consist of very ancient sandstone resting on gneiss. At Montreal, with beds of limestone, &c., we see, in the picturesque Mountain, a dark, massive or unstratified rock, a variety of the Trappean series, more or less closely allied to the lavas of volcanic regions; and rocks of a similar kind occur largely on the north shore of Lake Huron, and around Lake Superior, as well as in the Eastern Townships and other parts of Canada.

These examples are sufficient to shew the diversity which prevails with regard to the rock-matters of comparatively neighbouring localities. But if we look, not to the mineral characters of rocks, but to their general conditions of occurrence, by which their respective origins or modes of formation are indicated, we may refer them to two leading groups or sub-divisions, connected by an intermediate group, as in the following scheme :

SEDIMENTARY ROCKS—or Ordinary Stratified-Formations.

METAMORPHIC ROCKS—or Stratified Crystalline-Formations.

ERUPTIVE or UNSTRATIFIED ROCKS.

Sedimentary strata, comprising ordinary sandstones, limestones, &c., consist of detrital or other materials, collected, and arranged in more or less regular layers, by the action of water, as described below.

Metamorphic strata are regarded as consisting wholly or in great part of sedimentary deposits that have been altered or rendered crystalline by heat or chemical agencies. Eruptive rocks are known in many instances to have cooled down from a state of fusion, and are thought in others to have been consolidated from a plastic condition due to aqueo-igneous agencies. They have been formed, or have been brought into this condition, beneath, or deeply within, the Earth's crust, and have been forced upwards from time to time through fissures in the overlying rocks. In each of these divisions—Sedimentary, Metamorphic, and Eruptive—the included rocks belong to various periods of formation.

II. SEDIMENTARY ROCKS.

The rocks of this division make up by far the greater portion of the Earth's surface. Having been formed by the agency of water, they are often called *Aqueous Rocks*. They consist for the greater part of muddy, sandy, and other detrital sediments, collected by the mechanical action of water, and subsequently consolidated by natural processes, as described a few pages further on. Various limestones, however, and certain other rock matters of this division, have been deposited from waters in which their materials were chemically dissolved.

These sedimentary or aqueous rocks are characterized essentially by occurring in beds or strata; secondly, by exhibiting in many instances, a more or less clearly-marked detrital or sedimentary structure; and thirdly, by often containing organic remains. The latter, comprising shells, bones, leaf-impressions, &c. (see Part IV.), are the fossilized parts of animals and plants which lived upon the Earth, or in its waters, during the periods in which these rocks were under process of formation, as described below.

The sedimentary rocks may be conveniently discussed under the following heads: (1) *Composition or mineral characters ;* (2) *Modes of formation ;* (3) *Subsequent changes and effects produced by geological agencies.*

(1) COMPOSITION OF SEDIMENTARY ROCKS.

As regards composition, these rocks fall mainly under the following sub-divisions :

Sandstones, sands, and gravels—or arenaceous rocks.

Clays and clay-slates—or argillaceous rocks.

Limestones and Dolomites—or calcareous rocks.

Conglomerates and Breccias: rocks of variable composition (see below).

Trap tufas: stratified deposits formed out of materials derived from the denudation of trap and greenstone rocks.

Rock matters of carbonaceous origin, as the different kinds of coal.

To these may be added a few other substances of subordinate occur-rence, as gypsum, rock-salt, and bog-iron ore.

Sandstones are nothing more than beds of consolidated sand. They are of various colours, but chiefly present dull shades of yellow, red, brown, or green, and some are nearly pure white. The colouring matter is either sesquioxide of iron, or, in the case of the greenish varieties, a silicate of the protoxide. The harder and purer kinds, as some examples of our " Potsdam sandstone," are called *quartzose sandstones*. In other kinds, a certain amount of carbonate of lime is present, cementing together the component grains of sand, and thus forming *calcareous sandstones*. For special Canadian localities of these and other rocks mentioned under this division, consult Part V. Certain siliceous rocks, called "tripoli" and (erroneously) " infusorial marls," are formed almost entirely of remains of diatoms, microscopic vegetable forms of low organization. (See Part IV.)

Clay Slates are merely consolidated clays. They have a fissile structure, and are mostly of a grey, greenish, brown, or black colour —the dark tints being chiefly derived from the presence of finely dis-seminated carbonaceous or bituminous matter. Clays are also of various colours, as white, greenish, yellowish, blueish, black, and red. Those which contain little or no iron become white or pale yellow on ignition. Many clays are highly calcareous; others, bituminous, &c. The term *shale* is often applied to fissile consolidated clays; but this term, it must be remembered, is applied equally to fissile or slaty limestones and sandstones. When the term is used, therefore, the kind of shale should also be signified: as *an argillaceous shale, an arenaceous shale*, and so forth. *Bituminous shales*, as regards their mineral base, may be also arenaceous, calcareous, &c.

Limestones and *Dolomites* are principally, perhaps, of chemical for-mation. Water containing free carbonic acid (derived from decaying vegetable matter, &c.) dissolves a certain amount of carbonate of lime, but the bicarbonate, thus formed, is easily decomposed by various natural agencies, even by mere exposure to the atmosphere, and a precipitation of calcareous matter takes place. In this manner calcareous tufas (so common in many of our swamps, streams, &c.), together with stalactites and stalagmites, are produced; and similar processes, acting on a larger scale, may have given rise to extensive depositions of limestone strata in ancient seas and lakes. Some limestones, again, are formed almost wholly of the calcareous shells or tests of crinoids, foraminifera, and other organisms (see Part IV): but others are, undoubtedly, mechanical or rock deposits, derived from the wasting of coral reefs and other limestone formations. Limestones consist of carbonate of lime, more or less pure; dolomites, of carbonate of lime and carbonate of magnesia in equal atomic proportions; and dolomitic limestones of these two carbonates in other proportions, the lime carbonate generally predominating. Dolomites and dolomitic limestones appear in many cases to have been simple chemical precipitates, and, in others, to have originited from the alteration of limestone rocks by the action of soluble magnesian salts. These calcareous rocks are of various colours: grey, white, black, yellowish, &c. Their texture is sometimes very close and uniform. At other times, the stone is made up of small spherical concretions, when the texture is said to " oolitic." A bed of grey limestone of this structure occurs near the Chatte River in Gaspé. Oolitic limestones are of all geological ages. Some limestones, again, are of an earthy texture: the well-known chalk of Europe is an example; also our own " calcareous tufa," or "shell marl." Many of the dark limestones, as those of Niagara, &c., are more or less bituminous. Ordieary limestones dissolve in acids with strong effervescence; but dolomites as a rule produce merely a feeble or slightly perceptible effervescence unless the acid be heated.* Limestones which contain from 15 to 25 per cent. of argillaceous matter in intimate admixture, yield

---

*To determine the presence of magnesia in dolomitic limestones, a few grains of the rock may be dissolved in dilute hydrochloric acid. The solution is then boiled with a drop or two of nitric acid (to convert any FeO, that may be present, into Fe^2O^3), and ammonia is added carefully in slight excess. This will occasion a floculent precipitate if iron be present. Oxalate of ammonia is then added to precipitate the lime; this (after settling) is filtered off; the filtrate tested with another drop of oxalate of ammonia to make sure that all the lime has been thrown down; and finally the magnesia is precipitated by sodium phosphate or by solution of the blowpipe flux known as " microcosmic or phosphor salt."

hydraulic or water lime. Beds of this kind occur at Thorold, Cayuga, Loughboro', Kingston, Hull, Quebec, and other localities. See Part V.

Conglomerates consist of rounded pebbles or masses of quartz, sandstone, &c., cemented together, or imbedded in a paste of finer sandstone, or other rock substance. They are often known as "Pudding stones." Examples are not uncommon amongst our Silurian and other strata.

Breccias consist of angular masses or fragments of rocks, cemented together most commonly by calcareous matter. Whilst Conglomerates frequently contain imbedded water-worn materials derived from distant sources, true breccias are necessarily composed of detrital matters derived from neighbouring localities.

Trap-tufas are of comparatively rare or local occurrence. They are made up of materials derived from the wasting of trap or greenstone rocks, and are mostly of a green colour, weathering red. Their texture is generally more or less uniformly fine-grained ; but some occur as conglomerates and breccias, as on the north-eastern shore of Lake Superior, and elsewhere.

The other rock-substances enumerated above—Coals, Gypsum, Rock-salt, and Bog Iron Ore—occur only here and there as stratified rock deposits. For descriptions and Canadian localities, see Part I.

(2) FORMATION OF SEDIMENTARY ROCKS.

The manner in which the ordinary sedimentary rocks, sandstones, shales, &c., have been formed, or built up as it were, is rendered clear by the observation of certain natural processes still in action. We find, for example, at the present day, that sediments of various kinds are constantly carried down by streams and rivers into lakes and seas, and are there deposited. We find, moreover, that the cliffs of many sea and lake coasts are being continually abraded and washed away by the action of the waves. Observation shows also, that the sedimentary matters thus obtained, are always deposited or arranged in regular layers or beds, and that they frequently enclose shells and sea-weeds, together with bones and leaves drifted from the land, and other organic bodies. Hence it is now universally admitted, that, with the exception of certain limestones and dolomites, beds of rock-salt, gypsum, coal, and some other chemical or organic deposits of small extent, all the sedimentary rocks have been formed directly out of pr viosulyexisting rock-masses, by the wearing away or destruction of these; and secondly, that they have all been formed or deposited under water.

In pursuance of this inquiry, consequently, we have to consider, first, the origin or derivation of the sediments of which these rocks are made up; and, secondly, the processes by which the consolidation of the sediments into rock, properly so-called, was affected.

The sediments of which these rocks originally consisted, were derived from previously-existing rocks, by decomposing atmospheric agencies—rain, frost, and so forth; by the action of streams and rivers on their beds; and by the destructive action of the waves and breakers of the sea.

Action of the Atmosphere.—All rocks, even the most solid, are constantly undergoing decomposition and decay. The exposed face of a rock of any kind, for example, soon changes colour, and becomes in general more porous than the other portions of the rock. This effect is technically termed " weathering." Its action gives rise to the production of soils, and frequently causes the fossils contained in the rock to stand out in relief, these bodies being in many cases less easily destructible than the mass of rock itself. Every shower of rain that falls, takes part in this decomposing or disintegrating action, and carries off something, in s lution or suspension, to lower levels—*id est*, into streams, lakes, and seas. Frost, and, in certain localities, carbonic acid and other gases issuing through crevices in the rocks, assist this destructive process. Rain, acting on loosely coherent matters, is known in many districts to have excavated channels of considerable extent. These may become in course of time more or less permanent water courses, and the work of excavation be thus continously carried on.

Action of Streams and Rivers.—The action of streams and rivers, in wearing their channels, is both chemical and mechanical. Calcareous river-beds are wasted bit by bit by the dissolving power of the water, especially during the autumnal season, when dead leaves and other decaying vegetable matters yield the water a large supply of carbonic acid. On the other hand, a mechanical waste is also very generally taking place to a greater or less extent; and thus numerous rivers are continually cutting back their beds, and forming ravines. The Falls of the Niagara River have in this manner gradually receded from the face of the escarpment near Queenston to their present site;

and there is scarcely a river, or small stream indeed, in any part of Canada, that does not exhibit indications of having occupied at one period a wider bed and high level than at present. This erosive power of rivers has probably been assisted in many instances by a gradual elevation of the surrounding land. Some of the grandest examples of river erosion are exhibited by the cañons of the Colorado and other streams west of the Rocky Mountains. In some of these remarkable ravines, the stream has excavated its channel, within almost perpendicular walls of limestone and other rock, to a depth of a thousand feet or more.

The amount of detrital matters borne down by some rivers to the sea, is exceedingly abundant. This is well shown by the formation of deltas. The delta of the Mississippi on this continent, for example, like all other deltas, is derived essentially from the sandy and other matters brought down by the stream. On entering the sea, the velocity of the river is necessarily checked, and the sediments are thrown down. Much of the coarser matter is indeed deposited on the bed of the river itself, raising this, and compelling the formation of artificial banks, or levées, to prevent inundations. Finally, as a well-known illustration of the immense amount of sedimentary matters borne seawards by certain rivers, the case of the Ganges, as described so fully by Sir Charles Lyell, in his "Principles of Geology," may be here cited. That river, it has been demonstrated by actual observation and experiment, conveys annually to the sea an amount of matter that would outweigh sixty solid pyramids of granite, supposing each, like the largest of the Egyptian pyramids, to cover eleven acres at its base, and to stand 500 feet in height. The delta of the Ganges, composed of mud, &c., thus brought down by the river, extends for 200 miles along the coast, and commences far inland.

A considerable quantity of sediment is also produced by the slow movements of glaciers in Alpine and other districts in which these remarkable ice-rivers prevail. The glacier of the Aar, which covers with its tributaries an area of only six or seven square miles, thus furnishes daily, according to some recent researches of M. Collomb, at least 100 cubic yards of sand. This is carried off by its terminal stream or torrent.

Action of the Sea (and of large bodies of Water generally).—Vast in amount as are the sediments collected by rivers, they are far surpassed by the accumulation of detrital matters obtained by the waves and breakers of the sea. All who have resided for any length of time on an exposed and rocky coast, must be well aware of the des-tructive action of the waves. The cliffs subjected to this action, gradually become undermined and hollowed out; and thus large masses of rock are brought down by their own weight. These, sooner or later, are broken up, and spread in the form of sediment along the shore, or over the sea-bottom. On some coasts, the amount of land destroyed in this manner almost exceeds belief.* On some parts of the eastern shores of England, and the opposite or western shores of France, for example, the sea has thus carried off, within the present century, from fifty to over two hundred yards of coast—measured backwards from the shore-line—along a distance of many miles. Graveyards, shown by maps of no ancient date to have been located at considerable distances from the sea, have become ex-posed upon the cliff-face; and forts erected by the First Napoleon on the French coast, at two hundred metres and upwards from the edge of the cliff, now lie in ruins on the beach, or have altogether dis-appeared. These localities are mentioned as being more especially known to the writer; but in all parts of the world examples may be found of the same destructive process. In the clay and sandy bluffs of our own Lakes, as at Scarboro' Heights on Lake Ontario, and elsewhere, effects of this kind may be equally studied.

Confining our view at present to these results only, it must be evident to all that an enormous amount of sedimentary matter is annually, or even daily, under process of accumulation. The question then arises as to what becomes of this. The reply is obvious. The detrital matter thus obtained, is deposited in lakes or at river mouths or along the sea-shore, or over the sea-bed—contributing day by day to the formation of new rocks. In other words, existing rock masses, worn down by atmospheric agencies, by streams and rivers, and by the action of the sea, supply the material for other and of course newer rock deposits.

Deposition of Sediments.—All sediments diffused through deep or quiet water, arrange themselves under general conditions, in horizontal or nearly horizontal beds : the latter, if deposited on gently-sloping

---

* It would obviously be out of place in an Essay like the present to enlarge on this point. The reader unfamiliar with geological details of this character, should consult, more especially Lyell's *Principles of Geology*, and also the *Cours Elémentaire* of the late Alcide d'Orbigny.

shores.   Professor H. D. Rogers, in his Report on the Geology of
Pennsylvania, contests to some extent this usually-received view, and
maintains that certain inclined strata of mechanical formation were
originally of inclined deposition.   This may be true under local or
exceptional, but certainly not under general, conditions.   (See proofs,
further on.)   Where, however, sands and gravels are thrown down by
currents and running streams, an oblique arrangement commonly takes
place; but this is more or less confined to the subordinate layers of
which the larger beds consist, as shewn in the annexed figure.   The
inclined layers have sometimes different
degrees of inclination, and even dip (in different beds of the same strata) in opposite
directions, indicating changes in the tidal
or other currents by which they were
thrown down.   This inclined arrangement
is termed "false bedding," or "oblique

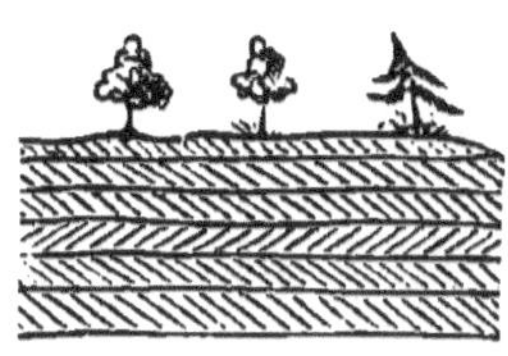

Fig. 84.

stratification."   It may be seen in some of the ancient, and also in
some of the more modern deposits of this continent, as in the Chazy
Sandstone of the south shore of Lake Superior, and in the Post-glacial
sands and gravels of many parts of Canada.

Consolidation of Sediments.—Having thus rapidly traced out the
formation of the mechanically-formed sedimentary rocks up to their
deposition in the state of detrital matter on the beds of seas, lakes, or
estuaries, we have now to inquire how these accumulations of mud,
sand, &c., become hardened into rock, properly so-called.

Most sediments hold within themselves the elements of their own
consolidation, in the form of particles of calcareous or ferruginous
matter, which act upon the other substances in the manner of a cement,
causing the whole to "set" or harden under water.   Frequently, also,
a large amount of calcareous matter is derived from the decomposition or solution of imbedded shells and other organic remains made
up of carbonate of lime.   In the majority of strata, and in sandstones
more especially, merely casts or impressions are thus left, in place of
the originally imbedded shells.   Masses of solid conglomerate are daily
under process of formation in places where springs containing calcareous or ferruginous matters infiltrate through the gravels and pebble-
beds of our Drift deposits.   Many thermal springs (as well as many

river-waters) also contain considerable quantities of silica in solution ; and there is reason to believe that in former periods of the Earth's history, springs of this kind must have prevailed to a very great extent. These, flowing into seas and lakes where sediments were under process of deposition, must also have lent their agency towards the consolidation of such deposits. Many of our Canadian limestones, it may be observed, are more or less siliceous.

The enormous pressure exerted upon low-lying sedimentary beds by masses of superincumbent strata, must likewise have been sufficient in many instances to have effected consolidation.

The heat transmitted in earlier periods from subterranean depths, or generated amongst low-lying sediments by chemical action, may also have been concerned in the work of consolidating the originally loose materials of stratified rocks. It may be remarked, likewise, that sediments occasionally become solidified by simple desication. The shell-marl, or calcareous tufa, of our swamps, &c., becomes thus hardened on exposure to the air.

(3) SUBSEQUENT ACTION OF NATURAL FORCES ON SEDIMENTARY ROCKS.

The more important effects produced on sedimentary rocks, from their first period of aggregation, are as follows :—(*a*) Elevation above the water-level, with Alternations of Upheaval and Depression ; (*b*) Denudation ; (*c*) Tilting up and Fracturing ; (*d*) Metamorphism. It is of course to be understood, that whilst certain strata may have experienced all of these effects, others may have been subjected to upheaval, or to upheaval accompanied by denudation, only.

(*a*) *Elevation above the Sea Level :*—The stratified rocks, it has been shown, must have been deposited originally, in the form of sediments, under water ; and from the marine remains which so many of these rocks contain, it is evident, that, as a general rule, they were laid down on the bed of the sea, either in deep or in shallow water. We find these rocks, however, now, at various heights above the sea-level, and frequently far inland. Hence of two things, one : either the sea must have gone down, or the land must have been elevated above the water.

The sinking of the sea would appear at first thought to be the more rational explanation of this phenomenon ; but if we look to existing Nature, we find no instance of the actual falling of the sea, whilst we have many well-proved examples of the actual rising and sinking of the land. In connection with this inquiry, it must be borne in mind that the sea cannot go down or change its level at one place without doing the same generally all over the world.

To afford a few brief illustrations, it may be observed that on several occasions within the present century, large portions of the Pacific coast of South America have been raised bodily above the sea, leaving beds of oysters, mussels, &c., exposed above high-water mark. The phenomenon, to the inhabitants of the coast, appeared naturally to be due rather to a sinking of the waters than to an actual elevation of the land ; but at a certain distance north and south of the raised districts, the relative levels of land and sea remain practically unaltered : and hence, if the sea had gone down within the intervening space, to the extent indicated, its surface must have presented an outline of this character ⌐⌐⌐⌐⌐⌐ : a manifest impossibility.

The land is also known to be slowly rising and sinking in countries far removed from centres of volcanic activity. Careful observations have shown, for example, that the northern parts of Sweden and Finland are slowly rising, and the south and south-eastern shores of the Scandinavian peninsular are slowly sinking : whilst around Stockholm there is no apparent change in the levels of land and sea. The whole of the western coast of Greenland is inferred to be slowly sinking : buildings erected on the shore by early missionaries, being now in places under water. A slow movement of depression is likewise taking place along the shores of Cape Breton and Nova Scotia generally ; and, probably also, to some extent, on the Atlantic sea-board of the United States. On the shores of Newfoundland, of Cornwall, and other districts, examples occur of sub-marine forests, or of the remains of modern trees,. in their normal positions of growth, below low-water mark ; whilst in neighbouring localities no change of level appears to have taken place. Besides which, without extending these inquiries further, we know that many fossiliferous strata are hundreds, and even thousands, of feet above the present sea-level. On the top of the Collingwood escarpment, for example, we find strata containing marine fossils at an elevation of over 1,500 feet above the sea; and on the Montreal mountain, shells of existing species occur at an elevation of about 500 feet. Hence, if these strata had been left dry land by the sinking of the oceanic waters in which they were deposited, an immense body of water, extending over the whole globe, must in some unaccountable manner have been caused wholly to disappear. It is therefore now universally admitted, that the sedimentary rocks, as a rule, have come into their present positions, not by the sinking and retiring of the sea, but by the actual elevation of the land.

Many strata afford proofs of having been elevated and depressed above and beneath the sea, successively, at different intervals. Many sandstones, for example, exhibit ripple-marked surfaces, and occasionally impressions of reptilian and other tracks, throughout their entire thickness. This indicates plainly that they were formed slowly in shallow water, and that they were left dry, or nearly so, between the tides. And it indicates, further, that the shore on which they were deposited, layer by layer, was undergoing a slow and continual movement of depression: otherwise the process of formation would necessarily have ceased, and the strata would present a thickness of a few inches only, or of a few feet at most. Afterwards a period of upheaval must have commenced, bringing up the rocks to their present level. In certain strata, also, the upright stems of fossil trees occur at various levels; and in some localities. beds containing marine fossils are over-laid by others holding lacustrine or fresh-water species; and these again by others with marine remains. Finally, to bring this section to a close, we have a striking example of alternations of land-upheaval and depression in the geology of Canada generally. Around Toronto, for example, we have certain strata of old date, belonging to the Lower Silurian Series, overlaid by deposits of clay, gravel, and sand of the Drift Epoch, a comparatively modern period. Between the two, a vast break in the geological scale occurs. Many intervening formations, indicating the lapse of long periods of time, are present in other parts of this continent; and hence, it is concluded that the Silurian deposits of this locality, after their elevation above the sea, remained dry land for many ages, whilst the intervening groups were under process of deposition in other spots; and that, finally, at the commencement of the Drift Period, the country was again depressed beneath the ocean, and covered with the clays, sands, and boulders of this latter time. Another period of elevation must then have succeeded, bringing up both the Silurian and the Drift formations to their present levels above the sea.

(b) *Denudation :*—This term, in its geological employment, signifies the removal or partial removal of rock masses by the agency of water. The abrading action of the sea, of rivers, &c., acting under ordinary conditions, has already been alluded to; but the erosive effects of water may be seen in numerous localities in which this action is no longer in force. Sections of the kind shewn in the accompanying figure, for instance, are met with almost everywhere, producing un-

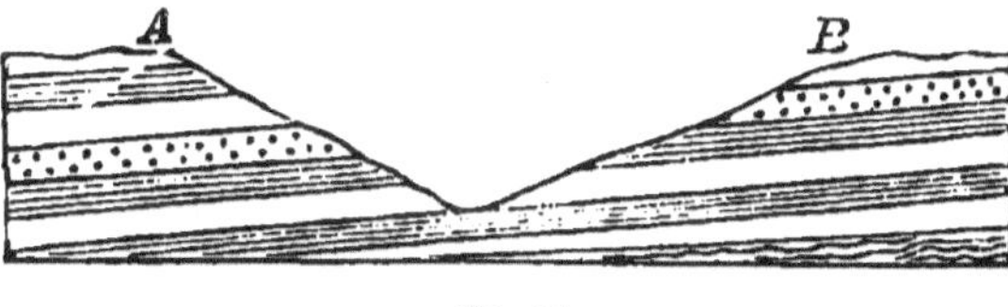

Fig. 85.

dulating or rolling countries. Here it is evident that the strata were once continuous in the space between *A* and *B*. Valleys which thus result from the removal of strata, are termed " valleys of denudation." Some of these valleys are many miles in breadth. Their excavation, consequently, could not, in the majority of instances, have been effected by atmospheric agencies, or by the streams which may now occupy their lower levels; but must have been caused essentially by the denuding action of the sea during the gradual uprise of the land, or during alternate movements of elevation and depression, in former geological epochs. If the bed of the Atlantic, for example, were now being raised from beneath, at the rate of a few inches in a year or series of years, an enormous valley would probably be scooped out along the course of the Gulf Stream; and in other places where currents prevail, more or less continuous valleys would also be formed. Isolated patches of strata have been frequently left by denudation at wide distances from the rocks of which they originally formed part. These are termed "outliers." Thus in Western Canada, small isolated areas, occupied by bituminous shales of the Devonian series, occur in the townships of Bosanquet and Warwick, and constitute outliers or outlying portions of the Chemung and Portage group (see Part V.) largely developed in the adjoining peninsula of Michigan. The

12

matter carried off in some districts by denudation, must have been
of enormous amount; and when it is considered that most of the inequalities on the earth's surface—those at least not immediately connected with mountain chains—have been thus produced, the part
played by the denuding agencies of former periods in providing the
materials of newer strata, may be readily appreciated.

(c) *Tilting up and Fracturing of Strata.*—Whilst some strata retain their original horizontality, others are more or less inclined, and
some few occupy a vertical and even a recurved position. That
strata were not originally inclined, at least to any extent, is proved
by the known arrangement of sediments when diffused through
water,—these (with the exceptional cases already pointed out)
always depositing themselves in horizontal, or nearly horizontal,
layers. The same fact is shown also by the frequent presence of

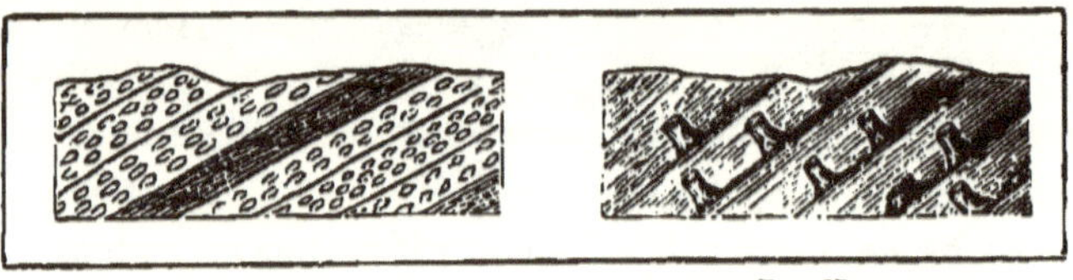

<table>
<tr><td>Fig. 86.</td><td>Fig. 87.</td></tr>
</table>

rows of pebbles, fossil shells, &c., parallel with the planes of stratification, as in Fig. 86; by the occasional presence of the fossilized
stems of trees (evidently in their positions of growth) standing at
right angles to these planes (Fig. 87); and sometimes by the presence
of stalactites suspended in a similar position. It is evident that
these bodies could not have been originally inclined in this manner
to the horizon.

The inclination of strata is technically termed the *dip;* and the
direction of the up-turned edges, the *strike.* The dip and strike are
always at right angles. In observing the dip, we have to notice
both its angle or amount, and its direction or bearing—as north,
north-east, N 10° E, and so forth. The direction of the dip is of
course ascertained by the compass; the rate of inclination, by the
eye, or by an instrument called a clinometer. The most convenient
instrument for both purposes, is a pocket compass, set in a square
bed, or attached to a square plate of metal, and furnished, in addition
to the needle and graduated limb, with a moveable index. The
latter hangs freely from the centre of the compass, and plays around

a graduated aic, as in the annexed figure. When the upper edge of the compass is held horizontally, the index cuts the zero point of the graduated arc. From each side of this point, the graduation is carried up to to 90°. If, consequently, the upper edge of the instrument be placed parallel with the inclined beds of any strata, the angle of the dip will be at once shewn by the index. A contrivance of

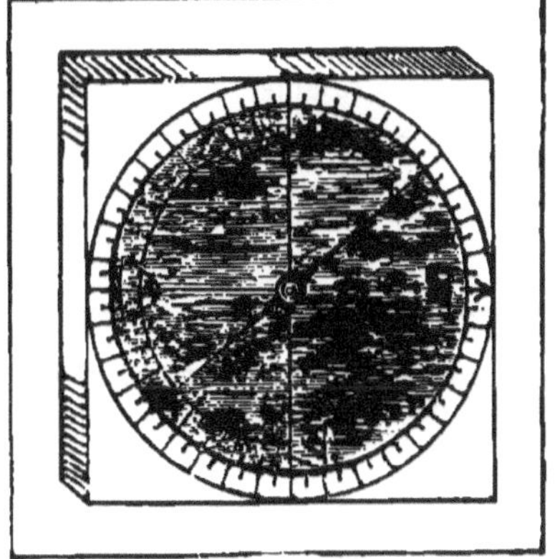

Fig. 88.

this kind, exclusive of the compass, may be easily made out of a semicircle of hard wood. The index may consist of a piece of twine extending below the graduated limb, and kept taut by a lead plumb or by a stone.

In a compass used for taking bearings, it is convenient to mark the *west* side EAST, and the *east* side WEST, as in the figure. If the *north side of the instrument* be then kept always in advance, and the angle be always taken from the *north end of the needle*—no matter what the actual direction of the line—the true magnetic bearing is obtained at once, and without risk of error. The compass is most readily held by passing the thumb through a short strap or loop, or through a hinged ring, attached to its under side. Where very accurate bearings are required, sights may be used, the instrument being fixed on a support; or a prismatic compass may be more conveniently employed.

When strata dip in two directions, as at *A*, in Fig. 89, the line along the culminating point of the strata is termed an *Anticlinal* or *Anticlinal Axis;* and the line from which the strata rise in oppcsite directions, as at *S* in the figure, is called a *Synclinal* or *Synclinal Axis.* Synclinals when of a certain magnitude, constitute " valleys of undulation." Anticlinals are often hollowed out by denudation, forming valleys or troughs called " valleys of elevation," as shewn at *E* in Fig. 89. The term "elevation" applies here, it should be observed, to the raised strata, and not to the actual position of the valley, as many of these so-called valleys of elevation lie in the beds of rivers, or occupy comparatively low ground.   The River Humber near Toronto, for example, flows at the lower part of its course over a denuded anticlinal of this character.*   Finally, it may be observed,

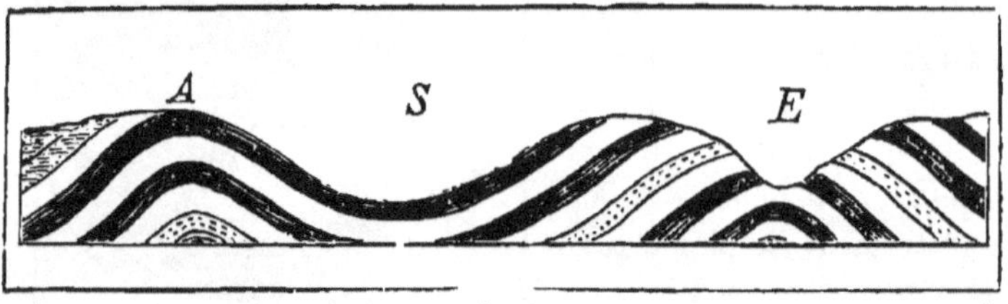

Fig. 89.

that when strata lie in parallel beds (as in Figs. 85 and 89), the stratification is said to be *conformable* or *concordant.*   When on the other hand, the beds are not parallel, the stratification is said to be

Fig. 90.

unconformable.   The accompanying section, in which the inclined beds belong to the Laurentian, and the overlying beds to the Lower Silurian Series (see Part V.), as shewn on Crow Lake, north of Marmora village, is an example of uncomformable stratification, or of want of concordance between these two series of rocks.   As explained further on, a want of conformability indicates almost invariably the commencement of a new geological period.

Both horizontal and inclined strata frequently exhibit fractures of greater or less extent.   Mineral veins, it may be mentioned, consist essentially of cracks or fractures formed at some more or less remote period in the surrounding rocks, and filled subsequently by various agencies, with sparry, earthy, and metallic matters.   The strata on one side of a fracture are often displaced, being thrown up or down as it were.   This peculiarity is technically termed a *fault.*   The

---

* Professor Robert Bell in his Report on the Manitoulin Islands, has pointed out the occurrence of fifteen anticlinals, crossing the Great Manitoulin in a general north and south direction.   These anticlinals give rise on the north shore of the island to deep indentations or bays, and inland to a series of parallel lakes.

levels occupied by a displaced bed are
sometimes only a few inches, and at
other times upwards of a thousand feet,
apart. At the first formation of a fault
or slip, an escarpment or terrace of
greater or less height must necessarily
have been produced; but in very few
cases (if in any case unconnected with
existing earthquake phenomena) is any-

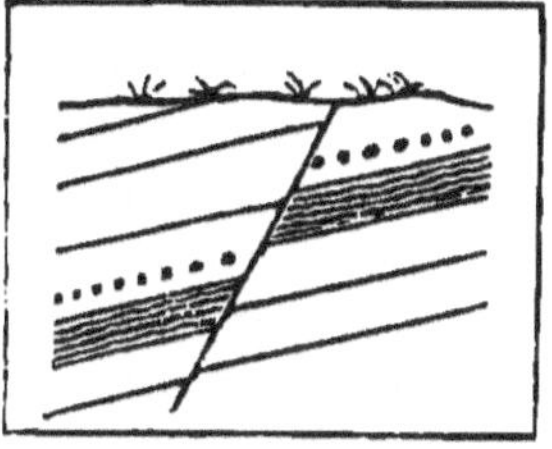

Fig. 91.

thing of this kind now observable, the ground having been levelled
down at some after period by the agency of denudation. In mountainous districts the fracturing of strata has sometimes given rise to
narrow gorges or so-called "valleys of dislocation," but most of these
have been subsequently enlarged by the atmospheric disintegration
of the surrounding rocks, and by the streams or torrents of which
they usually form the channels. In most faults the displaced beds
have slipped downwards, and have thus been brought into a lower
position than that which they originally occupied; but occasionally,
as in the so-called "reversed faults," they have been forced upwards,
so as in some instances to overlie the other beds.

(d) *Metamorphism.*—Many strata afford undeniable proofs of
having been greatly altered, as regards texture and other mineral
characters, from their original sedimentary condition. In many instances, indeed, the original composition of the rock appears to have
been changed. Strata thus affected, are commonly known as metamorphic or altered rocks. In some cases a passage can be traced from
the altered into the unaltered parts of the rock; but, frequently, where
rocks have been subjected to this action, the alteration has extended
over wide areas, and has been more or less complete. It consists most
commonly in the assumption of a crystalline structure, and is very
generally accompanied by the presence of crystallized minerals and
other indications of chemical action.

In numerous instances, metamorphism, on a limited scale, has
evidently resulted from the direct intrusion of eruptive rock matters
amongst sedimentary formations. Where trap dykes or masses of
granite, for example, have been thrust up through fissures in ordinary
strata, the latter are seen in many cases to have been more or less
altered around the points of contact, as though by the agency of
intense heat, or by that of steam or other gases acting under pressure.

Coal has been thus converted, within a certain distance of interpenetrating trap masses, into cinder and coke; earthy limestone, into crystalline marble; sandstone into quartz rock, and so forth; and somewhat analogous effects are occasionally produced in sandstone blocks that have been long exposed to heat and heated vapours in the interior of certain furnaces. These effects, however, have not always followed the intrusion of eruptive rocks; and in no case do they appear to have extended far into the mass of the surrounding strata. The alteration of extensive regions therefore, such as the wide area occupied by the Laurentian strata of Canada (see Part V.), points evidently to some more general although probably related cause, in explanation of the facts of metamorphism.* Whatever view be adopted respecting the internal condition of the earth, it is clear that immense spaces filled or partially filled with molten and vaporous matter must have existed through untold ages at certain depths beneath the surface rocks; and the chemical action going on within these spaces, and emanating from them, may be regarded as sufficient to produce the results in question, even if we cannot explain, to our thorough satisfaction, the actual processes involved in the production of these effects. See further under the METAMORPHIC ROCKS described below.

A special effect of metamorphism, developed more particularly in fine-grained argillaceous strata, is the production of *slaty cleavage.* Rocks thus affected, exhibit a more or less strongly-pronounced fissile texture, arising from the presence of numerous divisional planes running parallel with one another through the rock, and usually in a direction inclined to that of the planes of deposition. It is not always easy, in inclined strata, to distinguish the latter planes from planes of cleavage; but their direction is generally revealed by the presence of fossils, or by intercalated layers of a different shade of colour, degree of fineness, &c., across which the cleavage lines commonly pass without interruption. Cleavage in rocks, as shown by this latter condition, and by the fact that fossil bodies and imbedded stones are frequently drawn out or unnaturally elongated in the direction of the cleavage planes, is evidently a superinduced effect; but much obscurity still prevails with regard to its actual origin.

---

* We follow here the generally received view respecting the nature of these stratified, or at least laminated, crystalline rocks. But the assumed metamorphic character of these rocks is contested by some observers—in Canada, notably by Mr. Thomas Macfarlane—who regard them as original formations. Impartially considered, this view is not without strong grounds of probability.

It is usually attributed to mechanical pressure acting laterally upon the rock during elevations or depressions of contiguous areas; but it may be really due to the effect of long continued heat on confined masses of damp strata. Moist clay, for example, if exposed in a covered vessel to a certain degree of furnace heat, almost invariably assumes a fissile texture; and the same peculiarity is observable in ordinary biscuits, and more especially in those which have undergone an extra firing for ships' use. Oblique cleavage is exhibited by many of the clay-slates of the Eastern Townships, as those of Melbourne, Cleveland, Kingsey, &c.; but the clay slates of Lake Superior and other parts of the province, though more or less finely laminated, appear to be entirely destitute of true cleavage planes of this character.

III. METAMORPHIC OR CRYSTALLINE STRATIFORM ROCKS.

The rocks of this series are stratified or laminated rocks of a more or less crystalline aspect. In their mineral characters they frequently bear a great resemblance to eruptive rocks, to which indeed they are closely allied—almost every metamorphic rock having its representative in the eruptive series; but they differ from these latter by their general conditions of occurrence. As explained above, many sedimentary strata are seen to have assumed a crystalline texture, or to have lost more or less completely their normal sedimentary aspect, in the vicinity of intrusive masses of granite, greenstone, or other eruptive rocks. An alteration of this kind is known as local metamorphism. Earthy or ordinary limestones and dolomites are thus occasionally converted into hard crystalline marble, often veined with green and other coloured streaks and patches of serpentine, and filled in many cases with crystals and crystalline particles of graphite, pyroxene, amphibole, various micas, tourmaline, garnets, pyrites and other minerals, foreign to the rock in its sedimentary condition. In like manner. sandstones are changed in colour and texture, and are often converted into quartz-rock or some variety of gneiss; and clay-slates are transformed into mica-slate, talc-slate, hornblende-rock, and other so-called crystalline schists and gneissoid aggregations. These metamorphic results are probably due in part to the agency of various gases and heated vapours which accompanied the protusion of the eruptive mass. Alterations of a similar kind, but extending over wide areas, are assumed, on the other hand, to have taken place in many localities, without the direct intervention of eruptive rocks. This widely extended metamorphism has probably been effected by alkaline and other solutions acting on the heated rocks, or by the agency of superheated steam and other vapours on deeply-seated strata, or by other causes more or less immediately connected with the presence of subterranean heat. In many cases there can be no question as to these crystalline strata being really altered sedimentary deposits, and thus, by inference, a similar origin is generally attributed to all rocks of this character. Whilst sedimentary rocks, proper, are the products of surface action, and eruptive rocks—as regards their present condition, if not in all cases their actual origin—are products of internal or subterranean forces, metamorphic formations may be regarded as the result of both external and internal agencies.

The metamorphic rocks of Canada belong, as regards their geological position, to two essentially distinct series. The older series of Archæan age, comprises the rock formations of the Laurentian and Huronian periods, and occupies all the more northern and north-western portions of Quebec and Ontario, its strata consisting chiefly of enormous beds of gneiss, crystalline limestone, siliceous slates, and other rocks, enumerated below, and described more fully in Part V. The higher or less ancient series is apparently intermediate in position between the Huronian and Cambrian formations. Its strata are chiefly developed in the form of chloritic and talcose slates and beds of serpentine, throughout the Eastern townships and adjoining region south of the St. Lawrence, in the Province of Quebec.

The following are the more important metamorphic rocks of Canadian occurrence :—

Gneiss :—This rock is made up normally of three minerals—quartz feldspar and mica ; the two latter being generally the common potash species, orthoclase and muscovite (See Part I.). In some districts, however, the rock consists almost entirely of quartz and feldspar, mica being absent or very sparingly present. In coarsely crystalline varieties of the rock, the component minerals are easily recognized. The feldspar is usually white or red, and is present in distinctly cleavable grains or masses ; the mica is in leafy masses or small scales of a silvery white, brown or black colour ; and the quartz in

·colourless vitreous grains. The striped or banded aspect of the rock generally serves to distinguish it, in hand specimens, from granite; and when seen in position, its stratified structure is in most cases very apparent. Vast beds of gneiss, and strata of gneissoid rock in which the component minerals are more or less indistinct, occur throughout the wide area occupied by the Laurentian rocks of the more northern regions of Canada (see Part V.), and also here and there, in the less ancient crystalline district south of the St. Lawrence. Most of the boulders scattered so abundantly over the surface of Canada, consist of micaceous gneiss, or of the hornblendic variety described below. In some localities the mica of ordinary gneiss is partially replaced by scales of graphite.

Syenitic or *Hornblendic Gneiss:*—This rock only differs from ordinary gneiss by containing hornblende in place of mica; but the two rocks frequently merge into one another, both hornblende and mica being present in certain varieties. Normally, the hornblendic variety of gneiss is composed of red or grey feldspar, with quartz, and black or green hornblende. The three minerals are sometimes very distinct; but in other cases they are intimately blended, so as to form a dark-green rock, which passes, by the gradual diminution of the quartz, into hornblende-slate or amphibolite. Syenitic gneiss occurs abundantly, with ordinary or micaceous gneiss, in the Laurentian districts of Canada (see Part V.).

Mica Slate:—This is a foliated or schistose rock, composed essentially of quartz and mica. It is generally of a dark-grey, greyish-green, or silvery-white colour, and occasionally black; and some varieties are highly lustrous from the presence of intermixed graphite, as in many parts of the "Eastern Townships" of Quebec. It passes into clay-slate, and also into fine-grained gneiss and other rocks of this series. Mica-slate occurs here and there throughout the Laurentian area of Canada, and in the crystalline districts south of the St. Lawrence (see Part V.); but characteristic examples are rare—the rocks in question being rather micaceous slates than mica-slate as commonly defined.

Pyroxenite or *Augite-Rock:*—This rock, of subordinate occurrence, consists at times of almost pure augite or pyroxene, but in general it forms a granular compound of augite and some kind of feldspar, more ·or less intermixed with carbonate of lime. Frequently, also, it contains chlorite, with grains of magnetic iron ore and other minerals. The normal colour is dark green. Examples occur here and there in connection with beds of crystalline limestone and iron ores of the Laurentian Formation, as at Calumet Falls on the Ottawa, and parts of Madoc, Marmora, Tudor, and adjacent townships. In many cases pyroxenite cannot be distinguished from hornblende rock ; and it closely resembles also, in general character and composition, certain eruptive masses and dykes belonging to the trappean series. Many Laurentian pyroxenites, indeed, are probably of igneous origin.

Amphibolite or *Hornblende Rock :*—This metamorphic product is sometimes described as diorite, but the latter term is properly restricted to eruptive greenstones of similar composition. Hornblende Rock is composed normally of a mixture of hornblende and soda-feldspar, but at times it consists of almost pure hornblende. Many varieties are also more or less calcareous, and in some, both mica and quartz are occasionally present. These pass into syenitic gneiss. The texture of the rock is compact, granular, fibrous or slaty. The slaty varieties are commonly known as Hornblende Schist, and the fibrous as Actynolite Rock or Schist. Examples occur in some abundance among the Laurentian strata of Marmora, Madoc, Elzevir, Blythfield, and throughout the Laurentian country generally, between the Ottawa and Lake Huron; also at various places on Lake Superior, as at Point-aux-Mines, Goulais River, and elsewhere ; but many of the hornblende rocks of these districts may be really eruptive masses. Hornblendic rocks and slates form part also of the crystalline deposits of Beauce and other districts of the Eastern Townships.

Wollastonite-Rock :—The mineral Wollastonite (No. 64, Part II), mixed with feldspar, pyroxene, quartz, calcite, and other minerals, occasionally form beds in the Laurentian series, mostly in association with crystalline limestone. Where the Wollastonite predominates, the rock presents a granular-fibrous structure, and is white or pale-greenish in colour. Examples occur in the counties of Argenteuil, Terrebonne, Leeds, &c., but are comparatively unimportant.

Epidote-Rock :—This is also of subordinate occurrence. It consists of a mixture of quartz and epidote, and presents both granular and compact varieties, mostly of a pale-green colour. Examples have been recognized amongst the Shickshock Mountains of Gaspé; and others occur in Melbourne and other parts of the Eastern Townships.

Garnet-Rock :—Subordinate beds of this rock, composed essentially of granular red garnets and crystalline quartz, occur among the Laurentian strata of St. Jérome on the Ottawa, and Rawdon in Montcalm county ; and also, according to Mr. Richardson, in association with micaceous schists at Baie St. Paul. Dr. Sterry Hunt has likewise made known the occurrence of beds of more or less compact and light-coloured garnet amongst the metamorphic series of the Eastern Townships. See under " Garnet," Part II.

Quartzite or *Quartz-Rock :*—This rock consists normally of pure crystalline quartz, either colourless, or of pale shades of red, yellow, green, or smoky-brown. Coarse and more or less opaque varieties, passing into quartzose sandstone and chert, exhibit various colours, however ; and the rock is often green and greenish-grey from admixture with chlorite. Some cherts are black from the presence of anthracitic matter. Enormous beds of quartzite, frequently very pure, occur in the Laurentian series of strata, as on the River Rouge in the county of Argenteuil, and elsewhere ; and these rocks are still more characteristic of Huronian strata. Laurentian quartzose conglomerates occur in the townships of Bastard and Rawdon ; and a very remarkable conglomerate of the Huronian series, consisting of pebbles of colourless quartz and red jasper in a colourless, greenish-white or pale-yellowish quartzose base, is met with in the Bruce Mines district. These crystalline conglomerates show unmistakably the metamorphic origin of the rock. Beds of chert and jaspery quartz occur also in places on Lake Superior, and in the crystalline region south of the St. Lawrence (see Part V).

Siliceous Slate :—This rock is probably an altered clay-slate. It passes into impure quartz-rock or jasper ; consists essentially of a silicate of alumina ; is hard or more or less slaty, and usually of a greenish-grey colour, or dark green from intermixed chlorite, and occasionally striped or zoned with lines of black, green, or red. Examples of siliceous slates are of common occurrence on the north shore of Lake Huron, and amongst the Huronian strata of the Rive Doré and other localities on Lake Superior. In many places, these slates hold rounded pebbles or masses of gneiss, syenite, &c., and thus form " slate conglomerates."

Argillite :—This is one of the least altered rocks of the metamorphic series. It is simply a more or less indurated clay-slate, and commonly presents a black or bluish-black or dark-grey colour, but

some varieties are dull chocolate-red, and others greenish-grey—the
rock passing by insensible transitions into ordinary unaltered shales
on the one-hand, and into silicious and micaeous slates on the other.
Many argillites are highly lustrous from intermixed graphite, and
some contain small straw-like crystals of chiastolite or andalusite, as
described under that mineral in Part II. Dark and more or less
lustrous varieties are common in Huronian strata, and are still more
abundant in the higher Animikie series of Thunder Bay, Lake
Superior (see Part V), and in various parts of the altered region
south of the St. Lawrence. In the latter district, as in Beauce and
elsewhere, many of the green, purplish, and red agillites present a
nacreous talcose aspect, but, as shewn by Dr. Hunt's analysis, they
contain little or no magnesia.

Chlorite Slate :—This metamorphic rock in its normal aspect is a
compound of chlorite and quartz, possessing a distinct green colour
and a foliated or schistose structure ; but in many examples the
structure becomes fine-granular or almost compact. In Canada,
chlorite slates, passing into chloritic strata in which the typical
character of the rock is more or less obscured, occur sparingly in the
Laurentian series, mostly in connection with iron ores. A bed of a
dark green colour, filled with numerous small octahedrons of magnetic iron ore, occurs in the township of Galway. Other examples,
but often of·somewhat obscure schistose structure, are characteristic
of the Huronian rock throughout the vast region north of Lake
Huron and Lake Superior. In the crystalline region south of the St.
Lawrence, chloritic slates are also abundant, and most of the copper
ores of the Eastern Townships are associated with these strata.
Other beds contain intercalated scales and layers of specular or micaceous iron ore ; and in the Townships of Bolton and Broughton,
more or less compact or subfoliated beds of greenish-grey chlorite,
known as " potstone," form workable beds of good quality. (See
Part II, No. 80).

Steatite or *Soapstone-Rock* :—This rock consists of granular or
slaty talc, frequently intermingled with carbonate of lime or dolomite. It usually presents a greyish or greenish-white colour, and
when pure is very sectile. A bed of somewhat inferior quality, from
intermixture with calcareous matter, occurs in the Laurentian strata
of Elzevir. The closely related substance known as Pyrallolite (see

under No. 82, in Part II), also forms beds among Laurentian strata, as in Grenville, Ramsay, and elsewhere. Many deposits of more or less compact soapstone occur likewise in the higher crystalline series of the Eastern Townships, as in various parts of Bolton more especially, and also in Potton, Sutton, Stanstead, Leeds, and Vau_ dreuil.

Ophiolite or *Serpentine Rock :*—This rock consists essentially of the hydrated magnesian silicate, Serpentine, described fully in Part II. It usually presents a green, brown, greenish-grey, or pale yellowish colour, often veined or mottled with lines and patches of darker or lighter green, red or reddish brown ; and it forms more or less compact beds, frequently of great extent and thickness. Subordinate examples occur in the Laurentian strata of many localities, mostly associated with bands of crystalline limestone, as in the township of Grenville, and at Calumet Island on the Ottawa ; also in Burgess, and elsewhere ; but the crystalline districts south of the St. Lawrence contain the most abundant and important deposits of serpentine rock, as at Mount Albert in Gaspé, and in the Eastern Townships of Melbourne, Oxford, Brougham, Bolton, Ham, and Garthby, more especially. The serpentine of these districts is very commonly associated with beds of chromic iron ore ; and many examples are intermixed with crystalline calcite or dolomite, forming ornamental " serpentine-marbles " of green, chocolate-brown and other colours.

Crystalline Limestone :—This rock consists of carbonate of lime in a crystalline or semi-crystalline condition. It is usually white, light grey, or pale reddish, in colour, and is sometimes veined or spotted with yellow, green, blueish-grey and other tints. It presents most commonly a fine or coarse granular structure, much resembling that of loaf sugar, whence the name " saccharoidal limestone " by which this rock is often known ; but some varieties are more or less compact ; and others present in places a fibrous aspect, from intermingled tremolite or white hornblende, or light varieties of pyroxene. The finer kinds from the ordinary marbles of commerce. In Canada, large beds of crystalline limestone, often containing scales of graphite, and crystals of apatite, pyroxene, amphibole, mica and other minerals, occur among the Laurentian and Huronian series of strata in numerous localities (See Part V) ; and also among the crystalline strata south of the St. Lawrence. In the latter district, as already mentioned, some of these limestone beds are intermixed with green and other coloured serpentines, but many of the so-called serpentine marbles from the Eastern Townships are mixtures of serpentine with dolomite or magnesite.

Crystalline Dolomite :—This rock resembles crystalline limestone in colour and other external characters, but consists of carbonate of lime and carbonate of magnesia, and only effervesces when tested with heated acid (See Part I). It occurs, here and there, amongst the Laurentian strata, as at Lake Mazinaw in the County of Frontenac, and elsewhere. Also among the strata of the Eastern Townships, in which district beds of *crystalline magnesite* (See Part II), mixed with mica, serpentine, &c., are likewise present. These magnesian beds, as pointed out by Dr. Sterry Hunt, assume a yellowish or dull-red colour by weathering.

Crystalline Iron Ores :—Vast beds of Magnetic, Specular and Titaniferous Iron Ore, occur locally amongst the rocks of the Laurentian series, and should thus be referred to in connexion with the metamorphic formations of Canada. The strata of the metamorphic region south of the St. Lawrence are also especially characterized in certain localities by the presence of chromic iron ore in rock masses ; and many of the chloritic and other schistose strata of this region pass locally into "iron slates" or "specular schists," by the addition of micaceous hematite or specular ore. The distinctive characters of these Iron Ores, and their principal localities, are given in Part II.

IV.　MASSIVE OR UNSTRATIFIED ROCKS.

The rocks of this division are commonly known as *Igneous* or *Eruptive Rocks*. With regard to the igneous formation of certain members of the Eruptive Series, there can be no possible doubt ; but the actual mode of formation of other rocks of this group is involved in great obscurity. All agree, however, in being essentially devoid of true planes of stratification. They occur either in irregular unstratified masses ; or in sheets or apparent beds, intercalated amongst, or overflowing, stratified deposits ; or in the form of more or less tortuous veins ; or in broader and simpler veins, technically known as "dykes," which frequently terminate at their upper extremity in overlying step-like and columnar sheets of matter. And in these conditions, they are frequently seen to traverse older rocks of the same class, or to penetrate various stratified formations. They are thus essentially *intrusive rocks ;* and they are also, in the words of

Humboldt, essentially *endogenous* rocks—*i. e.,* they come from more or less deeply-seated sources within or beneath the Earth's crust, from whence they have been forced

FIG. 92.

up from time to time through cracks and fissures opened in the overlying or surrounding rock masses. From this it follows, as a general rule, that the intrusive rocks in question must have been at one time in a soft and plastic state, if not in an actually fluid condition. Certain trachytic and basaltic rocks—members of this group, described below—cannot be distinguished by chemical or mineralogical characters from ordinary lavas ; and the former existence of many basalts in a molten or highly heated condition is established by the effects produced by veins or dykes of these rocks on coal beds and other strata through which they have been erupted. Coal in contact with dykes of this kind, has been burnt into cinder, or converted into coke ; clays have been baked into brick-like masses ; sandstones rendered more or less vitreous ; and various limestones, to cite no further instances, have been hardened and altered into marbles of crystalline texture. Intrusive veins and masses of granite and syenite are also known to have produced metamorphic effects on the rocks which they traverse. But in many instances no alterations of this kind have followed the intrusion of a vein or mass of unstra-tified rock amongst sedimentary deposits. Hence it is clear that although the intrusive rock must have been in a soft or plastic condition, it could not in these latter cases have been in a molten or intensely heated state. Occasionally also, solid granitic masses appear to have been thrust up amongst overlying strata, the intrusion being followed necessarily by signs of great mechanical disturbance. The condition of the quartz in granite and syenite, is opposed to the view of igneous fusion; and yet quartz of the same character does occur sparingly in many trachytes, and under conditions not favourable to the idea that it may have been subsequently introduced by aqueous agencies. Through these trachytes, moreover, there is a gradual passage into actual lavas or known fusion-products; whilst, on the other hand, many syenites (containing free quartz) merge gradually into greenstone and basalt, products intimately related to augitic lavas. It is, of course, impossible to say in what form a rock belonging now to the eruptive class may actually have originated. It may have been produced from an earlier formed igneous or crystalline mass, or from a sedimentary deposit buried deeply under overlying beds. The endogenous or subterranean agencies, whatever they may have been, that rendered granite and syenite plastic and crystalline, also produced the crystalline texture and other related characters of gneiss, mica schist, hornblende rock, and other members of the metamorphic series. It is now very generally assumed that whilst ordinary lavas and most trachytes and trappean (or basaltic) rocks have solidified from a molten condition, other rocks of this class, the granites and syenites more especially, have been rendered plastic and crystalline by "hydro-igneous" agency. These rocks, in other words, are thought to have undergone a kind of aqueous fusion and subsequent crystallization, the water, originally present in them, having been retained for a time by the pressure exerted on the plastic mass at great depths. But this view, it must be understood, is entirely hypothetical, and in many respects is far from satisfactory. All that is really known may be thus expressed :—Two sets of forces are concerned, either alone or con jointly, in the production of rock masses generally. One set, entirely external, or consisting essentially in the action of the atmosphere and waters on the surface of the earth-mass, produces the sedimentary or stratified rocks proper. The other forces, of internal or subterranean origin, produce the unstratified rocks, as we now see these latter, and lead to the crystallization and metamorphism of sedimentary strata brought within their influence. But whether granites, syenites, traps, and trachytes, be igneous or non-igneous rocks, they are evidently related products, and members of a common class.

These rocks are arranged by Sir Charles Lyell in two broad divisions: Volcanic and Plutonic rocks; but it is impossible to draw a distinct line of demarcation between the two. Granite and syenite for example, are placed in the Plutonic series, and trachyte, greenstone, basalt, &c., in the Volcanic division; but certain granitic trachytes connect the granites with the volcanic rocks; and in like manner, certain greenstones merge on the one hand into syenite, and on the other (the distinction between augite and hornblende, except in a purely mineralogical or crystallographic point of view, being practically of little moment) they pass into augitic lavas. This equally affects the sub-division into Volcanic, Trappean, and Granite rocks, adopted by other observers. I would therefore propose, as an arrangement of convenience, the distribution of our Canadian Eruptive rocks into the following groups:—1. Granites and Syenites; 2. Anorthosites; 3. Traps and Greenstones; 4. Trachytes; 5. Obsidians; 6. Lavas.

1. *Granites:*—The rocks of this group possess, normally, a crystalline aspect and strongly-marked granular structure, the term granite being derived from the latter character. Granites are also especially characterized by the presence of free silica or quartz in a crystalline condition. They occur occasionally in broad, straight veins or dykes, but are most commonly seen in the form of complicated, ramifying veins, or in large irregular masses which have often broken through and tilted up the surrounding rocks. Where a granite mass lies in contact with another rock, the latter will necessarily be the older formation if it be tilted up or otherwise mechanically affected by the granite; or if it be chemically altered near the points of contact; or if portions of its substance (in a more or less altered state) be enclosed within the granite mass; or if the granite run into it in the form of veins (Fig. 93). On the other hand, if the adjacent rock rest in undisturbed position on the surface of the granite, and exhibit no chemical alteration, it may

FIG. 93.

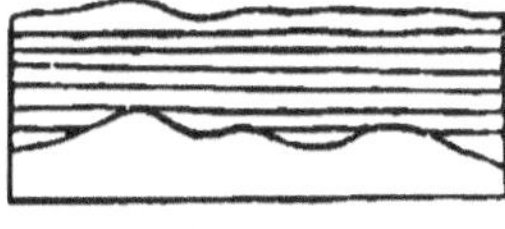

FIG. 94.

be inferred to be the more recent of the two (Fig. 94). Granitic veins frequently cross or intersect each other: intersected veins being necessarily older than those by which they are intersected. The diagram (Fig. 95) exhibits three veins of different ages. No. 1 is the oldest vein, as it is cut and also displaced or "faulted" by the other two. No. 3, again, is the most recent of the series, as it tra-

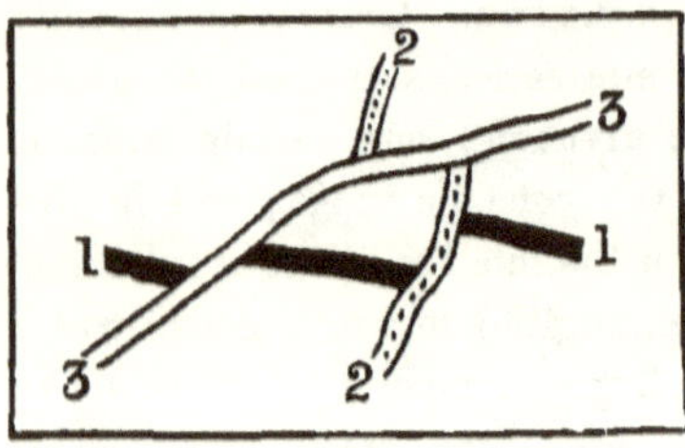

FIG. 95.

verses and displaces both Nos. 1 and 2. Granite rocks, by the decomposition of one of their essential components, feldspar, have become converted in some districts into white or light-coloured clays, largely used, under the name of kaolin, in the manufacture of porcelain.

Granite, properly so-called, is composed of three minerals : quartz, feldspar and mica. The feldspar is usually the potash species Orthoclase (see No. 57, Part II), but is occasionally represented by the Soda species Albite (Part II, No. 58), or by Oligoclase (No. 59). The mica is generally the common potash species Muscovite (Part II, No. 77), but is sometimes mixed with, or occasionally replaced by, one of the magnesian micas. As a general rule, the quartz, in granite, occurs in vitreous colourless grains ; the feldspar, in red, white, pink, or occasionally green or grey, lamellar masses, which exhibit smooth and somewhat pearly cleavage planes ; and the mica is mostly in small scales, or larger foliæ, of a pearly-metallic aspect, and silvery white, black, brown, pearl-grey, or greenish in colour. In coarse-grained granites, these component minerals are readily distinguishable ; but in rocks of fine-grain, they become blended into a common granitic mass. The mica frequently dies out, or is very sparingly present, in which case the rock is sometimes known as *Pegmatite*, but this name is applied by German lithologists to coarse-grained granites containing a small amount of silvery-white mica in comparatively large scales or leaves. Occasionally also in these quartzo-feldspathic granites, the quartz is arranged in the form of narrow, irregular crystals in more or less distinct bands, producing, in transverse sections, the appearance of a cuniform or Assyrian inscription : whence the term "graphic granite" sometimes bestowed on this variety. When, again, the quartz and feldspar become intimately blended, so as

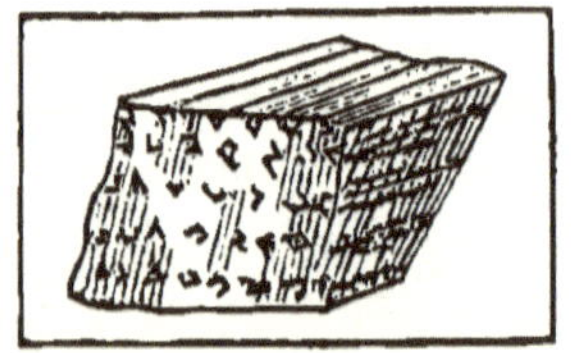

FIG. 96.

to possess more or less the appearance of a simple mineral, the rock has been termed *Felsite* or *Petrosilex*. Very frequently, through a base of this, or of ordinary granite, numerous crystals of feldspar are distributed, when the rock is known as *porphyry*, or, better, as *porphyritic gra ite* or *porphyritic felsite* (Fig. 97). The imbedded crys-

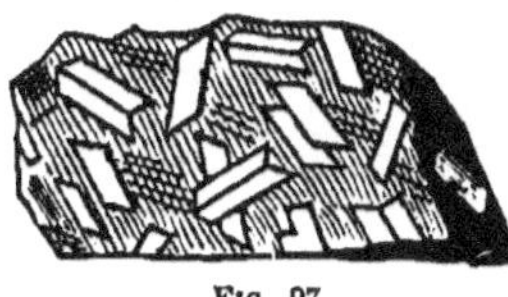
Fig. 97.

tals often show the twin or compound structure so common in feldspathic silicates. The term "porphyry" (from πορφυρα), as the name would indicate, was originally applied to rocks of this kind, in which either the base or the imbedded crystals presented a deep-red colour; but it is now bestowed conventionally on all rocks containing distinct crystals of feldspar or other minerals. We have thus porphyritic granites, porphyritic syenites, porphyritic trachytes, porphyritic greenstones (the original porphyry having been probably one of these latter), porphyritic lavas, &c. Finally, as regards other granite varieties (to many of which special names of uncertain or merely local application have been uselessly given), it may be observed that the mica of ordinary granite is occasionally replaced by talc, giving rise to *talcose granite* (the *Protogine* of some authors), or is accompanied at times by hornblende, the rocks in the latter case being known as *syenitic* or *hornblendic granite*. By the gradual diminution of quartz, the granites proper pass into granitic trachytes, described below; and they are represented in the metamorphic series by gneiss and gneissoid rocks generally, into which also they appear locally to merge.

Examples of intrusive granite occur in most parts of the large area occupied by the Laurentian rocks of Canada (see Part V.). Porphyritic felsite, in which the base is mostly dull-red or greenish-black, and the imbedded feldspar crystals red or pink, is seen in connection with a large mass of syenite in the Township of Grenville on the Ottawa. This variety, sometimes termed Orthophyre, is scarcely perhaps a true granite, but as it contains free quartz it must be referred conventionally to the granitic series. A broad dyke or vein of graphic granite (consisting of quartz and orthoclase-feldspar) is described in the Reports of the Geological Survey as occurring on Allumette Lake, north of Pembroke, and other examples of a similar character have been recognized in the neighbouring Township of Ross. Veins of both ordinary and quartzo-feldspathic granite, in some cases holding crystals of tourmaline or schorl, occur also more or less abundantly, in St.

Jerome, Escott, Lansdowne, Burgess, Madoc, Marmora, Galway, and indeed throughout the Laurentian region generally, lying between the Ottawa and Georgian Bay. In Laurentian strata, likewise, on the River Rouge, east of the Ottawa, and at Stony Lake, in the Township of Dummer or Burleigh, as well as in Bathurst and Burgess, granitic veins containing albite or soda-feldspar replacing or accompanying orthoclase, have long been known. The opalescent variety of Albite known as *Peristerite* (see Part II., No. 58) coems from a vein of this kind in Bathurst. Other veins and some considerable masses of granite occur on the north and north-east shores of Lake Superior, as in the vicinity of Michipicoten, at Point-aux-Mines, and here and there about Bachewahnung Bay, and elsewhere. A mass of red granite, inferred by Sir William Logan to be of Huronian age, is described as having broken through and tilted up Laurentian gneissoid strata south of Lake Pakowagaming on the north shore of Lake Huron; and granitic dykes and veins occur in the Bruce Mines District. A flesh-red granite underlies beds of Trenton limestone in the Township of Storrington, north of Kingston. Finally, intrusive masses and dykes of white or light-coloured granite occur on Lake Memphremagog in the Township of Stanstead, and others in the Townships of Hereford, Barnston, and Burford, of that district. Similar masses have been noticed on Lake St. Francis, Lake Megantic, and in the intervening townships. Some of these granitic masses, as described in the Revised Report of the Geological Survey (1863), cover areas of from six to twelve square miles.

A granite which contains hornblende in place of mica, was formerly defined by most geologists as *Syenite*, but this term is now generally restricted to a granitic greenstone, or mixture of orthoclase and hornblende. Keeping to the latter definition, we have in syenite a more or less distinctly granular or granite-like aggregate of potash-feldspar, and hornblende: the feldspar being usually red or white, and the hornblende green or black. Quartz in small amounts is also occasionally present. As in ordinary granite, both coarse and fine-grained varieties of syenite occur. In the latter, the component minerals are blended into a more or less uniform dark-green mass, and the rock resembles, and can rarely be distinguished from, an ordinary greenstone. From this trappean rock into well-defined syenite, indeed, an evident transition may be occasionally traced. On the other hand, syenite is represented in the Metamorphic Series by syenitic gneiss, and to some extent by amphibolite or hornblende-rock. Syenite, as already explained, is very frequently porphyritic—red or occasionally white crystals of feldspar appearing on a dark or black ground, or green or black crystals of hornblende being imbedded in a reddish granular mixture of the usual components.

In Canada, eruptive syenites appear to be confined mostly to Laurentian areas. The most remarkable example is the great syenitic mass described by Sir William Logan as covering a space of about thirty-six square miles in the Townships of Grenville, Chatham, and Wentworth, near the left bank of the Ottawa. It consists chiefly of red and white orthoclase, with black hornblende and a little quartz; mica being also present in one portion of the mass, which thus shows a transition into syenitic granite. Dykes pass from the main body of the syenite into the surrounding beds of crystalline limestone and gneiss. Two other series of dykes or eruptive masses occur in connection with the syenite of this locality. Some of these masses, consisting of a compact base of petro-silex, or intimate mixture of quartz and feldspar, with imbedded crystals of red orthoclase and fragments of gneiss and other rocks, traverse the syenite, and hence are of newer origin ; whilst others, consisting of trap or greenstone, are cut off, or interrupted in their course, by the syenite, and are therefore of anterior date. Dykes of syenite also occur, here and there, throughout the Laurentian country between the Ottawa and Lake Superior.

Anorthosites :—The term " anorthosite " was first applied by Dr. Sterry Hunt to a rock-mixture of various anorthic or triclinic feldspars, at that time regarded as a stratified crystalline formation or rock of the metamorphic series proper. Feldspathic rocks of this character occur in the counties of Argenteuil, Terrebonne and Montmorency, in the Province of Quebec, where they were thought to represent the so-called Upper Laurentian, Labrador or Norian formation, but they are now regarded by Dr. A. C. Selwyn, the present director of the Geological Survey of Canada, as eruptive products of Laurentian age, and this opinion, although not absolutely free from doubt, is probably the correct view. As stated in the previous edition of this work, their supposed stratification is exceedingly obscure.   They consist essentially of labradorite or albite, or of mixtures of these and other triclinic feldspars.   Their colour is mostly white, light-grey, pale lavender-blue or greenish-white; but some are pale-red or yellowish; and the cleavage planes occasionally show the green or greenish-blue reflected tints characteristic of labradorite.   All become opaque-white by weathering. Bronze-coloured or dark-green hypersthene, in foliated examples, is sometimes present in them, as at Château Richer and elsewhere.   This variety has been termed *Hyperite* or *Hypersthene rock.*

3. *Traps and Greenstones :*—The rocks of this series present a somewhat variable composition, but consist essentially of some kind of feldspar—usually Labradorite or Albite—or a mixture of feldspars, with augite, hornblende or chlorite.   Many also contain in addition, a mixture of zeolitic minerals, nepheline, magnetic and titaniferous iron ores, grains of olivine, scales of mica, carbonates of lime and iron, and other substances.   But free silica or quartz is altogether absent, or is present only as an accidental or inessential constituent.   The texture of these rocks is of two general or principal kinds : (1), compact or homogeneous; and (2), distinctly granular or granitic; but fine-grained examples offer a transition from the granitic to compact structure. In the latter, the component minerals are blended into a common or uniform mass, chiefly, unless weathered, of a grey, green or black colour. In each of these varieties of texture, a porphyritic structure (see Fig. 97, above) may also be present— the imbedded crystals consisting of albite, oligoclase, augite, hornblende or some other mineral. The compact varieties also frequently exhibit an amygdaloidal structure, Fig. 98, the rock being full of oval or irregularly-shaped cavities, usually of small size, and commonly lined or filled with amethyst, agate, or other varieties of quartz, or otherwise with calcspar, various zeolites, green-earth, &c.   These compact varieties, moreover, of both trap and greenstone very often assume a columnar or basaltiform structure, as in figure 99. In this case the rock exhibits a kind of rough crystallization, and contains numerous

Fig. 98.

joints or partings in the direction of
which it separates more or less
readily, forming prisms or prismatic
masses of from three to eight or
nine sides : and as these possess also
transverse joints at right angles to
the axis of a prism, a flat, tabular,
and step-like outline is very generally presented by columnar or sub-

Fig. 90.

columnar varieties of this kind.   Hence the term "trap" or "trappean rock," from *trappa*, a Swedish word signifying a set of steps—
attention having first been called to this peculiarity by Swedish
observers.   A good example is presented by the promontory of
Thunder Cape, Fig.
100, on the north-
west shore of Lake
Superior, in which
five very distinct
steps are observable,

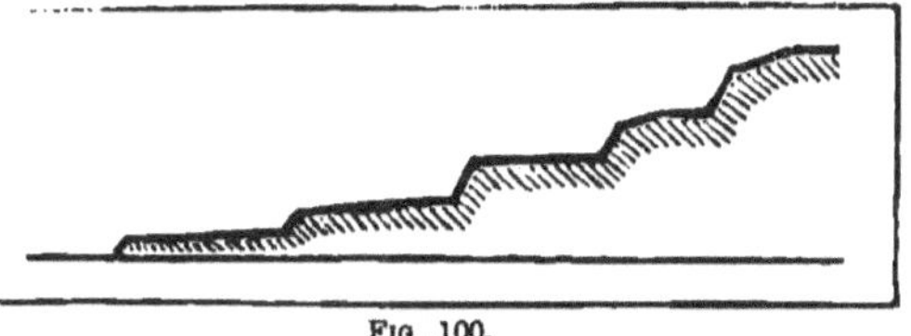

Fig. 100.

more especially when viewed from a certain distance.   The eruptive
mass of McKay's Mountain on the other side of Thunder Bay, as
well as similar rocks on Pie Island and elsewhere in that district,
exhibit also well-marked illustrations of this step-like outline, although most of these rocks present only a sub-columnar structure.
A similar step-like outline is exhibited by some large dykes of columnar dolerite in the township of Grenville on the Ottawa, as first
pointed out by the late Sir William Logan.

As regards their general conditions of occurrence, greenstones and
traps are seen very commonly in the form of more or less broad and
straight or simply-forking veins (Fig. 101), technically known as

01

dykes.   This term originates in
the fact that trappean veins
usually possess greater powers
of resistance to the decomposing influences of the atmosphere or the destructive action
of water than the rocks which
they traverse ;   in   consequence   of   which   they   often

project from the face of cliffs or hill-sides, or stand up above the general surface of the ground, and thus resemble in many cases the stone fences or walls known in certain localities as "dykes." The annexed figure exhibits a diagram-view—the surrounding foliage,

&c., being omitted—of a projecting dyke of this kind, as seen on Slate River, a small rocky stream which enters the Kaministiquia about twelve miles above Fort William on Thunder Bay, Lake Superior. The high cliffs of aluminous slate or shale on each side of the ravine through which the river flows, have been wasted by atmospheric action to a much greater extent than the

Fig. 102.

dyke; and the latter thus stands out from the face of the cliff on each side of the ravine, and presents the appearance of an old Gothic wall. On one side, it comes down close to the bank of the stream, as seen in the figure; and the arch, there shewn, must have been hollowed out, when the water flowed with fuller volume and at a somewhat higher level, during the gradual excavation of the valley. Most of the projecting points, reefs, and rocky islets on the shores of our northern lakes, consist of denuded portions of trappean dykes. Occasionally, however, trap and greenstones decompose more readily than the surrounding or encasing rock. Trench-like depressions in the ground, or clefts and open fissures on the face of the rock, are thus produced. Examples may be seen on some of the islands and parts of the coast of Lake Superior, near Neepigon Bay, and elsewhere.

Traps and greenstones occur also, in many districts, in the form of flat tabular masses, resting upon hill-tops. These are merely portions of ancient dykes, exposed and isolated by denudation. Finally, mountain-masses composed of trappean and greenstone rocks are of frequent occurrence, but these also may be regarded in most cases as the more salient portions of enormous dykes, several being often seen to lie in the same general direction, as though along an extended line of fissure. The picturesque mountain of Montreal, and the mountains of Belœil, Monnoir, Rougement, &c., are examples. These salient masses exhibit in places a distinctly conical or partly truncated form, as seen in the outline of the "Paps," (Fig. 103), on

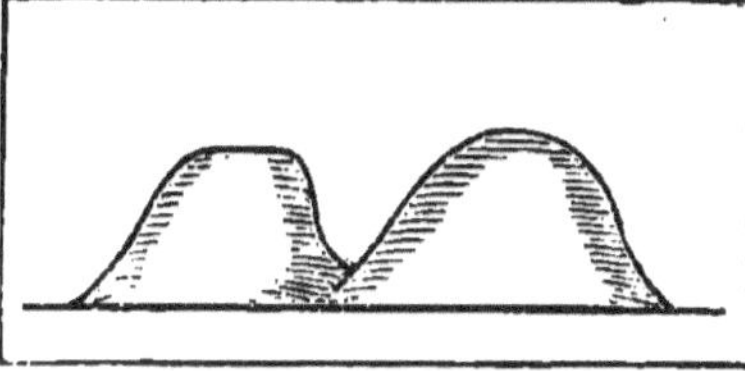

Fig. 103.

the east side of Black Bay, Lake Superior, and to some extent, also, in many of the greenstone hills of the Eastern Townships. A step-like and more or less tabular outline, as already remarked, is likewise very characteristic of rock masses of this group.

The variable composition and diversities of structure exhibited by trappean rocks have given rise on the part of lithologists to the formation of a great number of so-called species, each provided with a distinct name, usually of Greek derivation. But these attempted distinctions are in many instances of purely local application; and in very few cases can they be regarded as indicating definite admixtures of ready recognition. Names applied to particular varieties by one author, are applied quite differently by others. The terms melaphyre, porphyrite, diabase, &c., might be cited as examples. In many cases also, the same rock, if presenting slight differences of texture, or if assumed, without any possibility of proving the assumption, to contain augite in one case and hornblende in another, is described under different species. In this manner, fanciful distinctions which have no true foundation in Nature, and which cannot be rigorously or definitely applied, are attempted vainly to be carried out in many so-called systems of lithology. If minute chemical or mineralogical differences were regarded as essential, our Canadian varieties of this group of rocks might add many names to the already uselessly extended list. It is not possible however in the present state of the question, nor is it desirable in an elementary work of this character, to depart altogether from the beaten track. Retaining therefore some of the more generally recognised names and distinctions, whilst duly admitting the more or less arbitrary and uncertain character of these, we may refer our Canadian rocks of the Trappean series to the following varieties : (1) Trap or Basalt; (2) Dolerite or Granitoid Trap; (3) Greenstone or Aphanite; (4) Diorite or Granitoid Greenstone; (5) Diabase or Chloritic Trap. These varieties, it must be understood, merge more or less into each other, so that in many instances a rock might be referred with equal justice to two or more of their included types.

Trap or *Basalt* may be defined conventionally as a black, greenish-black or dark-grey rock of compact texture, composed of an intimately blended mixture of lime-feldspar (or lime and soda feldspars), augite, magnetic and titaniferous iron ores, zeolitic silicates, and carbonates of lime and iron. In some kinds, the feldspar is replaced by nepheline; and olivine, in visible grains of a green or greenish-brown colour, is very generally present in basaltic rocks. These component minerals. are deduced by calculation, it will of course be understood, from the separate ingredients, silica, magnesia, lime, &c., obtained in the analysis of the rock. Altered or weathered varieties of trap are frequently of a dull brick-red or brown colour, the change being caused by the higher oxidation of the iron. Unaltered basalts are always more or less strongly magnetic, easily fusible, and partially attacked by acids. Sp. gr. 3.0 to 3.1, but occasionally somewhat less from partial alteration of the rock.

The more common varieties or sub-varieties comprise :—(*a*) *Massive or amorphous trap*; (*b*) *Slaty trap*; (*c*) *Columnar and sub-columnar trap* (Fig. 99), the variety to which the term basalt is more generally applied; (*d*) *Amygdaloidal trap* (Fig. 98), containing oval or other shaped cavities mostly filled with agates, zeolites, calc-spar, green earth, &c., as explained on a preceding page; (*e*) *Porphyritic trap*, containing imbedded crystals of albite, augite, or other minerals. Examples of massive and porphyritic trap (consisting in places of almost pure augite or pyroxene, and hence termed "pyroxenite," by some observers) occur more especially in the Montreal Mountain, and in various parts of the Eastern Townships and lower St. Lawrence district. Columnar and sub-columnar trap is abundant around Thunder Bay; and amygdaloidal trap is of common occurrence along the northern and other shores of Lake Superior, the north shore of Lake Huron, and elsewhere. The agates of Michipicoten Island, Agate Island, and other spots on Lake Superior, are derived from the disintegration of these amygdaloidal rocks.

Dolerite is simply a trap or basalt of granitoid structure, in the coarse-grained varieties of which the component minerals are more or less perceptible, individually. Fine-grained varieties, which offer a transition into basalt proper, have been classed apart, by some lithologists, under the name of *Anamesite.* In these, the component minerals are scarcely, if at all, observable. Dolerites, as a general rule, are chiefly of a greyish colour, varying from light-grey with black specks and indistinct crystals of augite, to an almost uniform black or dark-grey tint. When much olivine is present, and occasionally in other cases, the rock assumes a greenish-grey or brownish-green colour; and weathered examples are frequently rusty-red, or otherwise dull white. Sp. gr. = 2.7. — 3.0.

This granitoid condition of trap presents, as in ordinary basalt, massive, slaty, columnar, amygdaloidal, and porphyritic varieties. Examples occur generally throughout Canadian districts in which trappean rocks prevail; especially in the Townships of Grenville, Chatham and Wentworth in the Ottawa region; also in parts of the Montreal Mountain and in the mountains of Montarville and Rougemont, and other parts of that district; abundantly also on the shores and islands of Lake Superior, (Gros Cap, Goulais Bay, Montreal River, &c.,) and throughout the northern lake region generally. The dyke on Slate River, shown in figure 102, consists of grey dolerite.

Greenstone or *Aphanite* is a compact trappean rock of a more or less decided green colour, passing into greenish-black. It is assumed from its general composition to be made up of an intimate mixture of lime and soda feldspars and hornblende, with very generally a certain amount of magnetic and titaniferous iron ore, and some carbonate of lime. Strictly, it cannot be distinguished, except conventionally by its green colour, from ordinary trap. It presents massive, slaty, columnar, anygdaloidal, and pophyritic varieties, and passes into diorite and diabase, the latter by the addition of chlorite, as well as into common trap. Dykes of this green variety of trap occur here and there on Lake Superior, but most of the so-called greenstones of that region are evidently chloritic, and hence would be regarded by systematists as compact and amygdaloidal varieties of diabase. Dykes of somewhat similar character occur also in the Madoc and Marmora region, and undoubtedly in other districts. The terms Greenstone and Aphanite, it should be observed, are applied by some authors to compact varieties of Hornblende Rock and other hornblendic examples of the Metamorphic Series.

Diorite is the name commonly given to a granitoid trappean rock made up of more or less distinctly visible grains or imperfect crystals of a soda-feldspar (or lime-feldspar) and hornblende, and containing very frequently, in addition, small grains of carbonate of lime, particles of magnetic iron ore, scales of mica, sphene, and other minerals. It passes into compact greenstone by almost insensible transitions ; and in many cases it cannot be distinguished readily, if at all, from varieties of dolerite or granitoid trap. Its feldspathic portion is usually white or grey, or sometimes reddish, and the hornblende black or green ; but fine-grained examples have very commonly a distinct green colour throughout. Massive, slaty, columnar, amygdaloidal, and porphyritic varieties occur, as in other kinds of trappean rock. The specific gravity varies from about 2.6 to 2.9.

Examples of diorite, of a more or less granitic aspect from the frequent presence of small scales of brown mica, occur in the eruptive masses of Belœil, Monnoir or Mt. Johnson, Rigaud, and Yamaska, of the Eastern Townships of Canada. Other examples, passing here and there into diabase, are seen at several spots on the shores of Lake Superior, as near Michipicoten Harbour and elsewhere in that neighbourhood, Batchewahmung Bay, &c. The term diorite, it must be remembered, has also been applied by certain authors to some of the stratified hornblendic rocks of the Metamorphic Series – these crystalline strata representing, as regards general composition, many diorites and other intrusive rocks containing hornblende, just as the gneissoid strata represent the granites and syenites. To avoid confusion, however, the term if employed at all, should be restricted, in accordance with common usage, to intrusive or eruptive rocks. If the same term is to be applied indefinitely to a stratified and eruptive form of rock, it follows logically that the term gneiss should be abandoned, and all the micaceous examples of gneiss should be known as granite, and the hornblendic varieties as syenite—a system, we presume, that few geologists, apart from those of a certain school, would be inclined to follow.

Diabase or Chloritic Trap—as defined by most authors—is an eruptive, feldspathic rock, containing augite or hornblende with a certain amount of chlorite : carbonate of lime being very generally present as an additional constituent. The term "diabase" is often applied, however, to chloritic and other varieties of hornblendic and augitic rocks belonging to the Metamorphic Series. Some kinds of eruptive diabase have also been described as melaphyre, but this term is also vaguely applied to many diorites and other granitoid rocks of the present group. Compact varieties, which are mostly of a decided green colour, pass into compact trap and greenstone by insensible transitions. Granitoid varieties merge also into dolerite and diorite. Both kinds offer amygdaloidal, prophyritic, and other examples. The feldspar in coarse-grained examples is either greyish-white, greenish, reddish, or brownish ; and the chlorite presents the form of small scales and particles of a green colour. Weathered examples are usually dull-brown or red. Varieties of diabase, as thus defined, occur both in the form of dykes and in intercalated bedded masses among the Huronian strata of Lake Superior, as in Michipicoten Island, as well as Cros Gap, Cape Maimanse, Point-aux-Mines, Goulais River, and elsewhere. The bedded examples may perhaps be really metamorphosed strata, but they consist most probably of portions of ancient trappean overflows formed during the gradual building up of the Huronian deposits. Some contain epidote ; others enclose well-defined crystals of augite ; and many are in the condition of calcareous amygdaloids.

Trachytes :—The rocks of this division are essentially feldspathic in composition, the more typical or charactaristic examples consisting almost wholly of orthoclase or potash-feldspar. Many of these are more or less porous or vesicular in texture, and are thus peculiarly harsh or dry to the touch, when the name "trachyte," from $\tau\rho\alpha\chi\grave{\upsilon}\varsigma$, rough. This character, however, will only apply to certain varieties. as many trachytes do not differ in this respect from other rocks, Most trachytes are white, light-grey, or pale reddish in colour ; but in the granitoid varieties the presence of scales of brown mica, small crystals or particles of green or black hornblende, and other accessory minerals, gives rise to a darker and variable tint. These trachyte rocks merge into members of the granitic and trappean series on the one hand, and into ordinary feldspathic lavas on the other. The substance known as pumice, for example, may be referred both to trachyte and to lava. Thus, many trachytes, occuring in connection

with active or extinct volcanic cones, are actual lavas in the common
sense of the term; but others, although undoubtedly of similar
origin, occur in localities to which the term volcanic has ceased to
apply.   Viewed generally, although no marked lines of demarcation
can be drawn between them, the Trachytes present the following
leading varieties :—Common or Porous Trachyte; Compact or
Massive Trachyte; Slaty Trachyte; Granitoid Trachyte.   Examples
of porphyritic structure occur in each of these varieties; and in the
trachytes of some localities the feldspar consists partially or wholly
of soda or lime species.

Common Trachyte is met with chiefly in regions in which active
or extinct volcanoes are distributed.   It is more or less porous, or of
an open granular texture, and is frequently porphyritic from enclosed
crystals of glassy feldspar.   Scales and specks of mica, &c., are sometimes scattered through it, and it contains occasionally some grains
of quartz.   The latter mineral is altogether of exceptional occurrence,
but its occasional presence serves to connect the trachytes with granitic
rocks.   *Compact Trachyte*, also known as " white trap" or "feldspar
trap," occurs in broad veins or dykes traversing both the older trap
or dolerite of the Montreal Mountain and the Lower Silurian limestones of that neighbourhood.   It contains at these localities a considerable amount of intermixed carbonate of lime ; whilst a related
variety of somewhat slaty structure, from near Lachine, is partly
zeolitic in its composition, and would thus be known by many lithologists as a *phonolite*.   These examples are partially in an earthy
state, a condition sometimes recognized by a special name, that of
Domite, a term applied to the earthy or semi-decomposed trachytes
of the Puy-de-Dôme in the ancient volcanic district of Central France.
A porphyritic variety of pale-red or yellowish trachyte, holding large
crystals of feldspar, occurs also at Chambly.   Examples of *Granitoid
Trachyte* are especially abundant in the Eastern Townships of Brome
and Shefford where they form eruptive masses of considerable extent
and elevation.   The trachytes of these mountains are both coarse
and fine granular, and are composed of orthoclas: or other feldspars
with intermixed scales of black or brown mica, grains of yellow
sphene and magnetic iron ore.   Some crystalline particles of black
hornblende are also occasionally present.   In the Yamaska Mountain of the same district, a micaceous rock of this character changes

somewhat in the composition of its feldspar, and becoming strongly hornblendic, passes into a variety of diorite; but the distinction between granitic trachyte and diorite is in many cases purely arti-ficial.

5. *Obsidians and Pitchstones :*—This division includes lavas and other rock matters of igneous origin which occur in a more or less vitreous or glassy state, and present an essentially feldspathic composition. The term obsidian is usually restricted to grey, green, brown, or black rocks of this character, occurring in actual connection with volcanoes, whether active or extinct. A variety containing small spherical secretions of a somewhat pearly aspect is known as Pearlstone. These rocks break with sharp edges, and the fractured surface shows conchoidal markings. Pitchstone occurs chiefly in the form of dykes in trappean districts. It is mostly of a black colour, and pitchlike or resinous aspect, but some varieties are dull-green, grey or red. A porphyritic variety, traversed by small veins of agate, occurs near the deserted copper workings on the Island of Michipicoten, Lake Superior; and some of the dykes and bedded traps near Michipicoten Harbour on the mainland, appear to be intermediate in character between pitchstone and ordinary basalt. Although not recognized in Ontario or Quebec, examples of obsidian are not uncommon in British Columbia.

6. *Lavas :*—These comprise the actual rock-matters which issue in a molten condition from volcanoes. They present vesicular, compact, columnar, porphyritic, and other varieties, and are of two general kinds as regards composition : feldspathic, and feldspatho-augitic. The first, and by far the more common of the two, are composed essentially of feldspar, and are mostly of a light or dark grey colour. They pass into trachytes. The second, composed essentially of feldspar and augite, are dark-green or black in colour and are undistinguishable, except by their actual conditions of occurrence, from many traps and greenstones. Examples of the group, as thus defined, are unknown within the limits of Ontario and Quebec, but occur in British Columbia.

V. MINERAL VEINS.

In a review of the characters and conditions of occurrence of rock-masses, the subject of mineral veins cannot be altogether passed over, but the scope of the present work admits only of a general reference to this subject.*

Mineral veins may be defined as cracks or fissures in the Earth's crust, filled or partially filled with stony and metallic matters. In some veins, stony or sparry matters, as quartz or calc spar, are alone present; but these matters are very generally accompanied by metallic sulphides, oxides, or other compounds, and occasionally by native metals. The sparry or stony substances are then known as *gangues* or *veinstones*. The more common veinstones comprise: quartz, calc spar, fluor spar, and heavy spar—two or more of these being frequently present together. In the higher part of a vein, frequently to a depth of several fathoms from the surface, the gangue and ores are often in a partially decomposed or earthy condition.

A mineral vein thus forms a more or less compressed sheet of mineral matter, extending often to unknown depths, and being frequently traceable for several miles across a line of country. Some veins are less than an inch broad, whilst others occasionally exceed twenty or even fifty feet in width. Many of the veins containing native silver in the district around Thunder Bay on Lake Superior, are at least twenty feet wide, and some are wider. A vein of calc spar, carrying galena, at the Frontenac Mining Location in the Township of Lough-borough (north of Kingston) is very nearly as wide, although the workable portion is limited to about twelve feet in breadth. As a general rule, however, few veins exhibit a greater average width than three or four feet; and in nearly all cases a vein contracts and expands more or less at different depths, or in different parts of its course. Many veins traverse the enclosing rocks, or "country," almost or quite vertically; others incline at a greater or less angle, the inclination being commonly termed the "underlie" or "hade;" and some again run almost horizontally, or like a narrow bed, for certain distances. The sides of a vein are known in mining language as the walls. These are very often separated from the enclosing rock by a band of brown ochreous matter or gossan, arising from the decomposition of pyrites, or by a layer of clay or other soft or earthy material. This is techni-

---

* Although true veins are of not uncommon occurrence among Canadian rock-formations, it should be premised that many of our metalliferous deposits, our iron ores especially, are chiefly present in the form of large irregular masses or "stocks."

·cally known as a "selvage." It usually facilitates the working of the vein. A broad selvage of this character lines the south wall of the Frontenac vein, referred to above. In inclined veins, the upper wall is generally termed the "hanging wall," and the lower, the "foot wall" or floor. A and B, in Fig. 104, illustrate these positions respectively.

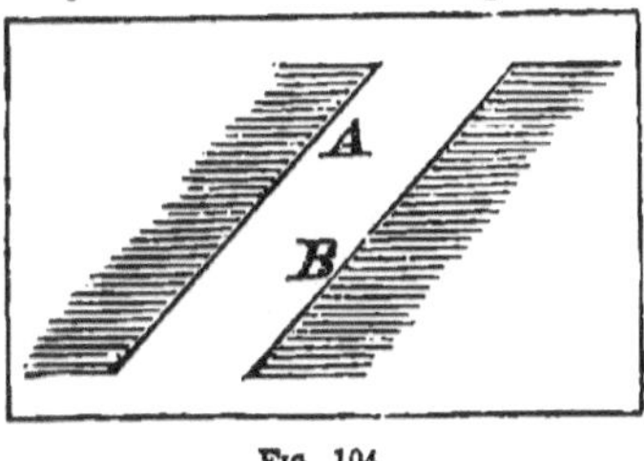

Fig. 104.

Mineral veins occur chiefly in mountainous or geologically-disturbed districts; and although present in certain localities among unaltered strata, they prevail mostly in metamorphic regions, especially where these are broken through by eruptive masses and dykes of granitic or trappean rock. In the Provinces of Ontario and Quebec, they occur chiefly in four districts:—First, in the Laurentian country lying between the Ottawa and Lake Huron, as, more especially, in the counties of Carleton, Lanark, Leeds, Frontenac, Hastings, Peterborough, and Victoria; secondly, in the allied Huronian strata on the north shore of Lake Huron; thirdly, in somewhat higher rocks on the shores and islands of Lake Superior; and fourthly, in metamorphic strata of apparently Pre-Cambrian age, in the Eastern Townships and adjoining region south of the St. Lawrence. These districts, as regards their geological relations, are described in Part V.

In reference to form and geological position, four different kinds of veins have been recognized. These comprise:—(1) *Independent* or *ordinary veins*, consisting of well-defined fissures which pass through rocks of various kinds, and generally hold a more or less straight course, whilst extending at the same time to great depths. In mining localities, several veins of this kind are commonly found to run in the same direction at greater or less distances apart. If crossed by another series of veins, the latter are usually found to carry ores of a different nature. The course of these veins may often be traced by trench-like depressions in the ground, arising from the atmospheric decomposition to which the surface of the vein has been subjected; but in some cases, especially when the gangue consists essentially of quartz, the vein has weathered to a less extent than the surrounding rocks, and thus stands

14

up in ridge-like form above the surface of the ground. (2) *Stock-werks.* This term, borrowed from German miners, is used to denote a series of usually narrow veins, ramifying amongst each other, and uniting occasionally into bunches or pockets of ore. (3) *Contact veins.* These are ordinary veins lying in immediate contact with eruptive masses, or between two different kinds of rock. Very frequently, for example, a band of metalliferous matter is found to lie between the edge of a mass of granite or trap and the enclosing stratified rock, in which case it is said to occur in the "contact country" of the two. (4) *Gash veins.* These are simply surface clefts or fissures of slight depth or extent. They are commonly filled with galena, and differ usually if not always from ordinary veins by the absence of veinstones properly so-called. In many cases they form mere strings of metalliferous matter. Attempts have been made to work deceptive veins of this character, in the townships of Eramosa, Clinton, and Mulmur.

Mineral veins may also be arranged to some extent as regards their structure or texture in five groups, as follows :—(1) *Compact veins.* In these, the fissure is filled entirely with a solid and more or less uniform mass of ore. (2) *Open ns.* The fissure, in these veins, is only partially occupied by mineral matter; open spaces occurring throughout the vein generally. Large cavities or "vugs," often lined with fine crystallizations, occur here and there in veins of various kinds ; but in these open veins, so-called, the insterstices or free spaces are especially numerous. (3) *Banded veins.* These are filled or lined with distinct bands or zones of different substances, the bands of the two walls corresponding in character, as in the annexed figure. The two outer bands, or those against the walls, may consist, for example, of brown ferruginous gossan, the two next of quartz, the two within these, of copper pyrites, succeeded by zinc blende, quartz, calcspar, galena, or other substances, in regular, banded alternations. Veins of this kind are

FIG. 105.

exceedingly abundant in many mining districts, but characteristic examples are rare in Canada. (4) *Spheroidal Veins.* In these the ore lies in the gangue in the form of spheroidal masses composed of concentric layers. Well-defined examples of Canadian occurrence do not appear to have been recorded, (5) *Brecciated Veins.* These form the great majority of mineral veins hitherto observed in Canada. The gangue contains angular and other fragments of well-rock, with the metalliferous portions of the vein arranged between and around these, occasionally in more or less distinct layers. The rock-frag ments are often traversed by thin strings of ore. When of large size, they form the so-called "horses" of the miners. These horses sometimes cause a good deal of trouble by coming in a direct line with the shaft, as happened at the Shuniah vein on Thunder Bay, Lake Superior, and in one of the shafts at the Ives Mine in the Eastern Township of Bolton.

Great obscurity prevails with regard to the processes by which vein-fissures have been filled with their contents ; but, in the majority of cases, several distinct agencies, acting both simultaneously and consecutively, have evidently been concerned in the repletion of these fissures. Some observers have sought to maintain that all the various matters found in veins were originally diffused through the mass of the surrounding rocks, and were drawn into the fissures by electrical currents passing through these : although they fail to explain how currents of this kind could possibly effect the operation in question. Others assume the mineral matters, in veins, to have been extracted from the surrounding rocks by the solvent power of water, and thus to have been gradually carried into the fissure. Many of the sparry, and some of the metallic matters, occurring in veins, may have been derived in this manner from the surrounding rocks ; but the supposed presence of diffused metallic matters in these rocks, considered generally, is, it must be remembered, entirely hypothetical, and open to many objections. On the other hand, we have undeniable proofs in volcanic and other districts, that metallic matters, in many respects similar to those found in veins, or capable of being converted into such by known chemical changes and decompositions, are actually brought from deeply-seated sources, both as sublimed products, and in solution in thermal springs. The weight of evidence, therefore, leads to the inference that the contents of veins generally, are due to endogenous action, rather than to surface forces; or that veins, in other words, have been filled essentially from below. In this connection, it must be remembered that many veins penetrate to unknown depths, and have yielded sulphurized or other ores, without being yet exhausted, to the amount of thousands of tons. Whilst many products found in veins are probably due in part or wholly to sublimation, the great majority of these products would certainly appear to have been deposited from solution : not necessarily in the condition in which they now appear, but in some other form from which their present condition has been derived. According to certain theorists, the whole of these bodies have been deposited from aqueous solution, but it is not easy to reconcile facts in all cases with this assumption. Such changes and decompositions as now take place in veins, lead to the conversion of many sulphurized ores into sulphates, carbonates, and other oxidized compounds ; but do not bring about, as the above hypothesis would require, the conversion of these latter on the large scale into vast bodies of galena, copper pyrites, arsenical pyrites, and other non-oxidized ores. But if these ores, now found in such vast quanities in mineral veins, really originated from soluble sulphates, chlorides, &c., the latter must undoubtedly have come from some deeply-seated source ; and their conversion into non-oxidized bodies could not have taken place on this enormous scale without the further collaboration of endogenous agencies.

Mineral veins are generally opened by shafts and adits, or by both of these methods combined. In the case of veins with considerable underlie, the shafts, or openings from the surface of the ground, are often carried down along the slope of the vein ; but in general, shafts are sunk vertically, and crosscuts are carried from the sides of the shaft at regular intervals to the intersection of the vein. Galleries are then driven along

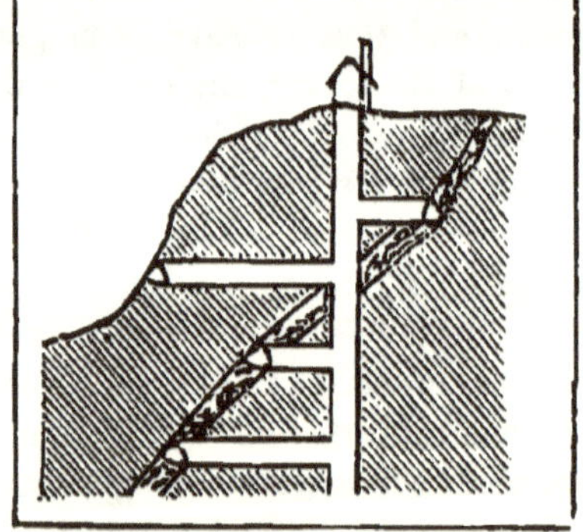

Fig. 106.

the course of the latter at these points, and the sheet of ore lying

between each pair of galleries or "levels" as these are commonly called,
is extracted by a system of step-like excavations, technically known as
"stoping." When a vein is nearly vertical in its position, a shaft
may of course be carried down to a great depth upon the substance
of the vein itself, and the material thus taken out of the shaft will
often pay for the sinking of the latter. Shafts are usually rectangular in form, and are not only strongly framed at the sides, at least
for a certain depth, but are commonly sub-divided vertically into two
or more compartments by brattice-work or planking; one of these
compartments being reserved for the pump rods and also for the
buckets or kibbles used for sending up the ore, or bringing it, in
technical phraseology, to grass; and another being fitted with ladders
or with a special lifting apparatus for the miners. An adit is a horizontal or nearly horizontal gallery driven from the side of sloping or
escarped ground, so as to strike the vein at a certain depth from the
surface outcrop of the latter. It serves in many cases, especially
where it opens out on a river bank, or on ground suitable for a tram-
way, &c., as a convenient roadway for bringing out the ore; and if
at a sufficiently low level it may greatly facilitate the drainage of
the mine, and assist in the ventilation of the works. Where two
shafts are sunk upon the vein, they should be located, if possible, on
high and low ground, respectively, in order to promote ventilation.
The ore, when brought to the surface, is usually "cobbed" or hand-
dressed by children, and the assorted portions, thus broken up by
hammers, are brought into the state of powder by subjection to
stamps or crushers. The powder is then agitated with water in long
narrow troughs or flat circular tubs called "buddles," the latter kind
being furnished with revolving arms or sweeps to which brushes are
attached; or it is shaken up with water in "jiggers" or tubs provided with movable sieves, until the metallic particles by reason of
their greater weight collect together, and so become separated, more
or less thoroughly, from the lighter earthy particles or refuse. commonly known as waste slimes or tailings. The dressed or concentrated ore is then ready for the furnace or reducing works.

Veins often cut or cross each other or are cut by eruptive dykes. In this case the intersected vein is very generally faulted or displaced. In mining language, a break of this kind in the continuity of the vein is commonly termed a "trouble," "heave," or "thrust," or an "upthrow" or "downthrow" as the case may be. The displacement may be very slight, or it may exceed

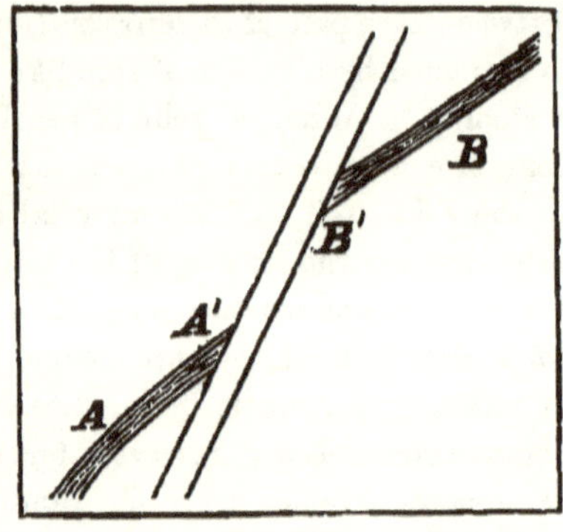

Fig. 107.

many fathoms; and great expense is often incurred in seeking for the displaced portion of a vein thus affected. As a general rule, if the intersecting vein or dyke be entered at its hanging-wall, as in working from A to A', Fig. 107, the continuation of the broken vein may be looked for "up-hill;" whereas, if the intersecting vein or dyke be entered at its foot-wall, or at B', the search for the displaced vein should be made "down-hill." This rule is not without its exceptions, but the exceptions are comparatively rare.

In order to ascertain the depth at which an inclined vein or bed of any kind may be reached by vertical sinking at a given depth from its outcrop, as at S, for example, in Fig. 108, we have the formula: $s = \tan i \times d$; or, Log $s = \log \tan i + \log d$; in which $s =$ the depth of the shaft; $i =$ the dip or inclination of the

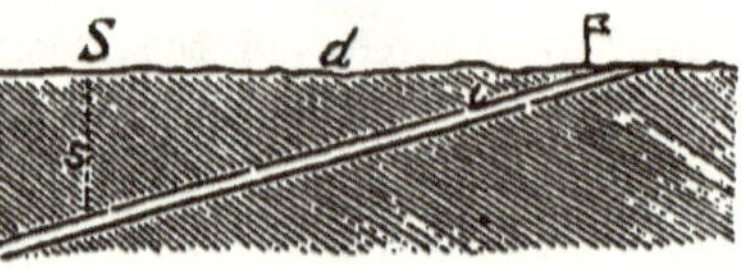

Fig. 108.

vein in degrees or minutes; and $d =$ the distance between the outcrop and the mouth of the shaft. If the ground at the proposed site of the shaft be higher or lower than at the outcrop of the vein or bed, the difference of level must of course be added to or deducted from s, as the case may require.

VI. CLASSIFICATION OF ROCK MASSES IN ACCORDANCE WITH THEIR RELATIVE PERIODS OF FORMATION.

Viewed in reference to their modes of derivation or general formative processes, rocks admit, as we have seen, of a distribution into three leading groups : comprising Sedimentary, Metamorphic, and Eruptive rocks ; but they admit also of another and far more interesting classification, based on their relative ages or periods of formation.

It is now universally conceded, on proofs the most unanswerable, that the various sedimentary and other rocks which make up the solid portion of our globe, were not formed during one brief or transitory period, but were gradually elaborated or built up during a long succession of ages. In areas of very limited extent, for example, even on the same cliff-face, or in excavations of moderate depth, we often find alternations of sandstones, limestones, clays, &c., lying one above another, and thus revealing the fact that the physical conditions prevailing around the spot in question must have been subjected to repeated changes. The same thing is also proved by alternations of marine and fresh-water strata in particular localities : and of deep-sea and shallow-sea deposits, in others. Again, the sedimentary rocks are frequently found in unconformable stratification, as explained above : horizontal beds resting upon the sloping surface or upturned edges of inclined strata. Here it is evident that the inclined beds must have been consolidated and thrown into their inclined positions before the deposition of the horizontal beds which rest upon them. In the absence of particular sets of strata in special localities, proving extensive denudation or long-continued periods of upheaval and depression—in the vast metamorphic changes effected throughout many districts—in the upward limitation of faults (Fig. 109), as sometimes seen— and, briefly, in the worn and denuded surface which a lower formation often presents in connexion with strata resting conformably upon it,—we have additional evidence of the lapse of long intervals of time during the elaboration of these rocks generally.

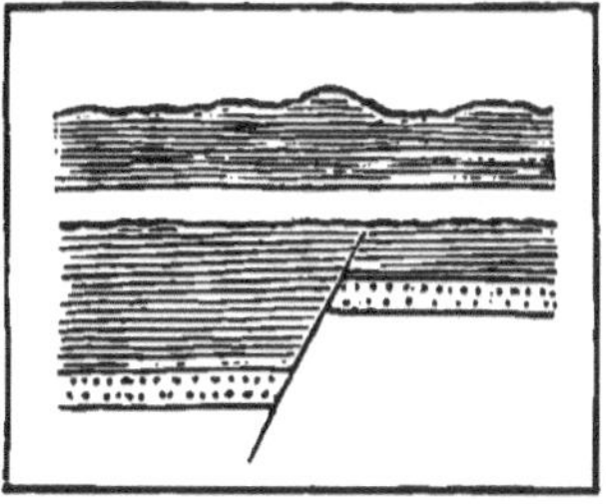

Fig. 109.

But a still more conclusive proof of this fact is to be found in the limited vertical distribution of fossil species of plants and animals, the remains of which are entombed in so many of the sedimentary rocks. The sediments now under process of deposition in our lakes, river-estuaries, and seas, frequently enclose, it will be remembered, the more durable parts, if not the entire forms, of various plants and animals, amongst which aquatic types necessarily preponderate. The sedimentary deposits of former geological periods have enclosed in like manner various organic forms peculiar to those periods. In the very lowest or earliest formed deposits, it is true, no traces of organic types have yet been met with, but above these beds, each group of strata holds its own characteristic fossils. Regarded broadly, the higher groups contain the higher organisms; and many structural conditions which are now embryonic or transitory, were manifested as adult or permanent forms of development in the periods represented by lower groups. Type after type lived through its allotted time, and then died out to be replaced by other and in general by higher forms of life. These facts are discussed more fully in Part IV. of this Essay, in which the leading questions connected with the subject of Organic Remains come under review. For present purposes, it will be sufficient to observe that by the careful study and comparison of these remains, geologists have sub-divided the series of rock-masses of which the Earth's crust or outer portion is composed, into a certain number of so-called Formations,—each Formation representing an interval or period in the ancient history of the Earth. These periods are thus made known to us by the various rocks produced by aqueous and other agencies during their continuance ; and by the organic remains, deriv d from the living forms of the periods in question, which are enclosed in these rocks. Each Formation, as already stated, holds its own organic types ; although, when viewed apart from local distinctions, consecutive Formations appear to merge into each other,—as an ordinary historic period blends insensibly with that which precedes and that which follows it. This is the case in natural groupings or classifications of all kinds : hard or sharply-defined lines being strictly unrecognized by Nature. The divisions however adopted by geologists, although overlapping as it were at their common boundaries, are distinct enough in the main ; and as some of these divisions are linked together more or less closely by the presence of certain related types of life, as well as by the general absence of other types, a grouping of Formations into larger divisions, representing longer geological periods or "ages," is conventionally adopted, as in the annexed tabular view.

FORMATIONS OF THE ANDROZOIC OR MODERN AGE.	Modern Deposits.
	Post-Glacial Formation.
	Drift or Glacial Formation.
FORMATIONS OF THE CAINOZOIC AGE.	Pliocene Formation.
	Miocene Formation.
	Eocene Formation.
FORMATIONS OF THE MESOZOIC AGE.	Cretaceous Formation.
	Jurassic Formation.
	Triassic Formation.
FORMATIONS OF THE PALÆOZOIC AGE.	Permian Formation.
	Carboniferous Formation.
	Devonian Formation.
	Silurian Formation.
	Cambrian Formation.
FORMATIONS OF THE ARCHÆAN AGE.	Huronian Formation.
	Laurentian Formation.

Notes on the above Table.

As the rock-formations enumerated in this table comprise a known thickness of many thousands of feet, it is evident that they can never exhibit a complete series at any one locality. But they are known to occur in this order, by a comparison of their relative positions at different places. Thus, in one district, we find (in ascending order) the Silurian and Devonian series; in another, the Devonian and Carboniferous, and so on.

(2) One or more of several consecutive formatious are often wanting or absent at a given spot. The Carboniferous rocks may thus, in certain districts, be found resting on the Silurian, without the intervention of the Devonian series. But the relative positions of these groups are never reversed. The Devonian beds are never found under the Silurian, for example, nor the Cretaceous under the Jurassic. The absence of particular strata, at a given locality, is accounted for by the elevation of the spot above the sea-level during the period to which the strata in question belong; or otherwise it is explained by denudation; or by the district having been situated

beyond the area of deposition to which the sediments extended.   (See some of the preceding observations under "Formation of Sedimentary Rocks," "Denudation," &c.)

(3) A formation of a given age may be represented in one place by a limestone ; in another, by a sandstone ; in a third, by argillaceous shales, and so on.   This will be easily understood, if we reflect that at the present day these different kinds of rock are being formed simultaneously at different places.   Many of our preceding observations have amply illustrated this, but the fact may be rendered still clearer by the accompanying diagram.   In this sketch, the dark outline is intended to represent a s o m e- what extended line of coast, with a river debouching into a deep bay.   In the latter, the argilla-

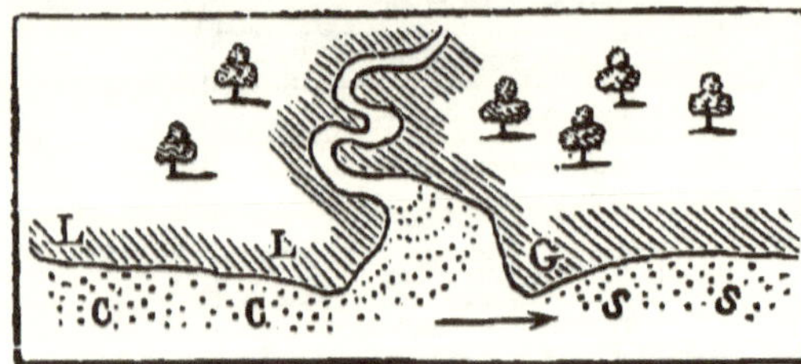

Fig. 110.

ceous or muddy sediments (*a*), brought down by the river, may be deposited.   At *G*, we may suppose a granitic headland.   The arenaceous or siliceous sediments (*s*) derived from the disintegration of this, will be arranged along the shore beyond it, by the set of the current.   Finally, at *L*, we may suppose the occurrence of exposed cliffs of limestone, yielding calcareous sediments (*c*).   These various sedimentary matters will be also in places more or less intermingled, producing rocks of intermediate or mixed composition.   But these rocks will be shown to be of the same period of formation, by the identity of some, at least, of the organic bodies contained in them : although many of the enclosed shells, &c., will necessarily be distinct, owing to the diverse nature of the sediments, the more or less exposed character of the coast, and the varying depths of water prevailing at different places.   We might expect, moreover, to find in one and all of these deposits, coins, pieces of pottery, and other objects of human workmanship, proving both their contemporaneous and their recent origin.   Hence, the age of a rock, it must be remembered, is in no way indicated by mineral composition : sandstones, limestones, &c., are of all geological periods.

(4) From time to time, during the gradual deposition of these sedimentary formations, various eruptive rocks were driven up

amongst them, producing (in general) chemical or mechanical alterations of greater or less extent. This action is still going on, as seen in volcanic phenomena.

(5) It is very generally assumed that the component matters of the earth and other cosmical bodies existed originally in a diffused nebulous condition, and gradually became condensed or solidified after passing through a state of igneous fluidity. And it is assumed, further, that this early molten condition of the earth is still retained in subterranean depths. This view—proposed by IMMANUEL KANT and maintained by LAPLACE—is necessarily hypothetical, but it is supported by many collateral facts, and has been very generally admitted.* If it be in the main correct, the rock-matters resulting from the first consolidation of the earth's surface must have been more or less akin to lavas and other volcanic products, although probably of a somewhat denser character from the greater density of the atmosphere then existing. No traces of these early lava-like rocks are now, however, visible. They must have been converted, long since, into other products, or have been re-melted and re-consolidated, probably time after time, beneath increasing thicknesses of superincumbent deposits. Sooner or later, after the first process of consolidation had set in, the continued radiation of terrestrial heat would allow the condensation of water to take place on the cooling surface of the earth. Then, a new set of phenomena would arise. The more exposed rock-surfaces would be worn down by aqueous and atmospheric agencies, and the materials thus obtained would form over the sea-bed a gradually increasing thickness of stratified deposits —most of which would undoubtedly be converted by subterranean agencies into crystalline or metamorphic formations. The earliest known rocks—those which underlie all other strata—are of this metamorphic character. They form vast beds of gneiss, mica-slate, hornblende-rock, chlorite-slate, quartzite, and other crystalline rocks, interstratified here and there with beds of crystalline limestone and dolomite, and are largely represented in Ontario and Quebec (see Part V.) In some of their strata, within the last few years, some

---

* "There are the strongest grounds for believing that during a certain period of its history the earth was not, nor was it fit to be, the theatre of life. Whether this was ever a nebulous period, or merely a molten period, does not much matter: and if we refer to the nebulous condition it is because the probabilities are really on its side."—PROFESSOR TYNDALL: *Address before the British Association: Liverpool, 1870.*

obscure and fragmentary examples of a supposed Protozoan, belonging or related to the Forminifera, have been discovered in Canada and at several European sites. A special genus has been framed for their reception under the name of Eozoon*; and the crystalline strata in which they occur, and which were formerly classed as formations of the Azoic age, are now generally known as " Archæan strata."

(6) Viewed broadly, the rock-representatives of the earlier portion of the Palæozoic age—comprising ordinary slates, sandstones, limestones, &c.—are characterized, more especially, by the presence of graptolites, cystideans, and trilobites, associated with tabulated corals, a great abundance of brachiopods (including species of the still surviving genus *lingula*), and many examples of orthoceras—an extinct cephalopod with straight shell, and simple, unlobed septa. These earlier Palæozoic formations are also distinguished, negatively, by the general absence or extreme rarity of land plants and vertebrated animal remains. The middle and higher portions of the Palæozoic series—including the Upper Devonian, Carboniferous, and Permian formations—contain, on the other hand, the remains of an abundant terrestrial vegetation, especially characterized by the presence of ferns and large equiseta (*calamites*), accompanied by peculiar types (*lepidodendra, sigillariæ,* &c.), some of which apparently indicate extinct transition groups between the higher acrogens and the gymnosperms of existing nature. In these higher Palæozoic strata also, the remains of cuirassed and other ganoid fishes, all of heterocercal type, occur more or less commonly; together with numerous tabulated corals, crinoids, brachiopods, and cephalopods, related generally to earlier or lower forms—whilst graptolites and cystideans are no longer met with, and trilobites die out at the base

---

* Examples of the *Eozoon Canadense* were first discovered in the Laurentian strata of Ontario by the late Dr. Wilson, of Perth, and others were subsequently found in the crystalline limestone of Grand Calumet Island on the Upper Ottawa, by the officers of the Canadian Survey. Although their organic origin was strongly suspected at the time by Sir William Logan, it was not definitely admitted until the publication of Dr. Dawson's microscopic researches, followed by those of Dr. Carpenter in England. By many observers, however, the organic nature of these remains is still contested, and it cannot certainly be regarded as fully proved. One point, and a great point, in its favour, is the undoubted resemblance of the better preserved Eozoon structures to examples of Palæozoic *Stromatopora*. On the other hand, it is somewhat remarkable that no other form of undoubted organic structure should as yet have been discovered in these Laurentian strata or in the less crystalline limestones of the succeeding Huronian series.

of the Carboniferous beds. Among the cephalopods, however, genera with angulated septa (*bactrites*, *goniatites*) make their appearance, and thus foreshadow the advent of the more elaborately-lobed types of the succeeding age. Of the higher vertebrates, mammalia and birds are entirely unknown, and reptiles all but absent—the only undoubted examples of the latter occuring in Permian strata. The remains of labyrinthodont amphibians, on the other hand, occur throughout the Carboniferous and Permian formations. Rock-representatives of the earlier Palæozoic periods occur widely throughout Ontario and Quebec. (See Parts V. and VI.)

(7) The succeeding strata, of Mesozoic age, are characterized essentially by the remains of numerous extinct types of reptilia of remarkable organization, including the paddle-footed Ichthyosaurians and Plesiosaurians, the winged Pterodactyls. the Dinosaurs, and many others. These strata are also especially distinguished by the ammonites and other genera of tetrabranchiate cephalopods with highly-lobed or foliated septa, which first appear in them, and which are altogether unknown in higher formations. Of these, the *ammonite* (as a genus) ranges throughout the entire Mesozoic series ; the *ceratite* is confined to Triassic strata ; and the *baculite, scaphite, ancyloceras, crioceras, turrilite,* &c., are essentially Cretaceous forms. The *belemnite*, a genus of the dibranchiate cephalopoda, known by its conical (internal) shell, is also especially characteristic of Mesozoic strata. Among the representatives of fish-life, homocercal forms appear ; and in the higher or Cretaceous portions of the series the rapidly-diminishing ganoids become almost entirely replaced by teleosteans of modern type. Vertebrates, higher than reptiles, are exceedingly rare ; but some bird remains of apparently reptilian character, referred to the extinct genus *Archæopteryx* (distinguished by an elongated series of caudal vertebræ: an embryonic type of structure as regards existing birds), have been found of late years in Jurassic strata ; and the Cretaceous beds of Western America have yielded some remarkable remains of toothed birds. A few small jaws and teeth, belonging most probably to marsupial mammals, are also known as Mesozoic forms. In the vegetation of the age, cycads and conifers stand out as dominant types—especially in its earlier and middle periods ; but monocotyledons are also represented throughout its series of strata, and dicotyledons appeared abundantly before

its close.   Mesozoic strata, although occurring in other parts of the Dominion, are unknown within the limits of Ontario and Quebec.*

(8).—The organic remains entombed in Cainozoic strata present, collectively, a marked resemblance to those of the existing epoch. Angiosperms, except under local conditions, evidently formed, as now, the prevailing vegetation of the age ; but among the dicotyledonous types, gamopetala, as compared with polypetala and apetala, were apparently less numerous than at present. In the animal world, the crinoids, brachiopods, chambered cephalopods, ganoid fishes and other types eminently characteristic of Mesozoic and Palæozoic periods were no longer predominating forms, but were reduced to a few comparatively unimportant representatives. The reptilian characteristics of the Mesozoic age had also given place to higher modifications of vertebrate structure and organization. Placental mammals formed the leading or more characteristic types, and were abundantly represented.   All Cainozoic orders still offer representatives ; but many Cainozoic genera, and practically all the species, have become extinct ; and two orders, at least, the Edentata and Proboscidea, exhibit marked decadency.   Cainozoic strata underlie an immense extent of country throughout the north-western portions of the Dominion, but have not been recognized in Ontario or Quebec.

(9).—During the Palæozoic and Mesozoic ages, and the earlier epochs of the Cainozoic age, the entire surface of the globe appears to have possessed a warm and comparatively uniform temperature, resembling that which now prevails in intertropical regions. This view is amply sustained by fossil evidence. The remarkable plant remains, so abundant in the middle or higher Palæozoic strata, exhibit in widely separated regions the same aspect and character. Those of Mesozoic formations, and also the characteristic reptilian and other types of the Mesozoic periods, exhibit the same law. The broad distinctions, due to geographical position, which now prevail, were absent also throughout the greater portion of the Cainozoic age, or only became indicated towards its close. Not only do we find, for example, in the Cainozoic strata of Northern Europe, and other comparatively high latitudes, the shells of conulariæ, nautili and

---

* Certain marls and sandstones associated with the trappean overflows of Thunder Bay and the Nipigon district, Lake Superior, have been thought by some observers to represent Triassic strata, but the weight of evidence is opposed to that view.

similar warm-sea mollusca, but these strata contain also many palm-fruits, with the remains of large ophidians, and the teeth and bones of large mammals allied to the existing tapir, rhinoceros, hippopotamus and other genera, now limited, or nearly so, to intertropical habitation. As time passed on, however, a great climatic change appears to have crept slowly over all the northern portions of both the eastern and western continents, and to have been experienced also in the extreme south. Under its influence, the once warm climate gradually gave place to all the rigours of an Arctic winter. This remarkable change was evidently accompanied, and probably in chief part produced, by great alterations in the previously-existing levels of land and sea. All the higher lands appear to have been covered by broadly-extended glaciers, whilst the seas were traversed by floating icebergs, bearing southwards the boulders and detritus of northern rocks. This condition of things was undoubtedly of long duration. Between its close and the actual commencement of the natural conditions now existing, no strict demarcation can be drawn. The one period merged slowly into the other, as the glacial manifestations were gradually beaten back to within the higher latitudes and alpine elevations in which they still prevail. Deposits of this glacial period are present almost everywhere throughout Ontario and Quebec. (See Part V.)

(10).—At some remote epoch of this later time, Man appeared upon the scene. His advent preceded the extinction of several important species and even genera of mammals; a fact which in itself goes far to prove the high antiquity of our race. The extinction of these types cannot be supposed to have taken place in any sudden manner, but was undoubtedly the result of slow organic and physical changes, proceeding continuously throughout long intervals of time. All the earlier tokens of man's presence on the earth—the rude flint and bone implements found in certain gravel-beds, associated in places with extinct mammalian remains—indicate a low state of civilization; a deduction sustained to some extent by the characters of certain human crania found in similar deposits. In later accumulations of detrital matter, as well as in many peat-morasses, etc., somewhat more finished types of stone utensils and weapons make their appearance; and these become replaced, in higher portions of the same deposits, by characteristic memorials of the so-called bronze and iron periods.

PART IV.

FOSSILIZED ORGANIC BODIES.

In sedimentary accumulations now forming at river-mouths and in lakes and seas, the shells of mollusca, bones of vertebrates, and hard parts generally of other animals, are being constantly enclosed— together with many sea-weeds, and the leaves and stems of various land plants. In like manner, the remains of many of the plants and animals which lived upon the earth or in its waters in former ages, became entombed in the sedimentary rock-matters then being deposited ; and they have thus been fossilized and preserved, although very commonly in the form of casts or impressions only. Many stratified rocks are thus found to contain fossilized organic bodies, either entire or in a fragmentary condition. In these fossil bodies, therefore, we see representatives of some, at least, of the life-forms of former geological periods ; but we must guard against the supposition that more than the merest vestiges of the floras and faunas of the Past are thus brought before us. Thousands of living forms, it is clear from their conditions of life and general surroundings, have little chance of being preserved in a fossilized state : and equally, in former periods, must the great majority of the types, then living, have passed away without leaving behind them any record of their existence. Imperfect, however, as must necessarily be our knowledge of the life-forms of past ages, the vestiges thus preserved to us reveal many points of great physical and biological interest. Their study enables us to distinguish the formations of one geological age or period from those of other periods—each holding its own proper and separate forms. These fossil forms, moreover, fill up many breaks or gaps in the existing life-series, shewing connections between living forms apparently remote in structural affinities, and indicating a gradual passage from earlier and comparatively generalized types, into higher and more specialized forms of existence. Their study reveals also in many instances the great changes in the distribution of land and sea that have taken place both in early and in comparatively recent times.

Certain fossil forms belonging to the most recent geological deposits are identical with existing species; others are akin to these generically, although specifically distinct; and many are wholly without representatives in existing nature. In some cases the fossils found in rocks were evidently entombed as living forms; whilst, in other cases, they have been drifted after death to a greater or less distance with the sediments of which they now form part. The process of fossilization is a gradual replacement (as in the case of many mineral pseudomorphs) of the original substance of the body by mineral matter. The fossilizing material is either the general substance of the enclosing sediments, or some special substance— mostly silica, calcic carbonate, or ferrous carbonate. The latter is frequently converted by after changes into hydrated ferric oxide, or occasionally into iron pyrites.

PLANT REMAINS.

The relations of fossil plants to existing vegetable types—except (and that not always) in the case of sea-weeds, ferns, and the plant remains of the less ancient rock-formations—are generally more or less obscure. This arises from the comparatively imperfect state of preservation of these bodies, their more characteristic structural parts being commonly destroyed; whilst, at the same time, the fragmentary remains of distinct species and genera are often mixed up together, and much uncertainty in their determination is thus occasioned. Plant remains in the fossil state are much less common than the fossilized remains of animals. The reason is obvious: aquatic types, literal and lacustrine forms especially, are those most favourably situated for fossil preservation, and these, in the plant world, are for the greater part composed of easily-perishable cellular tissue,—whereas most aquatic animals possess shells, bones, or other hard secretions, composed largely of mineral and comparatively indestructible matter. The shells of most mollusca, for instance, as well as most coral structures, contain more than 90 per cent. of calcic carbonate; in teeth, the inorganic matter varies from 60 to 70 per cent.; and in ordinary bones, from 50 to about 60 per cent.* Land plants, when preserved as fossils, are generally found in

---

* The inorganic matter in ordinary fish-scales exceeds 50 or 55 per cent.; whilst in the scales of reptiles (as in the nails, horns, and hair of mammals) it is under 5 or 6 per cent. This explains the rare occurrence of reptilian scales, even as impressions, in strata, whilst those of fishes are comparatively abundant.

argillaceous shales and in certain sandstones, usually in the form of casts or impressions, with the surface coated with a thin layer of black carbonaceous matter.

The vegetable forms of existing Nature admit of a separation into two primary subdivisions : *Cryptogams* and *Phanerogams*, or flowerless and flower-bearing plants, respectively. In the latter—although in many of their types the floral organs are quite inconspicuous—the ovules or embryo seeds are produced and fertilized into seeds, properly so-called, within the flower; whilst in cryptogamic forms there are neither flowers nor ovules, and consequently no true seeds, but, in place of these, certain seed-like bodies, termed spores, of a peculiar character. Some of the higher cryptogams, however, notwithstanding these and other points of difference, make a close approach in many respects to the lower phanerogamic types ; and it is very probable that certain extinct forms of Palæozoic vegetation may have constituted connecting links between the two divisions.

In each of these primary series, various subordinate groups are recognized—as shewn, in condensed form, in the following tabular view :—

I. CRYPTOGAMS :

  1. THALLOGENS :
      (i.) Land Thallogens.
      (ii.) Aquatic Thallogens.

  2. ACROGENS :
      (i.) Cellular Acrogens.
      (ii.) Vascular Acrogens.

II. PHANEROGAMS :

  1. GYMNOSPERMS :
      (i.) Cycads.
      (ii.) Conifers.

  2. ANGIOSPERMS :
      (i.) Monocotyledons.
      (ii.) Dicotyledons.

CRYPTOGAMS.

CRYPTOGAMS, as already explained, comprise all flowerless forms of vegetation, as lichens, sea-weeds, mosses, ferns, and the like. They may be arranged broadly under two series : *Thallogens* and *Acrogens*.

Thallogens are composed of cellular tissue only, and in them there is no proper (although sometimes a seeming) differentiation of stem and leaf. They may be classed as *Land Thallogens*, growing in the air; and *Aquatic Thallogens*, growing in the water of seas, lakes and streams.

Land Thallogens comprise *Fungi*, which have no chlorophyll (green colouring matter) in their tissues, and which require organized matter for their nutrition; and *Lichens*, most of which form incrustations on stones and the bark of trees. Neither are of palæontological interest.

Aquatic Thallogens may be regarded broadly as consisting of *Algæ* (ordinary sea-weeds and related fresh water confervæ), and *Microphytes*—separating, under the latter designation, a group of very minute or microscopic forms usually placed with the algæ, and known as diatomaceæ, desmidiæ, and volvocineæ.

Algæ occur as fossil casts and impressions in rocks of all ages from the Cambrian upwards; but in the majority of instances fossil forms can only be referred doubtfully to the special groups into which modern algæ are divided. They are commonly termed "fucoids." and are abundant in our Silurian and other strata. Some of the more common examples are shewn in the annexed figures: 111-115.

Fig. 111.
Lithrophycus Ottawaensis (Billings).
Trenton Formation.

Fig. 112.
Bythotrephis tenuis (Hall).
Medina and Clinton Formations.

The remarkable impressions, commonly regarded as the tracks of crustacea, which occur in the Potsdam (Cambrian) formation of the vicinity of Perth in Ontario, and in Beauharnois, Quebec, should probably be referred to algæ. These are known as *Climactichnites* and *Protichnites*, figures 116 and 117. In *Climactichnites*, the form is that of a long ribbon, five or six inches wide and several feet in length, mostly with a kind of beaded border and central groove, and

Fig. 113.
Arthrophycus Harlani (Hall).
Medina and Clinton Formations.

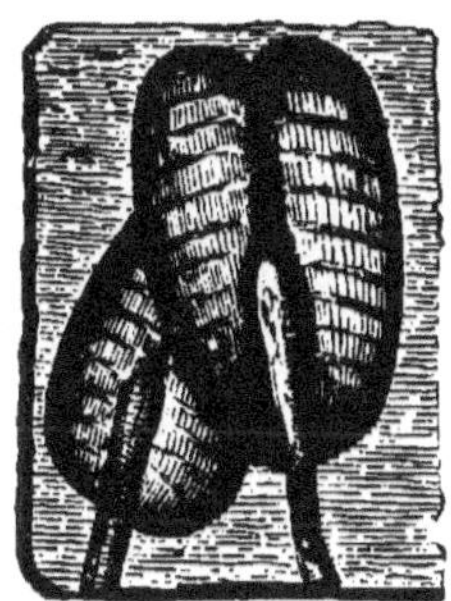

Fig. 114.
Rusophycus bilobatus (Hall).
Medina and Clinton Formns.

Fig. 115.
Fucoides (=Spirophyton) cauda-galli.
Portage-Chemung Formation.

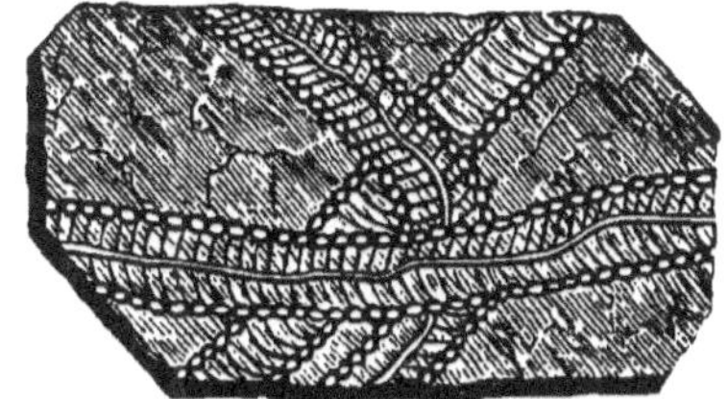

Fig. 116.
Climactichnites Wilsoni (Logan).
Potsdam Formation.

with numerous transverse furrows. *Protichnites* consists of a central groove, more or less interrupted, with on each side, at nearly regular

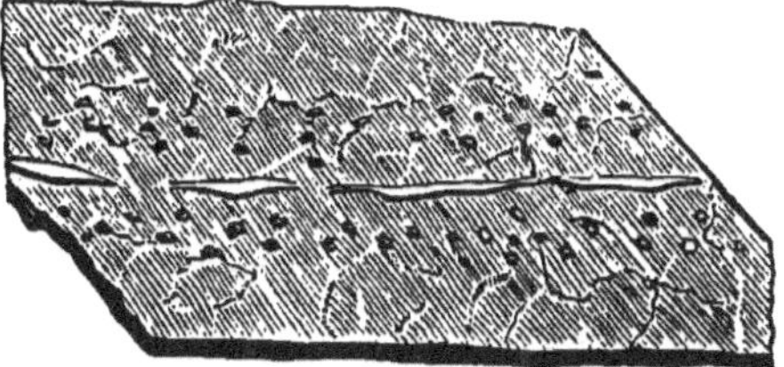

Fig. 117.

distances apart, a series of small pits or indentations, varying in number in different impressions, or being occasionally absent.*

___

* The central groove, by those who consider these impressions to be crustacean tracks, is supposed to have been made by the caudal spine of the creature, and the lateral pit marks by its claws. But these pit marks differ in number in different impressions, and as the number of the feet in the crustacea is a very important and constant character, Professor Owen has

Of the microscopic forms, here separated under the name of *Micro-phytes* from the *Algæ proper*, the diatoms only are of palæontological interest—representatives of the other groups being unknown, or only doubtfully known, in the fossil state. The diatoms, which abound in modern seas and in most fresh waters, secrete a siliceous test or shell. Under the microscope, they present circular, stellate, triangular, sigmoid, and other shapes. Many siliceous deposits of cainozoic

---

been forced to refer the impressions to four or five distinct species—one impression being without the lateral indentations. As stated by the writer more than ten years ago (*Canadian Journal*, 1877; also, *Annals of Nat. History*, of the same year), the association of so many different species, if the supposed tracks be really of animal origin, is at least a very remarkable circumstance; one, indeed, that might cause doubt in unprejudiced minds as to real nature of these impressions. On the other hand, there is really nothing in them to conflict with the view that they may be simply the impressions of large fucoids. Many of the existing *Melano-spermæ* grow to a great length: and in many genera with flattened or riband-like fronds there is a well-defined midrib, sufficiently hard to make a distinct impression when the frond is pressed upon damp sand. The lateral indentations of our Potsdam examples may have been made by groups of spores or sporangia arranged (as seen in many existing sea-weeds) along the sides of the fronds. Even apart from these, the air-bladders in many algæ are capable of making very distinct impressions on moist sand: and it would not be more unreasonable to infer that in these ancient sea-weeds the spores were of a somewhat harder or denser nature, than to have to admit with Professor Owen that the crustaceans by whose feet the indentations are commonly supposed to have been made, were "wholly distinct from the crustacean forms of later geological periods or of the present day."

If the impressions be fucoidal, the otherwise remarkable character of these lateral pit-marks, in differing in number and grouping in different impressions, becomes easily explained without the necessity of having recourse to imaginary specific distinctions. In the impressions in which they do not appear, it may be inferred that the fucoid had already scattered its spores, or that the development of the latter had not taken place, when the frond was cast upon the ripple-marked shore of the old Potsdam sea.

The supposed fucoidal origin of these impressions would not, I confess, however, have been thus advanced, were it not for their association or connection in at least one locality—the vicinity of Perth, in Eastern Ontario—with impressions of an analogous character to which an animal origin can scarcely be attributed on any rational grounds. These are the impressions known as *Climactichnites*. It is probable that the supposed animal origin of these latter impressions would never have been conceived, but for their general relations to the *Protich-nites* impressions. They may be described, generally, as being in the form of a band, several feet in length, although clearly fragmentary, with a width of from five to six inches. In their general dimensions they agree, therefore, very closely with the *Protichnites* impressions. But they differ from the latter in being traversed transversely by a series of narrow parallel ridges, about an inch and three-quarters apart, and by having a kind of beaded edge or border —the impression, as remarked by Sir William Logan, thus somewhat resembling a rope-ladder, whence the name *Climactichnites*. In some examples there is a central groove or ridge running roughly parallel with the length of the impression.

The points, here, to which attention should be chiefly directed, are, first, the presence of these numerous transverse ridges; secondly, their constancy, and the uniform clearness of their outline, throughout the impression; and thirdly, the unbroken continuity of the impression throughout its entire length. It must be evident that there are only two ways—both exceedingly improbable—by which these impressions could by any possibility have been made by any animal. If the impression be really a track, the animal must either have had, or have been able to assume, the form of a complete sphere or cylinder with ribbed surface, and it must have possessed sufficient internal force to roll itself over and over throughout a length of

and recent age consist almost entirely of these minute forms. The well-known "Tripoli," used as a polishing material, is of this charac-

ter. Diatomaceous deposits were formerly called "infusorial marls," diatoms having been at first regarded as
animal infusoria. Fig. 118 shews a few
of these forms, highly magnified. In
Canada, diatomaceous beds are all but
unknown. The only recorded example
(Rep. Geol. Survey, 1863) is in the valley of the Petewahweb, in the Upper
Ottawa region ; but a bed of this nature
is said to occur at Westbury in Compton

FIG. 118.
Recent Diatoms.
Greatly magnified.

County, Quebec. Thin sections of chert nodules from our corniferous
(Devonian) limestone, have also shewn the presence of diatoms.

Acrogens :—Whilst in Thallogens, the plant, so to say, is practically all leaf or all stem, and growth takes place from no definite
point, in acrogens there is a distinct differentiation of stem and leaf,
and the growth is acrogenous or terminal. The division falls into
two subdivisions : *Cellular Acrogens* and *Vascular Acrogens.*

In cellular acrogens, the plant, as in the thallogens, is composed of
cellular tissue only. The sub-division comprises : *Characeæ, Mosses,*
and *Hepaticaceæ.* Fossil representatives of these (with the exception
of the "nucules," of certain charæ) are exceedingly rare, and of no
special interest. The nucules of charæ (fresh-water plants) are
minute seed-like organs, encased in five spirally-twisted filaments, the

---

many feet ; or otherwise the creature must have moved forward by a series of spasmodic jerks
or jumps, alighting always in an exact line with the end of the trail, so as to avoid the *slightest
overlap* or other *break of symmetry* in the entire impression. Any other mode of progression
would unavoidably have effaced or smudged the transverse grooves or ridges as the body of
the animal passed over them. There is also another point which appears to be in complete
opposition to the assumed track-origin of these impressions. In places, two, or even three, of
these supposed tracks cross one another, but at the crossing points there is no sign of disturbance or smearing, so to say, such as must inevitably have occurred if one trail had been carried
across another. As shown especially in Sir William Logan's original figure, representing a
group of several "tracks" (Geol. of Canada, 1863, p. 107), the one impression simply conceals
or lies over the other at these points, as would happen if two fucoid-fronds, or other similar
bodies, were drifted together to a sandy shore, and were there covered simultaneously with
sediment.

In attributing these impressions to large fucoids, we encounter, on the other hand, no real
difficulty. Many algæ, it is well known, present transverse furrows ; and a salient example of
this character may be seen in our *Arthrophycus Harlani,* so abundant in many of the Medina
and Clinton beds.

free ends of which form a kind of coronal on the top of the nucule. These little bodies, when first found as fossils, were taken for foraminifera and called "gyrogonites." They occur in Triassic and many higher (especially Cainozoic) fresh-water deposits.

Vascular acrogens comprise the more typical acrogenous forms, those in which vascular tissue is present. They include : *Equisetaceæ*, *Ferns* and *Ophioglossaceæ*, *Hydropteridæ* or *Rhizocarps*, *Lycopodiaceæ*, and *Lepidodendraceæ*. The latter are usually placed with the lycopodiaceæ, but although more or less closely allied to these, their true affinities are still uncertain ; and their comparatively large size and other characters warrant their separation as a distinct and higher group. All are extinct ; and their remains are apparently confined to Palæozoic strata.

The *Equisetaceæ* comprise only one living genus, *Equisetum*, common species of which are familiarly known as "Dutch rushes," horsetails, &c. The *Equiseti* form hollow-jointed stems, arising from creeping rhizomes or root-stalks, with, in fertile examples, a terminal cone or spike containing the sporangia. The stem in most cases is longitudinally striated, and the joints (in which, more especially, silica is deposited) are surrounded by a toothed sheath of united

FIG. 119.
Calamites inornatus (Dawson). Portage-chemung Formation.

leaves or scales, and also in some examples by a whorl of slender branchlets. Fossil forms are chiefly represented by *Calamites*, although undoubted equiseti are known in carboniferous and higher strata. *Calamites* are abundant in many Devonian and Carboniferous beds. They occur usually in the form of stem-fragments (or impressions of these), transversely jointed and longitudinally furrowed, and as a rule more or less flattened or compressed. These stems vary from about an inch, or less, to more than a foot in diameter, the average width being from two or three to about six inches. Fig. 119 represents an example of a calamite from the dark bituminous shales (Devonian) of Cape Ipperwash or Kettle Point in the township of Bosanquet on Lake Huron. Other examples occur in the Devonian rocks of Gaspé. At the latter locality, some impressions of radiating leaves (annularia laxa, Dawson) belonging probably to a related type of plant, have also been found. The narrow radiating leaves (often attached to furrowed stems) known as *Asterophyllites,* and which are so abundant in many Carboniferous and in some Devonian strata, have not yet been recognized within the area of Ontario and Quebec. They are commonly regarded as calamite leaves, but on very uncertain evidence.

Ferns (Filices), although so abundant in the higher Devonian, Carboniferous and other strata, mostly in the form of leaf or frond impressions, have not as yet been discovered in a fossil condition within the limits of the Provinces referred to in the present work.

The *Hydropteridæ* or *Rhizocarps,* sometimes known as water-ferns, comprise merely a few fresh-water forms (Marsilea, Piluria, Salvinia, &c.) without fossil representatives in our strata.

The *Lycopodiaceæ* of existing Nature comprise a small number of inconspicuous forms belonging to both land and fresh-water types. The former (lycopodium, selaginella) are small, moss-like cryptogams, dichotomously branching, and with narrow, more or less clasping or imbricating leaves. The aquatic types (Isoetes) are rush-like forms. True lycopods occur in Devonian and higher strata; and some apparently related forms from the Devonian rocks of Gaspé have been referred by Sir William Dawson to this division. The principal of these form the genus *Psilophyton* (Fig. 120), represented by fragmentary impressions of narrow, stem-like plants, circinate (as in ordinary ferns) at their terminal points. The leaves are very small and thorn-like.

The *Lepidodendraceæ,* represented typically by the fossil genera, *Lepidodendron* and *Sigillaria,* are entirely Palæozoic. They consist, in most cases, of casts or impressions of tree-stems, usually fragmentary, but found occasionally in lengths of more than thirty or forty feet. The lepidodendraceæ, proper, are regarded as closely allied to the lycopods, whilst the sigillariæ are thought by some authorities to constitute a higher type of vegetation, more nearly allied to the cycads. The two, however, are very closely alike,* and

---

* In their dichotomous branching, their supposed leaves, their leaf-scars, woody structure, roots, &c. See a comparative tabular view of these homologies by Professor Schimpfer, in Zittel's *Handbuch der Palaeontologie* (1880), p.p. 209, 210.

they appear in certain intermediate forms (*Lycopodendron vasculare*, *Sigillaria Defrancii*, *S. tesselata*, &c.) to merge into each other. Typically, the sigillariæ have the outer surface marked with strongly-

Fig. 120.

Psilophytum princeps (Dawson). **P. elegans.*(Id.) Devonian : Gaspé.

defined, longitudinal ribs and furrows, whilst these are absent in the lepidodendra; but ribs are also absent in many sigillariæ, at least upon the outer surface : whence the two groups *costatæ* and *acostatæ* of the latter, as commonly adopted. The oval, rhombic, or other shaped impressions on the stem-surface of both types, indicate the original sites of leaves, and are thus known as "leaf-scars." Within these, generally towards the upper part, lie in most cases, three small indentations known as "vascular scars," but in some genera only one is present. Typically, in lepidodendra the central vascular scar is more pronounced than the lateral scars ; whilst in sigillariæ the reverse of this occurs, or the central scar may be altogether wanting. Very probably these extinct types represent connecting links between the higher cryptogams and the gymnosperms of existing Nature.

Remains of lepidodendroids are unknown in Ontario, but a species of lepidodendron (Fig. 121) occurs in the Devonian rocks of Gaspé,

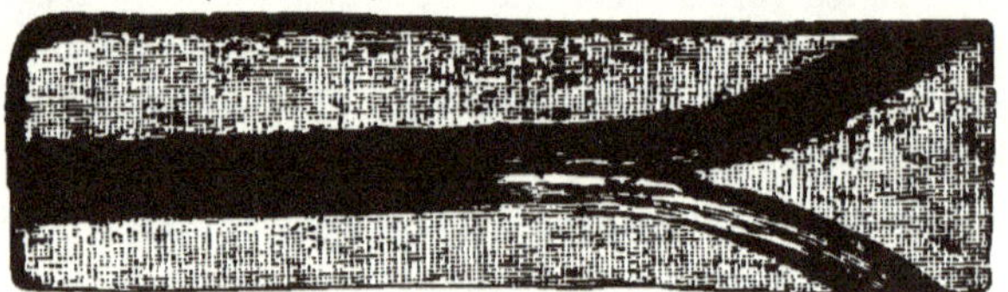

Fig. 121.

Lepidodendron Gaspianum (Dawson).   Devonian : Gaspé.

together with impressions of long, narrow, parallel-veined leaves referred to *Cordaites** (Fig. 122).

---

* Some of the more characteristic lepidodendroid and sigillarioid fossils of Devonian and Carboniferous rocks, are comprised in the following synopsis :

1. *Lepidodendron :*—Stem-forms, often of great length, bifurcating, the surface marked with oval, cordiform, or rhomboidal leaf-scars.   Upper Silurian (?), Devonian, Carboniferous, Permian.

2. *Lepidostrobus :*—Oval or cylindrical bodies with more or less distinctly hexagonal surface-markings.   Supposed " fruit cones " of lepidodendroids, Dev., Carb.

3. *Cordaites :*—Long-pointed leaves with comparatively broad base and parallel venation· Supposed leaves of lepidodendroids.

PHANEROGAMS.

All flower-bearing vegetable forms belong to this division ; but in many, the flowers are exceedingly inconspicuous, being destitute of corolla and other parts of the floral enve-lope or perianth, and thus reduced to parts directly concerned in the production of seed. The phanerogams fall into two principal series : *Gymnosperms* and *Angiosperms.* In the first, the seeds are "naked," that is, they are not developed within a special ovary ; and these types present in other re-spects certain peculiarities of organization which, notwithstanding the exogenous struc-ture of the wood, render them more or less akin to the higher cryptogams. In the an-giosperms, on the other hand, the ovules are developed and con-verted into seed within a special receptacle or so-called ovary.

Fig. 122.

Cordaites angustifolia (Dawson).
Devonian : Gaspé.
See preceding paragraph

Gymnosperms :—These, distinguished essentially by their naked seeds, are commonly classed under three orders : *Cycads, Conifers, Gnetaceæ ;* but the latter, some of which are known popularly as "jointed firs," are of no palæontological importance.

Cycads, represented essentially by the genera cycas and zamia, are intermediate in aspect between tree-ferns and palms ; and those of existing Nature are entirely confined to intertropical regions. Fossil forms are cited from Devonian and higher strata, and are especially characteristic of Mesozoic formations.   (See the table on page 201.) Examples are unknown within the area of Central Canada.

Conifers comprise firs, pines, cypresses, and other (typically) cone-bearing, resin-secreting types, with acicular or narrow leaves. Vertical sections of the wood shew under the microscope rows of circular discs, regarded practically, as a typical character ; but these discs are shewn also by cycads, and likewise, although far less pro-minently, by certain dicotyledons. Conifers occur in all latitudes

---

4. *Sigillaria :*—Stem-forms, with (typically) longitudinal ribs and grooves on the surface, and with more or less hexagonal or oval leaf-scars within the furrows. The scars in some examples are comparatively far apart; in others, close together. Some examples, also, are without longitudinal ribs and furrows. Dev., Carb.

5. *Stigmaria :*—Stem-like casts with irregular, more or less rounded, surface-markings. Supposed roots of sigillariæ and lepidodendra.

within the limits of arboreal vegetation, although certain types are confined to special localities. Fossil examples date from the Devo-

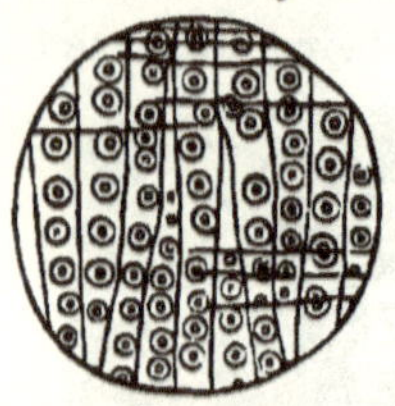

Fig. 123.
Circular discs of conifer-
ous wood.

nian (or perhaps Upper Silurian period). Those of Carboniferous and Jurassic strata are thought to have been closely allied to the *Araucariæ*, now limited to Australia and the more southern portions of South America. Pines and firs, proper, first appear in Lower Cretaceousbeds, and are largely present in the Upper Cretaceous and Lower Cainozoic brown-coal deposits throughout the North-West Territories and British Columbia; whilst junipers, yews, and gnetaceæ are comparatively modern types.

In the Devonian rocks of Gaspé some casts of comparatively large stems have been referred by Sir J. W. Dawson to coniferæ, under the name of *Prototaxites*, but this view is disputed by other authorities.

Angiosperms :—The plants of this subdivision, as explained above, comprise all flowering types in which the seeds are enclosed in an ovary. They fall into two leading series: *Monocotyledons* and *Dicotyledons.* In the first, as the name implies, the embryo-plant has but one cotyledon or seed-leaf; whilst in the second, the embryo has two cotyledons.

Monocotyledons :—These (formerly known as endogens) comprise grasses, lilies,. palms, and other representatives with (typically) straight-veined leaves, flowers composed of parts in threes or sixes, and wood made up of irregularly disposed vascular bundles. Obscure examples are cited from Carboniferous strata, but the earliest undoubted examples are Mesozoic. No fossil examples occur in the strata of Ontario or Quebec.

Dicotyledons :—In these plants, the leaves are typically net-veined, the flowers composed of parts in fives or fours, or multiples of these numbers, and the woody stem made up of rings of vascular bundles traversed by medullary rays. The greater number of the flowering plants, and all the trees (conifers excepted) of temperate regions belong to this subdivision. Fossil examples appear first in Lower Cretaceous strata. In Canada, so far as regards the Provinces referred to in this book, the only fossil examples consist of modern leaves, &c., as those of *populus balsamnifera*, and our common species of maple, oak, and other trees—impressions of which occur in many Post-Cainozoic clays, shell-marls, calcareons tufas, &c., as at Green's Creek on the Ottawa, and elsewhere, at numerous localities, in both Ontario and Quebec.

ANIMAL REMAINS.

The forms of the Animal Kingdom may be classed under nine leading divisions or so-called sub-kingdoms, namely : 1. *Protozoa ;* 2. *Polystomata ;* 3. *Cœlenterata ;* 4. *Echinodermata ;* 5. *Vermes ;* 6. *Arthropoda;* 7. *Mollusca ;* 8. *Tunicata ;* 9. *Vertebrata.*

SUB-KINGDOM I.

PROTOZOA.

This sub-division—apart from a few fossil representatives—comprises a number of minute, and in great part microscopic, types, consisting of gelatinous sarcode-matter, destitute of true tissues and special organs, and either naked, or protected by an external test or shell of a horny, calcareous, or siliceous nature. Nearly all are aquatic, but some few are internal parasites. They have no true body-cavity, the sarcode-matter absorbing nutriment through its entire substance, although in some of the higher forms (Infusoria) certain portions of the body are more permeable than other parts— an approach towards an alimentary canal being thus indicated.* The Protozoa admit of a sub-division into three natural groups :— *Pseudopodifera*, in which the body substance is extensible into long or short pseudopodia (see preceding note) ; *Gregarina*, entozoic types, pseudopodous only at an early stage of existence; and *Ciliata*, including the ordinary infusorial forms, furnished with long or short cilia, or, in one section, with retractile tubular suckers.

Of these three groups, that of the *Pseudopodifera* alone presents fossil representatives. This group includes four classes : 1 and 2, *Monera* (?) and *Amœbina*, both soft-bodied and without fossil forms ; 3. *Foraminifera*, mostly with calcareous shell, and long, anastomos-

---

* The *amœba* of our ponds and ditches will convey a good idea of a typical protozoan. Under the microscope, this is seen to consist of a small gelatinous mass which possesses the power of extending itself into short irregular projections, technically known as *pseudopodia*. By the aid of these it moves along and captures passing infusoria or particles of nutriceous matter, over which the body closes until digestion is effected. The creature thus improvises a a temporary stomach. In *actinophrys*, a related form, often found in rain puddles, gutters, &c., the pseudopodia are long and thin, and regularly radiated.

ing pseudopodia* ; and 4, *Radiolaria* with siliceous, highly foramin-ated test as regards the more typical forms.

Remains of *Foraminifera*, all belonging to living species, were detected some years ago in the leda clay formation (immediately

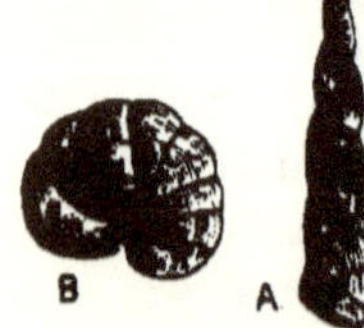

FIG. 124.

B *Polystemella* (or No-nionina?) *umbilicata.*
A *Textularia (varia-bilis?).*

above the true drift deposits, see Part V.) by Sir J. W. Dawson in the vicinity of Montreal, and at Beauport, near Quebec. The most com-mon form is the *Polystomella* (or Nonionina?), shown at B in the following highly magnified figure. Another, but much less common form, from Beauport, observed by the writer in some sandy matter in the interior of a fossil balanus, is shewn at A. It is a species of *Textularia* : a living genus, dating from the Carboniferous epoch.

In addition to these essentially microscopic forms found in our post-cainozoic deposits, some comparatively gigantic types, referred rightly or wrongly to the foraminifera, have been discovered in our Archæan and Palæozoic (Cambrian and Lower Silurian) rocks. These comprise, chiefly, the *Eozoon* of Dawson, the *Archæocyathus* and some related forms together with the *Pasceolus* of Billings, and the *Receptaculites* of Ferdinand Roemer. The true nature of these fossil forms, however, is exceedingly obscure. The *Eozoon* is regarded by Möbius, King, and other high authorities, as entirely of inorganic origin, notwithstanding the able memoirs of Dawson and Carpenter in defence of its assumed foraminiferous character. It occurs, with us, in the crystalline Laurentian strata of North

---

The Foraminifera may be sub-divided, practically, as in the following synopsis :—

GROUP I.—*Imperforata :* Body-covering or shell with single external opening for passage of pseudopodia.

     § 1. *Chitonosa :* With chitonous (or indistinct) body-covering :
         Fam. 1. *Gromidæ (e.g.,* Gromia, Lieberkühnia).
     § 2. *Arenacea :* With body-covering made up of agglutinated sand-grains, &c.
         Fam. 2. *Lituolidæ (e.g.,* Lituola, Saccamina).
     § 3. *Porcellanea :* With calcareous, non-foraminated, porcelain-like shell.
         Fam. 3. *Miliolidæ (e.g.,* Cornuspira, Miliola. &c.).
GROUP II.—*Perforata :* with distinctly foraminated shell :
     § 4. *Vitrea :* Shell more or less distinctly hyaline ; calcareous ; foraminated :
         Fam. 4. *Lagenidæ :* Shell with very minnte foramina *(e.g.,* Lagena, Cristel-laria, &c.).
         Fam. 5. *Globigerinidæ :* foramina comparatively large ; shell thin *(e.g.,* Textilaria, Globigerina, &c.).
         Fam. 6. *Nummulinidæ :* foramina comparatively large ; shell solid *(e.g.,* Nummulina, Fusulina, &c.).

Hastings, and elsewhere, in the form of concentric, wavy, partially constricted or irregular layers, made up of calcite or dolomite with intervening layers or areas of serpentine. The more perfect examples are of circular shape, and vary in diameter from three or

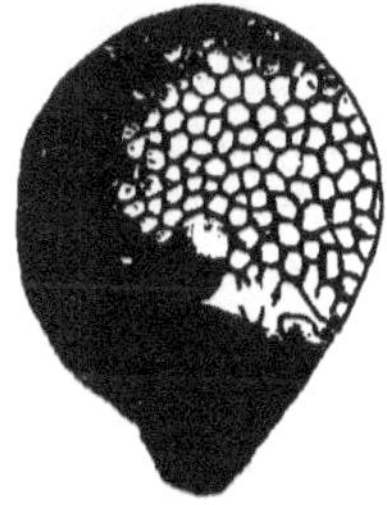

Fig. 125.
Pasceolus.
Trenton Formn.

four inches to nearly a foot. *Archæocyathus* is cyathiform in shape, much resembling many corals and some sponges, to the latter of which groups it was at first (and probably correctly) referred. It occurs in the Potsdam and Calciferous formations (see Part V.) of Eastern Quebec. *Pasceolus* (Fig. 125) forms oval or small globular masses, an inch or two in diameter, with the surface covered with hexagonal markings. It occurs in the Trenton (Lower Silurian) formation of Ottawa. *Receptaculites* presents shallow, saucer-like or circular forms, often a foot or more in diameter. The surface, as shewn in Fig. 125, is divided into small, rhombic areas by fine (usually somewhat dotted) lines, curving in opposite directions, like the lines on "engine-turned" watch-cases, from a central root-like nucleus. It was at first placed with *Dactylopora*, now regarded as a calcareous fucoid. It occurs in our

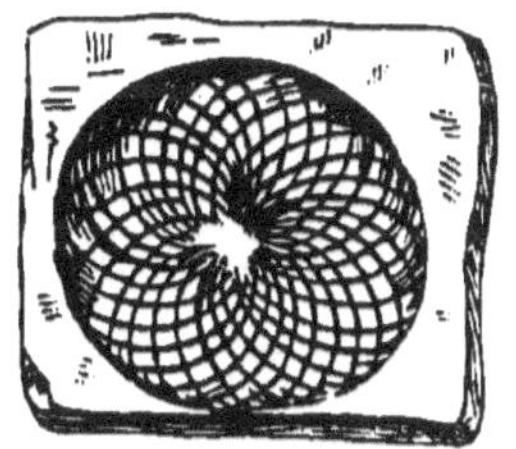

Fig. 126.
Receptaculites.
Lower Silurian.

Lower Silurian strata, both in Central Canada and in Manitoba.

The fourth group of the *Pseudopodiferous Protozoa*, the siliceousshelled *Radiolaria*, also known as *Polycystina*, have not as yet been recognized in Canadian strata, although their remains in a fragmentary state probably occur with sponge spicules, &c., in some of our post-cainozoic deposits.

SUB-KINGDOM II.

POLYSTOMATA OR SPONGIDA.

The representatives of this division are aquatic and mostly marine organisms, consisting of ciliated gelatinous matter, with internal cavity traversed by numerous inhalent pores or canals, and having one or several outlets or oscula. In the great majority of sponges, the gelatinous matter is strengthened by a fibrous, horny framework (the sponge of commerce), or by spicules or a spicular-skeleton of silica or of carbonate of lime. The spicules are of various forms—mostly needle-like, or three-pointed, anchor-like, irregularly-branching, or stellate; and the shape, and in some cases the arrangement, of these spicules is found to be a more or less constant character, whilst the outer form of the sponge is exceedingly variable. Hence, the modern classification of sponges is based essentially on spicular characters; but these, in fossil examples, are as a rule of somewhat difficult observation. Occasionally, they may be made out if the sponge be dissolved in dilute acid; but this method of observation is very frequently inapplicable from the entire body of the sponge having become silicified by fossilization. The internal structures can then only be detected by the microscopic examination of thin slices or chippings (ground down with emery powder on a cast-iron plate) under an object glass of tolerably high power.

Sponges are thus commonly classified as in the annexed table:

I. *Myxospongiæ*—gelatinous, only.

II. *Fibrospongiæ*—with horny framework, or separate or united siliceons spicules, or both.

III. *Calcispongiæ*—with calcareous spicules.

Classes II. and III. are sub-divided further into families—as (in Class II.) *Ceraospongidæ* (with horny framework. sometimes containing simple spicules); *Monactinellidæ* (with simple, unbranched siliceons spicules); *Tetractinellidæ* (with siliceous four-pronged spicules); *Lithistidæ* (with branching, often united, spicules); and *Hexactinellidæ* (with six-armed siliceons spicules). The *Calcispongiæ* fall into: *Ascones* (with thin "wall," and regularly-arranged three-rayed and other calcareous spicules); *Leucones* and *Pharetrones* (with thick wall, traversed by irregularly-branching canals, and with scattered or united spicules); and *Sycones* (with thick wall, traversed by radiating canals, and with regularly-arranged spicules).

The fossil sponges, or bodies regarded as sponges, hitherto found in the strata of Ontario and Quebec, are very few in number, and all are more or less obscure in character. The principal comprise: *Stromatopora; Archeocyathus* (already mentioned under the foraminifera, but which is probably a calcareous sponge of the order *Sycones*); *Eospongia;* and *Astylospongia.*

Stromatopora is of not uncommon occurrence in our Silurian formations. It has been referred to the *Foraminifera*, the *Sponges*, the *Zoantharia* and the *Hydrozoa*. It forms hemispherical or more or less irregular masses, often many inches in diameter, made up of numerous concentric, wavy lamellæ. Our most common species is the S. rugosa, found especially in the Trenton (including the Black River) formation of various parts of Ontario and Quebec. Another closely related species *S. concentrica* occurs in the Niagara formation.

Fig. 127.
*Stromatopora
rugosa.*
Trenton Formation.

Archeocyathus occurs in expanding, beaker-like or horn-shaped forms, with deep central cavity, the sides of which are marked with what appear to be the openings of radiating canals. The form has thus a general resemblance to a *Zaphrentis* or other cyathiform coral. Species have only been found, at present, in the Potsdam and Calciferous formations of the Straits of Belle Isle and the Mingan Islands of Eastern Quebec. *Eospongia* is mostly pyriform or subglobular in aspect, with central depression and radiating pores. Species occur in the Chazy formation of the Mingan Islands. *Astylospongia* occurs in small, globular or sub-globular forms, without a central cavity, or with only indications of this; but with radiating lines or pores at the somewhat flattened upper portion, and without any signs of a stem-attachment at the under side. Species have been found in the Trenton and Niagara formations, but are of comparatively rare occurrence.

SUB-KINGDOM III.

COELENTERATA.

The typical cœlenterates are distinguished from lower forms by the possession of a distinct body-cavity with single mouth-opening; and from higher forms, by the absence of a distinctly separated stomach —the body-cavity and stomach being practically identical. The mouth-opening is surrounded by tentacles. All cœlenterates are aquatic types. They may be classified conveniently, as shewn in the annexed tabular view:

15

A. Without natatory cilia :
    A.[1] Stomach-cavity completely identical with body-cavity :
        (*i.*) Without internal calcareous corallum :
            Class I.　*Hydrozoa.*
        (*ii.*) With internal calcareous corallum :
            Class II.　*Hydrocoralla.*
    A.[2] Stomach partially separated from body-cavity :
        (*i.*) Mouth-opening with eight fringed tentacles :
            Class III.　*Crossocoralla* or *Alcyonaria.*
        (*ii.*) Mouth-opening with numerous simple tentacles .
            Class IV.　*Anthocoralla* or *Zoantharia.*
B. With natatory cilia :
            Class V.　*Ctenophora.**

I.

HYDROZOA.

This class, as here defined, is composed of soft-bodied aquatic types, without internal stony "corallum."[†]　In some of its representatives, however, a chitonous or horny, cellular support is present.　The class may be sub-divided broadly into the following orders :—1. *Hydrida* (*e.g.*, the fresh-water *Hydra*); 2. *Hydromedusæ* (*e.g.*, *Tubularia, Sertularia,* &c., and extinct *Graptolites*); 3. *Discomedusæ* (*e.g.*, the true *Medusæ, Rhizostoma,* &c.); 4. *Lucernarida* (*e.g.*, *Lucernaria,* &c.); and 5. *Siphonophora* (*e.g.*, *Physalia, Velella,* &c.). Of these, the *Hydrida, Lucernarida,* and *Siphonophora,* have no fossil representatives.　The *Discomedusæ,* being entirely soft-bodied, gelatinous types, are most rare in the fossil state ; but their impressions have been occasionally found in Mesozoic rocks, as in the lithographic slates of Solenhofen in Bavaria.　The remaining Order, that of the *Hydromedusæ,* represented by the living sertularians, &c., contains, on the other hand, an extinct group of forms, the Graptolites, of great palæontological interest.　These forms are exclusively of lower palæozoic age, and are typically characteristic of Silurian strata.

Graptolites :—The extinct forms, thus known, were apparently

---

* An aberrant group, forming a passage-group into the echinodermata.　Fossil representatives are unknown.

† The *Milleporidæ* are Hydrozoids with secreted calcareous corallum, and should thus b placed with the Hydrocoralla, as in the present classification.　See page 229.

free-floating marine types, living in colonies of individuals which secreted in common a horny or chitonous cellular support. The latter, in a more or less fragmentary condition, has alone been preserved, forming markings or impressions, mostly in argillaceous slates. It is technically known as the " stipe." Most commonly it presents a narrow, linear shape, " toothed " or serrated along one or both of its edges. Frequently, these linear stipes bifurcate, and in some forms (*Rastrites*, &c.) become partially enrolled or even spiral, and assume in others a leaf-like form. Occasionally, the lateral serratures are obliterated by transverse compression. These serratures are the mouths or openings of minute cells, and thus much resemble those of modern sertularians. They are pointed or even mucronate in some genera, and obtuse in others. A somewhat prominent thread-like line runs up the centre of the stipe (or along the outer edge in the forms with one row of serratures) and often projects beyond the stipe. This thread-like line is known as the axis. Where its projecting extremity, or in bifurcating forms the united extremities of two axes, forms a sharp or blunt point, this is known as the " radicle " or " sicula." In some examples from the Quebec slates of Point Levis, several bifurcating stipes radiate from a common centre around which there appears to be a thin connecting membrane or supposed " float." In the leaf-like forms from this locality, as first pointed out by Professor Hall, of Albany, two, or more properly four, stipes were united originally in a cruciform mode of structure, although now generally separated.

Graptolites may be arranged under five groups. These comprise : (1) *Monoprionidians*,* with single stipe celled on one margin only (*e.g.*, Monograptus, Spirograptus, Rastrites) ; (2) *Dichoprionidians,* with dichotomously-branched stipe, celled on one margin, only (*e.g.*, Didymograptus, Tetragraptus, Loganograptus, &c.,—all Lower Silurian) ; (3) *Metaprionidians,*† stipe bifurcating, with single row of cells on the separated portions of the forks, and a row on each margin where the forks come together (*e.g.*, Dicranograptus, Lower Silurian) ; (4) *Diprionidians*, with cells on each side of stipe (*e.g.*, Diplograptus, Climacograptus, Phyllograptus) ; and (5) *Retioprionidians*, with comparatively broad, bi-serrated, stipe, net-veined or dotted on the surface (*e.g.*, Retiolites, Retiograptus).

---

* πρίονωτὸς, serrated, saw-like.
† Μετὰ, between, intermediate—as regards the group.

The annexed figures shew some of our more common or characteristic forms:

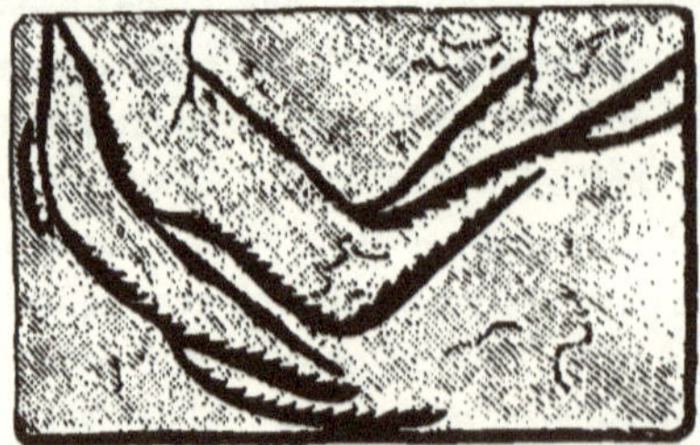

Fig. 128.
Graptolithus (Didymograptus) flexilis : Hall.

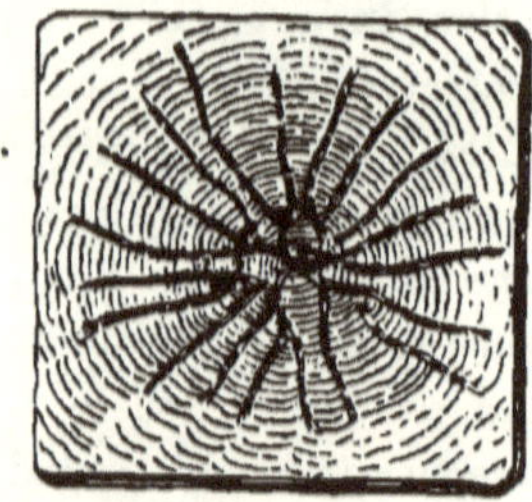

Fig. 129.
G. (Loganograptus) Logani : Hall.

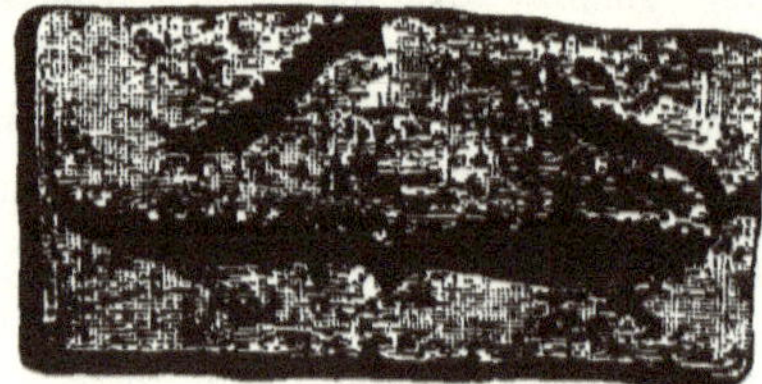

Fig. 130.
G. (Didymograptus) pennatulus : Hall.

Fig. 131.
G. (Dicranograptus) ramosus : Hall.

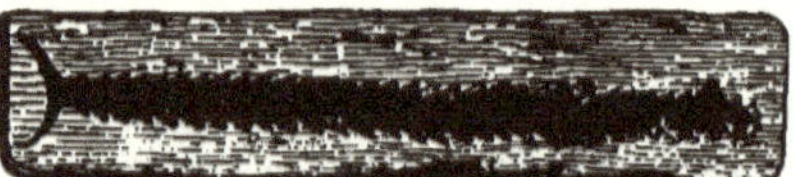

Fig. 132.
G. (Climacograptus) bicornis : Hall.

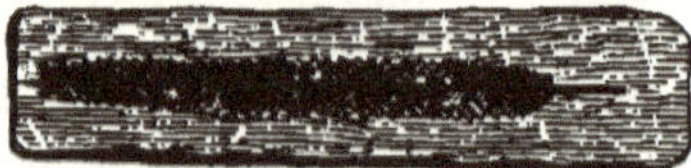

Fig. 132.*
G. (Diplograptus) pristis . Hisinger.

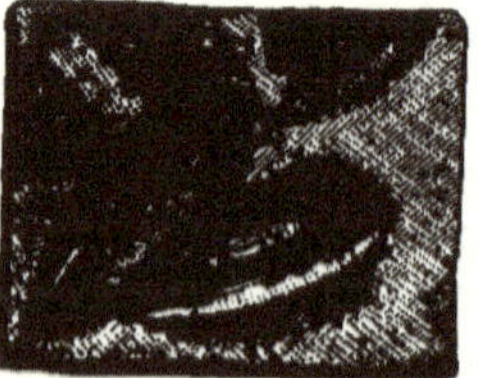

Fig. 133.
G. (Phyllograptus) typus : Hall.

In addition to these forms, several more or less related fossil-types are commonly referred to the graptolites. As regards Canada, the *Dictyonema* is the princi_pal representative of these doubtful types. It forms dark, usually carbonaceous, undulating, recticulated markings, as shewn in figure 134, which represents a common species, *D. retiformis*, of our Niagara and Clinton Formations. By some palæontologists it is regarded as a Bryozoon, by others as a sea weed.

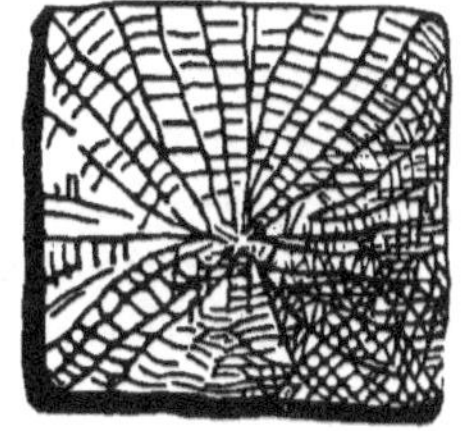

FIG. 134.
Dictyonema retiformis : Hall
Upper Silurian.

II.

HYDROCORALLA.

This sub-division is to some extent a group of convenience, rendered necessary by our still uncertain knowledge of its included forms. Some of these are undoubted *Hydrozoa* with tabulated calcareous corallum. The tabulated character of this corallum—with other features, as the absence (or rudimentary nature) of radiating septa, &c.—compels the collocation in the same group, as maintained by Agassiz, of many of the "Tabulata" of older groupings—thus separating these latter from the typical "Hexamerous Corals " with which they are still so commonly placed, although from the general absence of septa it is impossible to tell whether the tentacles of the living animal were hexamerous or not. These tabulata again, offer a complete transition into the allied—although in classifications usually widely separated—Rugosa, in many of which the assumed tetramerous character of the septa is not recognizable. Some of the forms placed under the present division may perhaps be Bryozoa, and others, again, Alcyonarians : but it is not possible to determine this. On the other hand, the tabulated structure which they possess in common, serves to unite them conveniently—and, in default of negative evidence, naturally also, into a common group, under the name of *Hydrocorolla*. This name indicates their affinities to the Hydrozoa, on one side, and to the corals proper, on the other.*

---

* The classification of corals into *Hexacoralla* and *Octocoralla*, although definite enough in certain cases, is on the whole entirely conventional. In many forms there are no visible septa ; and in many in which well-developed septa are present, the actual number is exceedingly variable. The genera *Stylina*, Lam., *Stylocœnia*, Edw. and Haime, *Heterophyllia*, McCoy, and numerous others, are examples.

Before referring to the more typical forms of Canadian occurrence, it will be necessary to explain briefly a few common terms employed in the description of the Hydrocoralla and corals generally. The animal substance of corals consists of soft gelatinous matter containing a digestive sack or stomach (or a number of these, united) with a series of tentacles around the opening or so-called mouth. The gelatinous matter possesses in the great majority of cases the power of secreting amongst its tissues a calcareous or horny framework, technically known as a corallum, the " coral " of popular language. This corallum is either " sclerodermal," or " sclerobasal." A sclerodermal corallum is secreted within the tissues, and reveals more or less the form of the polyp or fleshy part of the animal. It is always composed of a single cell or tube,—round, oval, hexagonal or polygonal in shape—or of a number of these in close juxtaposition or otherwise united : (1) by two sides only of the cell *(Halysites)*; or (2) by short lateral tubes *(Syringopora)*; or (3) by a mass of more or less porous, or occasionally compact tissue, known as " cœnenchyme ". The sclerobasal-corallum, on the other hand, is never cellular. Is is secreted, as a common support, by the entire fleshy mass of the compound coral, the separate polyps being developed only within this fleshy mass, as seen for example, in the well-known " red coral " of the Mediterranean.

In the sclerodermal or cellular corallum, the interior of the cell is in some cases quite smooth or open; in others, the walls are marked

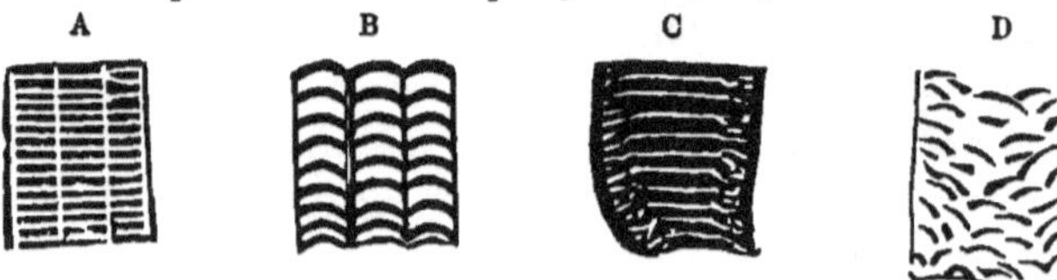

Fig. 135.

by vertical striæ ; in others, vertical projections radiate from the walls towards the interior of the cell, and often unite in its centre. These projections, known as " septa " or " radiating septa," present in some groups of forms a very distinct hexamerous character, and in others a tetramerous system of arrangement, but these characters in fossil examples cannot always be made out, and in many genera, the septa are quite variable in number. Sometimes, in place of one of the primary or more strongly pronounced septa, there is a narrow empty space or " septal fossette " as in *Zaphrentis*, etc., but this in badly preserved examples is not always observable. Its presence is usually marked by a ridge upon the outside of the cup or cell.

Whether radiating septa are present or absent, the interior of the cell in many genera—typically, in all the HYDROCORALLA—is divided transversely by a series of plates or "tabulæ," also called "diaphragms," which extend more or less regularly across the cell, and are flat or arched in form (Fig 135 *a.* & *b.*). In other cases, these tabulæ, when present, are confined to the more central part of the cup, or cell, the sides of the latter being filled with short, irregular, plates, technically known as "vescicular tissue" (Fig. 135*c*). Occasionally, again, as in the genus *Cystiphyllum*, this vesicular tissue occupies the entire cell (Fig. 135*d*).

Viewed broadly, the HYDROCORALLA* may be classed under six sections, as in the following table :

§ 1. *Inornata.*

§ 2. *Tabulo-Stellata.*

§ 3. *Vesiculo-Stellata.*

§ 4. *Vesiculosa.*

§ 5. *Operculata.*

§ 6. *Integri-Stellata.*

These sections are defined as follows :

§ 1. *Inornata :*—Tabulæ well developed, extending entirely across the cell. Radiating septa absent, or quite rudimentary.

All the genera of this section are compound forms ; and all, with the exception of *Millepora* (a living type, dating only from the Eocene period) are extinct Palæozoic types. The more common Canadian genera comprise : (1) *Fistulipora*, mostly in irregular, encrusting masses with very small cells of two sizes, the smaller forming a kind of pseudo cœnenchyme. (2) *Monticulipora* (including in part, *Stenopora* and *Chætetes*), with narrow, capilliform cell-tubes, in branching and rounded or hemispherical examples, Fig. 136. (3) *Favosites*, in irregular and pyriform, sometimes branching, masses, composed of polygonal cell-tubes with perforated walls and straight tabulæ, Fig. 137. (4) *Alveolites* Fig. 137 *a.* with obliquely-opening cell-mouths and perforated walls : (5) *Michelinea,* with short, wide

---

·cells, and convex tabulæ, Fig 138.   (6) *Halysites* (the "chain corals"),
with oval or round cell-tubes united in chain-like groupings, Fig. 139.
Syringopora, with round, reed-like cell-tubes, united by short transverse tubes, Fig. 140.

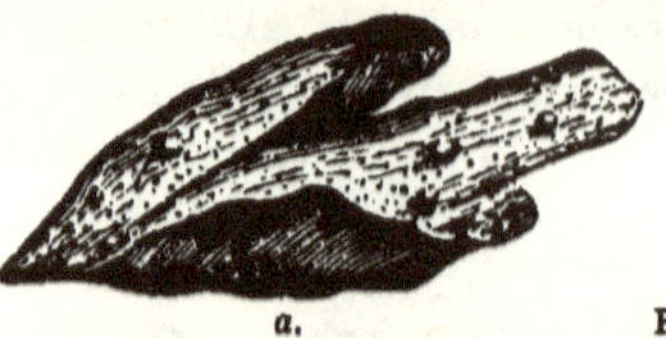
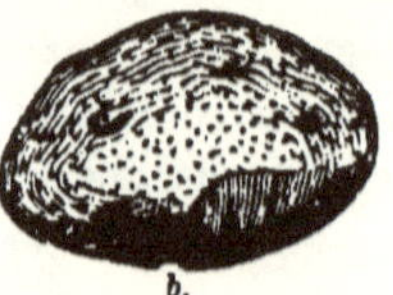

a.   FIG. 136.   *b.*

a. *Monticulipora (Stenopora* or *Chætetes) fibrosa;*   b. *M. petropolitana.*   Trenton
and Hudson River Formations.

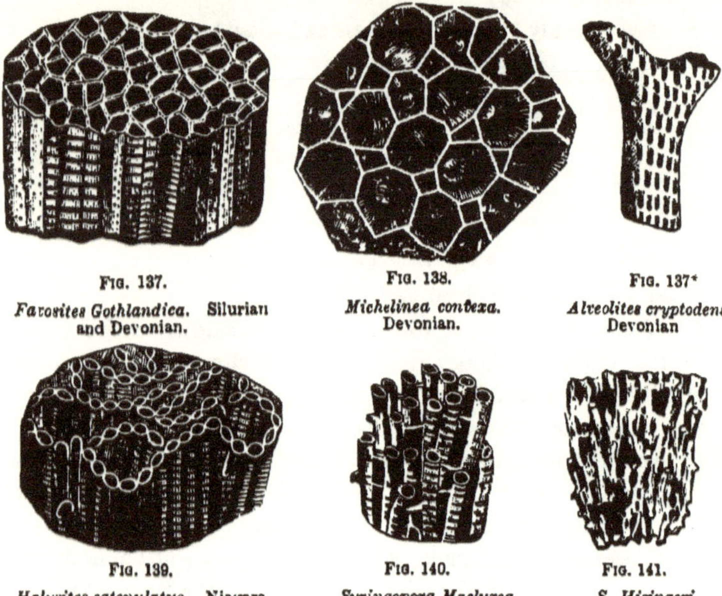

FIG. 137.   FIG. 138.   FIG. 137*

Favosites Gothlandica.   Silurian   *Michelinea convexa.*   *Alveolites cryptodens.*
and Devonian.   Devonian.   Devonian

FIG. 139.   FIG. 140.   FIG. 141.

Halysites catenulatus.   Niagara   *Syringopora Maclurea.*   *S. Hisingeri.*
Formation.   Devonian.   Devonian.

§ 2. *Tabulo-Stellata :*—Tabulæ and radiating septa both present
No vesicular tissue, or traces merely.

Both compound and simple forms are included in this section.
The more common Canadian genera comprise : (1) *Columnaria* Fig.
142, with compound corallum made up of hexagonal or polygonal
cell-tubes, with short septa and horizontal tabulæ.   Distinguish from
Favosites, which it much resembles in general aspect, by the border
of short radiating-septa at the cell-walls and the absence of pores ;

Favistella is identical or closely allied, but with longer septa, none of which however reach the centre of the cell. (2), *Amplexus*, Fig. 143, mostly with simple corallum, in the form of a round, often more or less contorted tube. bordered with short septa and divided transversely by horizontal tabulæ. (3), *Zaphrentis*, (Fig. 144,) a genus of common occurrence in our Devonian strata, in simple horn-like forms with well-developed radiating septa, and a septal fossette. One species, *Z. gigantea* is of comparatively large size, but the smaller species *Z. prolifica*, figured below, is more abundant. In both, the septa reach the centre of the cup. *Streptelasma*, a silurian genus, is of very similar conformation, but some of its septa form in the centre of the cup a kind of twisted axis or " pseudo-columella."

FIG. 142.
Columnaria alveolata
(*Goldfuss*) Black River
(Trenton) Formation.

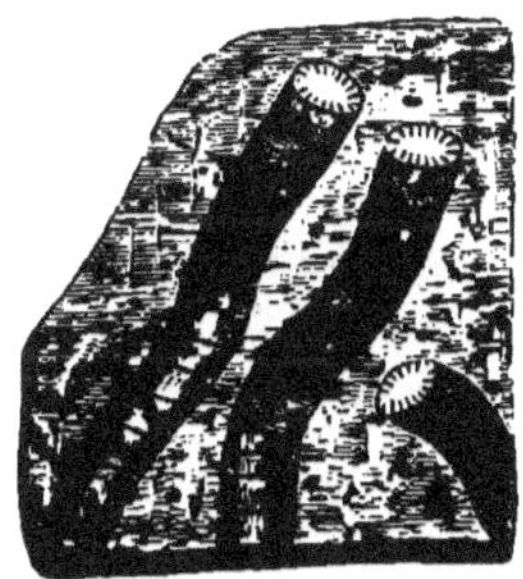

FIG. 143.
Amplexus laxatus (Billings).
Devonian.

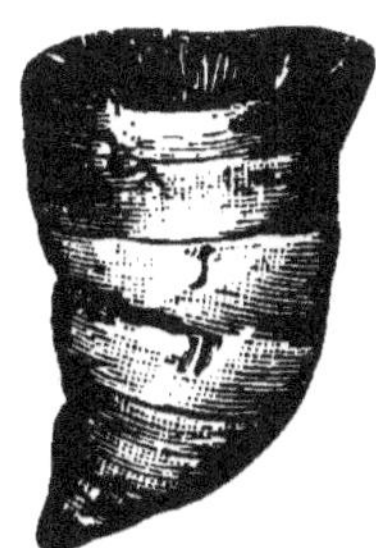

FIG. 144.
Zaphrentis prolifica (Billings). Devonian.

§ 3. *Vesiculo-Stellata :*—Tabulæ confined to central or inner part of cell, the outer part filled with vesicular tissue. Radiating septa always present.

This section includes a large number of both simple and compound forms, belonging in part to somewhat ill-defined genera. The more common examples of Canadian occurrence, comprise : (1) *Cyathophyllum*, simple and compound, septa with smooth sides and edges ; (2) *Heliophyllum*, Fig. 145, like *Cyathophyllum*, but with ridges or projections on the sides of the septa ; (3) *Clisiophyllum*, simple, horn shaped, with conical elevation in centre of the cup or cell,* and (4), *Phillipsastrea*, Fig. 146, compound, astræiform, with radiating septa prolonged beyond the outer walls of the cells.

---

* A vertical section shews in *Clisiophyllum* three areas : a central area indicated by the raised ends of the united septa ; an outer or marginal area of fine vesicular tissue ; and an intermediate area represented by more or less irregular tabulæ or diaphragms.

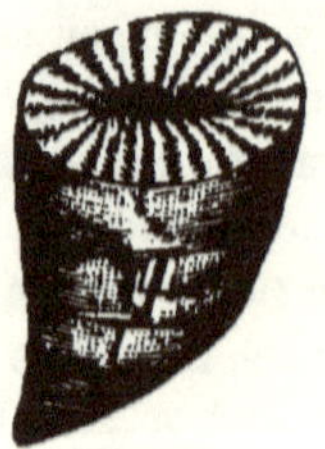

Fio. 145.

Heliophyllum Halli Devonian.

Fio. 146.

Phillipsastræa. Devonian.

§ 4. *Vesiculosa :*—Tabulæ replaced by irregular vesicular-tissue. Radiating septa absent or quite rudimentary.

The only Canadian genus referrible to this section, is *Cystiphyllum*, in which the entire cell is filled with short vesicular tissue.   Examples, one of which is shewn in figure 147, are mostly simple and more or lesss horn-shaped; but at least one compound species, *C. aggregatum*, is known.

§ 5. *Operculat i :*—cup or cell furnished with an operculum composed of a single valve or of several pieces.   Septa more or less rudimentary.

Canadian representatives of this section have not as yet been discovered.   Among other types it includes the curious slipper-like, triangular form, *Calceola*, until lately regarded as a brachiopod.   This genus is especiallv characteristic of Devonian rocks in Europe.

§ 6. *Integri-Stellata :*—Radiating septa well developed.   Tabulæ and vesicular-tissue entirely absent, or the latter present only in mere traces.

Fio. 147.

Cystiphyllum Senecaense (Billings.) Devonian.

Petraia, Fig. 148, with deep cup, in simple horn-shaped or turbinate forms, appears to be the only example of this group hitherto recognised in Canadian Strata.   Species, however, are often confounded with *Zaphrentis*, owing to the difficulty of determin_ ing the internal structure without obtaining artificial sections of the fossil.   The typical representative of the group is the genus *Cyathaxonia*, in which a central columella is present; but of this form we have no examples.

Fio. 148.

Petraia. Lower Silurian.

III.

CROSSOCORALLA OR ALCYONARIA.

This division is composed largely of living forms. In these, the polyps possess eight fringed tentacles, and there is a partial separation of the stomach from the general body-cavity. The corallum is sclerodermal or thecal in some forms, and sclerobasal in others.* The Crossocoralla may be arranged under four Sections, (1), *Tubulifera ;* (2), *Spiculosa ;* (3), *Incellata ;* (4),*Pinnigera.*

§ 1. *Tubulifera :*—Corallum sclerodermal, tubular, without septa or other internal structures.

This section includes the living *Tubipora* or " Organ Corals," and most probably the extinct (palæozoic) *Aulopora.* The latter genus is of not uncommon occurrence in Canadian strata. Figure 149 represents a Devonian form.

§ 2. *Spiculosa :*—Corallum slerodermal, coriaceous, with imbedded calcareous, branching, spicula; fixed. Includes the *Alcyonidæ,* doubtfully represented in the fossil state.

Fio. 149.

Aulopora cornuta (Billings). Devonlan.

§ 3. *Incellata :*—Corallum sclerobasal, horny or calcareous ; fixed. Includes the *Gorgonidæ* or " sea fans," the *Isidaceæ,* and the *Corallidæ* —the latter represented by the well-known " Red Coral " of the Mediterranean and Red Sea. No fossil representatives in Canadian strata.

§ 4. *Pinnigera :*—Corallum sclerobasal, horny ; free. Includes the *Pennatulidæ* or " sea pens "—Pennatula, Renilla, Virgularia. No fossil Canadian representatives, unless the Graptolites, as inferred by some palæontologists, belong to this section.

IV.

ANTHOCORALLA OR ZOANTHARIA.

The general absence of tabulæ, and the typically hexamerous character of the radiating septa, are the leading characters of this class. The Anthocoralla include a great number of existing corals and many Cainozoic and Mesozoic genera and species ; but Palæo-

----

* See explanation of these terms in the lutroductory remarks prefixed to the *Hydrocoralla* on a preceding page.

zoic types are exceedingly rare, and fossil examples in our strata are of very doubtful occurrence.   Viewed broadly, the Class may be subdivided into four sections : *Aporosa ; Perforata ; Sclerobasica ;* and *Malacodermata.*

§ 1. *Aporosa :*—Tissues of the corallum ("sclerenchyme") comparatively or essentially solid and compact.   Includes the families of the *Turbinolidæ, Astræidæ, Oculinidæ,* and *Fungidæ.*   No fossils in strata of Central Canada.

§ 2. *Perforata :*—Substance of the corallum essentially porous ; cell-walls perforated.   Includes the families of the *Eupsammidæ* and *Poritidæ* (placing the *Madreporidæ* with the latter).   No fossils in rocks of Central Canada.

§ 3. *Sclerobasica :*—Corallum sclerobasal, horny or spicular ; Polyptentacles simple, 6, 12, 18, or 24 in number.   Includes properly only one Family, that of the *Antipathidæ* or so-called " black corals. " No fossils.

§ 4. *Malacodermata :*—No-corallum : entirely soft-bodied.   Includes the Families of the *Actinidæ* or " sea anemonies," *Ilyanthidæ,* and *Zoanthidæ.*   No fossils.

———

SUB-KINGDOM IV.

ECHINODERMATA.

The representatives of this division are marine, and, in the adult state, typically radiated forms, with stomach distinctly separated from the general body-cavity.   The latter (with its system of appendages, when present) is protected by an external calcareous test, composed of numerous plates ; or otherwise, by a coriaceous integument strengthened by calcareous plates, tubercles, or spicula.   In some forms, the body is attached to the sea bottom, either permanently or during the earlier period of life, by a long or short stem made up of numerous calcareous plates, mostly round or pentagonal in shape, and perforated through the centre by a circular, stellate, pentagonal or quadrate, orifice.   The structural parts are almost invariably in fives or multiples of five.   In the more typical forms, the test or skin carries numerous movable spines, whence the name

Echinodermata. The more common existing forms comprise "star-fishes," "brittle stars," "sea-urchins," and holothurians. Pedunculated forms were especially characteristic of the earlier periods of the Earth's history, but are now comparatively rare, or, as regards the greater part, entirely extinct.

The Echinoderms are distributed under the following classes :

1. *Crinoidea.*   5. *Ophiuroidea.*
2. *Cystidea.*   6. *Asteroidea.*
3. *Blastoidea.*   7. *Echinoidea.*
4: *Edrioasterida.*   8. *Holothuroidea.*

I.

CRINOIDEA.

This class comprises the various so-called "sea-lilies," now nearly extinct. The Crinoids are attached, typically, to the sea-floor by a comparatively long, flexible stem ; but some become free in the adult condition. They consist essentially of three parts : the body or "calyx, the "arms" or tentacles, and the stalk or stem, as indicated in the annexed sketch, figure 150. The body, oval or cyathiform in shape, is protected externally by a number of calcareous plates, meeting, and in some cases partially interlocking, at their edges. These plates comprise : (1) a series of "basals," usually three or five in number, immediately above the stem, but often forming two horizontal rows, known respectively as lower and upper basals; (2) a series of "radials" often in more than one zone ; (3) a series of "inter-radials," more or less numerous, but sometimes absent; and (4) a series of "brachials," from which the arms or tentacles immediately spring, (see figure 151). The upper part of the calyx is covered in many genera by numerous small, more or less irregularly arranged plates, and is termed the "vault" or "roof," but

Fig. 150.

this in most of the more modern forms is simply coriaceous, or destitute of plates, properly so called. The calyx has a central opening, usually regarded as the mouth, and, in living forms, an excentric

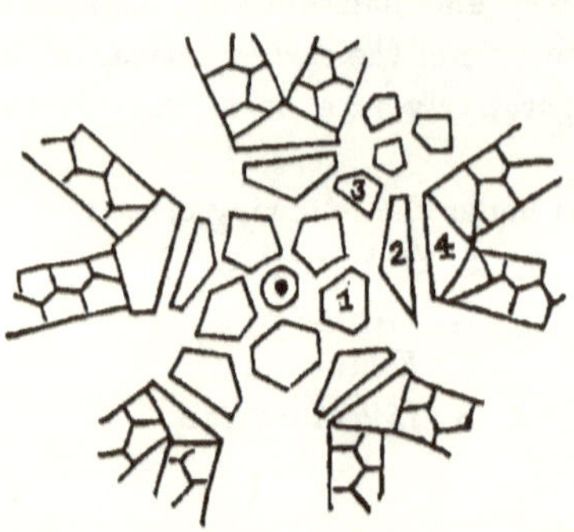

Fig. 151.

Simplified dissection of a crinoidal calyx,
viewed from below.

anal opening; but in most fossil genera only one opening is present. This is situated centrally or sub-centrally, and is frequently placed at the summit of a so-called "proboscis" or elevated, tubular portion of the calyx roof. The arms are in some cases short and simple; in others, long and dichotomously ·branched. As a rule, they are free or separate, but sometimes they are more or less united. Commonly, also, they are provided with attached pinnulæ, among which in living forms generation-products are developed. The plates which protect these arms externally, form either a single regular series, as in *a*, figure 152; or a single alternating series as at *b*; or a double interlocking series as at *c*. The same arm, however, sometimes presents two of these conditions. The plates which compose the stem, are circular or pentagonal (more rarely tetragonal)· in form, and usually shew a radiately-striated surface on their planes of junction.

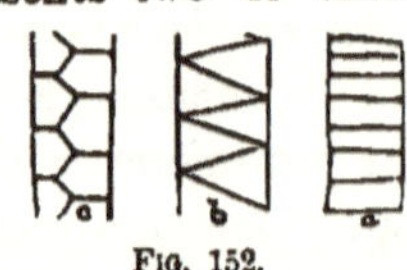

Fig. 152.

They are either of one diameter throughout the stem, or of alternating diameters ; and they have always a central perforation, round, pentagonal, five-rayed, or rarely quadrate, in shape. Occasionally, additional orifices are also present.

The Crinoids are usually sub-divided by palæontologists into two leading groups, named respectively, *Tesselata* and *Articulata*. The *Tesselata* are distinguished essentially by the calyx-roof being closed in with calcareous plates ; and by the calyx-plates, generally, being comparatively thin and but loosely attached to each other : whilst in the *Articulata* the roof has merely a coriaceous covering (strengthened in some forms by small plates or tubercles), and the calyx-plates, as a rule, are comparatively thick, and in a measure locked together. All the *Tesselata* are extinct ; and their remains (with the exception of two Cretaceous genera) are exclusively confined to Palæozoic strata.

All the Articulata, on the other hand, are Post-palæozoic types, fossil genera being found especially in Triassic, Jurassic, and Cretaceous formations; and about eight living genera are known.*

Nmerous examples of crinoid stems in a more or less fragmentary

FIG. 153.
Crinoid stem-fragments.

condition occur in our Silurian and Devonian rocks. Figure 153 represents a piece of shale, from the vicinity of Toronto, covered with portions of Crinoid stems, some seen in transverse section, and others longitudinally; and similar examples are abundant in the Chazy, Trenton, Niagara, and Corniferous limestones, of other parts of Central Canada. Well preserved or entire examples of crinoids, and especially of the calyx (on which, distinctive characters, as regards genera and species, are chiefly based) are, on the other hand, comparatively rare. The stems, unfortunately, are of little use in the determination of genera, as the character of the stem differs frequently in different species of the same genus; and occasionally the same stem is found to vary in different parts of its own length. Next to the stems, fragments of crinoid arms are of most frequent occurrence. In the following enumeration, therefore, of some of our more commonly occurring forms, the genera are arranged after the more easily recognised arm-characters :

§ 1. *Arms with pinnulæ.*

Glyptocrinus :—Pinnulæ very fine; arm-plates in single row; calyx-plates radiately ridged; stem generally round, the plates alternating in diameter, (sometimes pentangular); stem-orifice, five angled. Figure 154.

Thysanocrinus :—arms long and thin, bifurcating; arm plates in a double row, otherwise much like *Glyptocrinus :* Silurian and Devonian,

Dendrocrinus :—arms thin, long, much branching; calyx-plates, large; stem five-angled. Lower Silurian.

Heterocrinus :—arms long, simple or bifercating, with strong pinnulæ; arm plates in single row; stem variable. Lower Silurian.

FIG. 154.
Glyptocrinus decadactylus. Hall. Silurian.

---

* In 1882, the Author proposed a new classification of the Crinoids in three groups and twenty-

§ 2. *arms without pinnulæ.*

Palæocrinus :—arms long and thin, of equal size, without pinnulæ, bifurcating. Lower Silurian.

Hybocrinus :—arms very long and thin, not bifurcating, and without pinnulæ ; Lower Silurian.

Cheirocrinus (including *Calcecrinus*, &c.):—arms very long, decumbent, unequal in size, with single row of plates. Silurian, Devonian.

Ichthyocrinus :—arms short, without pinnulæ, more or less in close contact throughout their length, gradually merging into the calyx ; arm plates in one row ; stem round, with small circular orifice. Roof of calyx composed of small, imbricating plates. Upper Silurian to Carboniferous. Figure 155 represents our principle species.

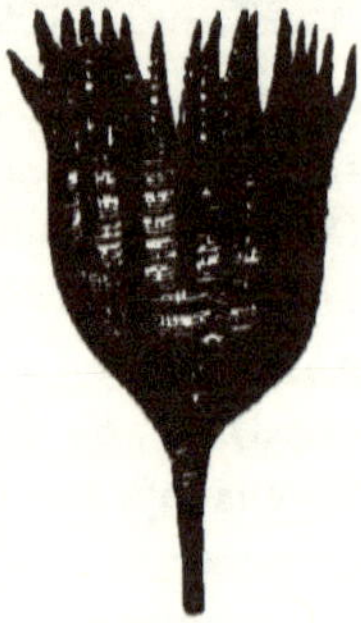

Fig. 155.
Ichthyocrinus lævis.
Niagara Formation.

Lecanocrinus :—Closely allied to *Ichthyocrinus* (if not really identical), but calyx plates larger and less numerous. *L. elegans*, Trenton Formation, is our best known species.

Other typical genera of the tesselated crinoids, comprise: *Pisocrinus*, *Marsupites* (a stemless form), *Actinocrinus*, *Crotalocrinus*, &c. The *Articulata* (see above) comprise, more especially, *Encrinus* (to which the "lily encrinite," E. liliiformis, of Triassic strata, belongs) ; *Apiocrinus* (including the "pear encrinite" of Jurassic strata) ; *Pentacrinus*, *Antedon or Comatula*, &c, but no examples of this group occur in Central Canada.

II.

CYSTIDEA.

The Cystideans from an entirely extinct group of Cambrian and Silurian age. They present relations, in some cases obscure, in others very marked, to both crinoids and blastoids. The typical cystidean may be described as an oval or spherical body, covered

two distinctive sections, the groups based on the presence or absence of a canaliculated structure in the plates of the calyx and arms. A synopsis is published in the *Transactions of the Royal Society of Canada* : vol. 1, pp. 113-116. This grouping is not followed here—as in a little work of this unpretending character (dealing only with local forms, necessarily limited in number) it would be out of place to enter into the classification details which it involves.

with calcareous plates united at their edges, without, or with merely rudimentary arms, and attached to the sea-floor by a short stem. Some, however, appear to have had no stem ; and in one or two genera closely related to the crinoids *(Porocrinus, Caryocrinus)* a well-developed system of arms was present.   Two other salient characters are commonly present, also, in all typical cystideans.   These comprise a so-called " pyramidal orifice," and a system of pores or minute fissures, by which some or all of the plates are traversed.   The " pyramidal orifice " is an opening, usually near the summit of the body closed by several triagular plates, forming a five or six-sided elevation.   It was most probably the oval orifice*.   A second opening is generally present at the upper part of the body, and occasionally a third opening is seen.   An enlarged view of a pyramidal orifice is shown in figure 156.   The pores so commonly present in cystideans, have not been recognised in all genera.   When present, they are either scattered irregularly over the surface of the body ; or are grouped in twos in small oval areas, on some or most of the plates ; or otherwise are arranged in fine lines forming lozenge-shaped areas, the so-called

Fig. 156.

"pectinated rhombs," or "hydrospires," which extend across the sutures into adjacent plates, as shown in enlarged form in figure 157*B*.   Figure *A*. shows a series of paired or double pores.

No very satisfactory classification of cystideans has yet been proposed, but these forms may be arranged conveniently under five sections, as follows :—

§ 1. *Occultiperforata :* —Without visible or distinct pores ; but pores

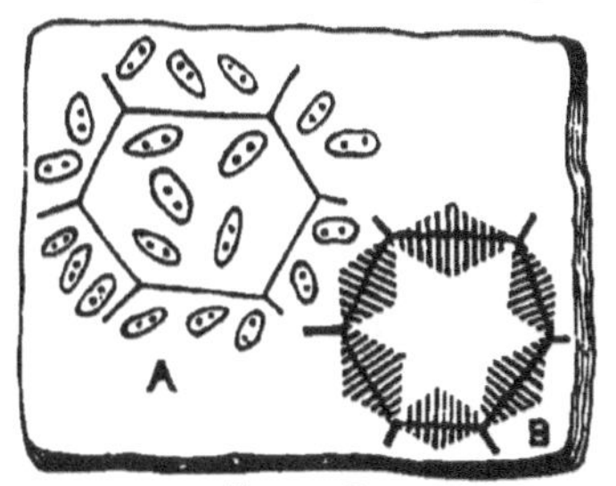

Fig. 157*B*.

may perhaps open on the sides of the plates between the sutures.

Canadian genera include : *Amygdalocystites* (Billings) : with few body-plates and two recumbent arms : Trenton formation ; *Ateleocystites* (Billings) with plane and convex sides, respectively, and two free arms : Silurian, Devonian ; and *Malocystites*, with numerous

---

* It is regarded by some palæontologists as the mouth, and by others as an ovarian aperture. The system of valves, closing from without, would appear to indicate that it was an orifice of emission, not of entrance.

16

body plates and several recumbent arms or pseudo-ambulacra : Lower Silurian.

§ 2. *Biperforata* : With double pores, in small separate areas, on the body plates.

The section includes :—*Gomphocystites* (Hall) with somewhat club-shaped body covered with numerous plates : Niagara formation ; *Holocystites*, etc.

§ 3. *Rhombiperforata* :—With numerous pores (or fine linear grooves) in rhombic areas which extend into adjacent plates.

Includes *Porocrinus* (Billings), a small crinoid-resembling form, with few body-plates, distinct arms, and pore-areas at the angles of the plates : Trenton formation ; *Caryocrinus*, also a crinoid-like form, but of larger size, with radiately ornamented body-plates, distinct arms, and long stem : Niagara formation ; and *Echinosphœrites* (not yet found in Canada), with globular body, covered with numerous, small, hexagonal plates : Lower Silurian.

§ 4. *Pauciperforata :* With pectinated rhombs more or less apart and comparatively few in number.

Fig. 158.
Glyptocystides Lo-gani (Billings). Trenton formation.

Includes *Pleurocystites* (Billings) with few body plates on one side of the body, and many smaller ones on the other side, two arms and short stem, the latter made up of round plates alternating in diameter : Chazy to Hudson River formations ; *Glyptocystites* (Fig. 158.); *Callocystites*, etc.

§ 5. *Tectiperforata :*—With hydrospire at upper surface of body.

This section includes, with probably a few other imperfectly known types, the blastoid-resembling *Codonaster*, in which the form is more or less conical or top-shaped, with flat or truncated upper surface carrying a five-rayed pseudo-ambulacral star, and intervening narrow plates, between which occur the minutely punctured lines of the hydrospire. Apart from the presence of the latter, the form and general arrangement of the body-plates are those of a typical blastoid : Devonian, Carboniferous.

III

BLASTOIDEA.

This class, like that of the Cystoidea to which it is more or less closely related, is entirely extinct, and chiefly characteristic of Devonian and Carboniferous strata. The typical Blastoid has an oval or bud-shaped body, covered with comparatively large, regularly arranged plates, with a five-rayed "pseudo-ambulacral star" at the summit. The pseudo-ambulacral plates carried, it is supposed, small pinnulæ during the life of the animal. At the underside of the body there is usually a short stem.

The known or supposed genera (for some are of very doubtful position) may be arranged as follows :

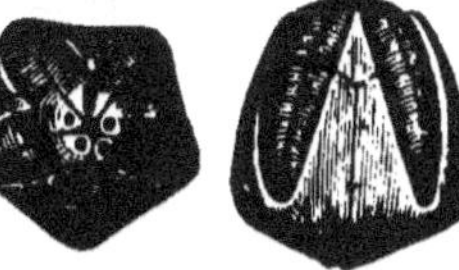

Fig. 159.

§ 1. *Without any inter-ambulacral apertures :* This section includes *Pentremites* (with broad-ambulacral areas, and comparatively large basal plates *i.e.* those immediately above the stem), Upper Silurian, Devonian, Carboniferous; *Granatocrinus* (with long and narrow pseud-ambulacral areas and very small basal plates) Carboniferous ; and *Eleuthocrinus* (stemless, with four linear and one short pseud-ambulacral area) Devonian.

§ 2. *With (anal) aperture in one of the inter-ambulacral spaces :* *Orophocrinus* (with general aspect like that of *Pentremites*), Carboniferous. *Nucleocrinus* (with very minute based plates, and long pseudambulacral areas) Devonian, Carboniferous. *Stephanocrinus?* (with long basals, and five sharp points at the summit of the calyx) . Upper Silurian : Niagara formation.

Blastoids are rare among Canadian fossils, but examples of *Nucleocrinus* and *Stephanocrinus* are occasionally met with. Fig. 160 shows the upper surface (about twice enlarged) of *Nucleocrinns Canadensis* (Montgomery), perhaps identical with *N. lucina* (Hall), from the Hamilton formation of Bosanquet township in south-western Ontario.* Fig. 160 *bis.* is a figure of *Stephanocrinus angulatus* of the Niagara formation.

Fig. 160. *Nucleocrinus Canadensis.* Devonian.

Fig. 160 *bis.* *Stephanocrinus angulatus,*

---

* A full description of this species by its discoverer, Prof. H. Montgomery, an old student and graduate of Toronto University, will be found in the *Canadian Naturalist,* vol. x, No. 2

IV

EDRIOASTERIDA.

This class, entirely extinct and Palæozoic, comprises a very limited number of representatives. These apparently connect the cystideans with the star fishes. They are stemless, circular, depressed forms, varying from about a quarter of an inch to a little over an inch in diameter. The under side of the body is unknown. The upper side carries a five-rayed ambulacroid star, composed of a double series of interlocking plates. The rays are curved in some forms, and straight in others ; and are entirely confined to the upper surface of the body. The margin of the disciform body is covered with very small imbricating or partially-overlapping, scale-like plates. The other parts or inter-ambulacroid spaces are protected also by imbricating plates, but of somewhat larger size ; and a "pyramidal orifice," resembling that of a cystidean, is situated in one of these inter-radial spaces.

The principal genera comprise : (1) *Agelacrinites*, with curved rays, like the "arms" of many ophiurian star-fishes ; and (2) *Hemicystites*, with short, straight rays.*   Species of both genera occur in our Silu-

---

* The generic name *Agelacrinus* (now more commonly. written *Agelacrinites*) was given by Vanuxem to these forms, from the Greek, αγελη, a herd or crowd—the first found examples consisting of several individuals heaped or crowded together.   But this condition of occurrence is purely accidental.   It was also thought that these forms, although without a stem, were always attached to shells or other submarine objects by their broad base ; and this idea is still retained by many writers.   Hence the term *Edrioasterida* of Billings, from εδραιοσ fixed, sessile.   Examples are found now and then attached to fossil shells, but that condition is by no means general.   Out of fourteen or fifteen examples belonging to several species examined by the writer, only one occurred in contact with a brachipod shell, and accidental contacts of that kind are common among fossil bodies generally.   The structural characters of the Edrioasterians are still very imperfectly known.   The mouth for instance, is almost universally regarded as lying in the centre of the ambulacroid area at the summit of the disc.   In a communication published in the *Canadian Journal*, and in the *Annals of Natural History*, so long ago as 1860, the writer strove to maintain that this was not its true position, but that it was to be looked for, as in ordinary asterians, &c., in the centre of the disc, below.   No little support seems to be lent to this view, by the subsequent discovery by Wyville Thompson of his *Echinocystites*, regarded by him as a transitional type between cystideans and echinida.   The body is covered (apart from the ambulacra, &c.) with *imbricating*, irregularly arranged plates ; the mouth is central, on the under side of the body ; and the anal opening, with *protecting pyramid of plates*, is interradial in position (*Edinburgh New Philosophical Journal*, 1861).

A magnificent specimen of *Agelacrinites*—probably the best in Canada, if not on this Continent—is in the collection of Dr. Grant of Ottawa.

rian, Trenton Formation. but are comparatively rare. The genus *Edrioaster* of Billings only differs from *Agelacrinites* in shewing pores in the ambulacroid spaces; but pores (obliterated by fossilization) were probably present originally in all forms. Figure 161 represents an example of *Hemicystites Billingsii* from the Trenton Limestone of Peterborough in Ontario, Other species, but commonly of smaller size, occur in the same formation at Ottawa and elsewhere.

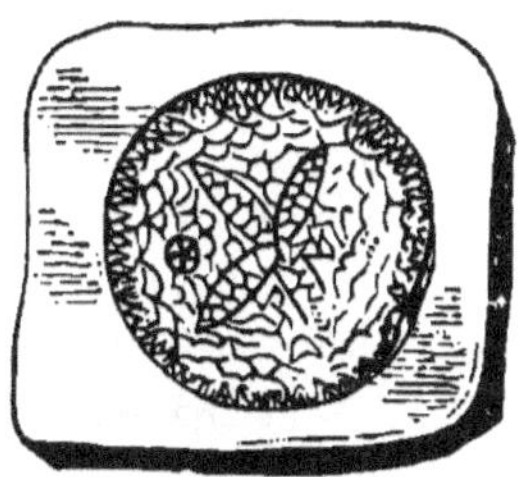

Fig. 161.

Hemicystites (Agelacrinites) Billingsii ; Chapman. Trenton Formation.

V.

OPHIUROIDEA.

This class comprises the so-called " Brittle Stars," or star-fishes with typically small central body, and five long, thin, more or less serpentiform arms or rays, entirely distinct from the central stomach-cavity, and without open ambulacral groove. Both disc and arms are generally plate-covered, but in some forms they are coriaceous only, or partially tuberculated. The month is in the centre of the underside of the disc, and there is no separate anal orifice.

Two orders are generally recognized : *Euryalida,* with coriaceous integument, and mostly with branching arms which are capable of being curled towards the mouth ; and *Ophiuridu,* the true brittle stars, with plate-covered disc and arms. Examples of both orders date from the Silurian period ; and some few genera (*Protaster, Eugaster,* &c.) are exclusively Palæozoic. *Tœniaster cylindricus* (Billings), with narrow arms and partially overlapping, spinous plates, from the Trenton Formation, is our best known representative.

VI.

ASTEROIDEA.

The representatives of this class comprise the starfishes proper, in which the stomach cavity is continued into the so-called arms or rays. In most, there is an anal orifice ; and, in all, the ambulacral groove is open or uncovered. The mouth is in the centre of the under-side, and the body-covering is chiefly coriaceous, or partly tuberculated or plate-covered. The star-fishes date from the Silurian period, but fossil examples in the lower rocks are comparatively rare. The class is usually subdivided into two orders: *Brisingida*, connecting the class with that of the Ophiuroidea, the arms being distinct to some extent from the body cavity; and *Stellerida*, comprising the true star-fishes.

The *Brisingida* are unknown in the fossil state. The *Stellerida* may be conveniently arranged under three sections as follows: §1. *Multiradiata*, with more than five, usually from 13 to over 20 arms or rays (e.g. *Lepidaster*, Upper Silurian; *Solaster*, a living type, dating from the Jurassic period; *Luidia*, &c.) § 2. *Curtiradiata*, with very short or in some cases almost suppressed rays, the shape being then pentagonal (e.g. *Palasterinus*, Lower Silurian; *Palæocoma*, Upper Silurian; *Goniaster*, the " cushion-stars," first appearing in Liassic strata; *Asterodiscus*, &c). § 3. *Quinqueradiata*, with five well-developed rays (e.g. *Palæaster*, *Stenaster*, Cambrian and Lower Silurian to Carboniferous; *Asterias*, *Oreaster*, &c.).

The genera *Palasterina*, *Palæaster* and *Stenaster* are occasionally represented in our Lower Silurian strata, more especially in the limestones of the Trenton formation. In *Palasterina*, the rays extend a very little way beyond the central part of the body: *P. Stellata*, a pentagonal form, is our best-known species. In *Palæaster* (including *Petraster*) the form is distinctly five-rayed, and the ambulacral furrows are bordered by two rows of small plates; whilst in *Stenaster* (= *Urasterella*), also a distinctly five-rayed form, the ambulacral grooves are margined by a single row. Our principal species comprise, *Palæaster rigidus* from the Trenton, and *P. bellulus* from the Niagara formation; *Stenaster pulchellus*, with very narrow rays, and

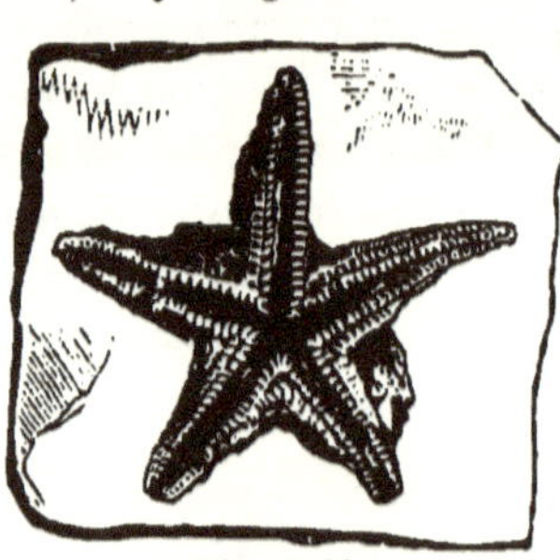

Fig. 161 *bis.*
Petraster Bellulus, after Billings. Niagara Formation. Grimsby, Ont.

S. Salteri with comparatively broad rays, both from Trenton strata. Full descriptions of these species, by the late Mr. Billings, will be found in Decade III. of "Canadian Organic Remains," issued in 1861 by the Geological Survey.

VII.

ECHINOIDEA.

The representatives of this class are the "sea-urchins" or "sea-eggs" of popular parlance. They present a globular, disciform, oval, heart-shaped, or other form of body, entirely enclosed in calcareous plates. In a small number of both ancient and living genera, the plates are overlapping, but in the great majority of echinids, they are joined at their edges. These plates are of two general kinds : *ambulacral* and *interambulacral*, respectively. The ambulacral plates, in all echinida, form five separate, converging, linear or petaloidal areas, each composed of two rows or zones of perforated plates. The interambulacral plates, lie also in five separate areas ; but in all Palæozoic types, with a single exception (*Bothriocidaris*), each area contains more than two rows, commonly five or six ; whilst in all succeeding with only two known exceptions (*Anaulocidaris?* and *Tetracidaris*) these interambulacral areas contain two rows only.* The perforations in the ambulacral plates give passage to delicate sucker-feet ; and movable spines—in some cases hair-like, in others club shaped or cylindrical and comparatively large—are borne on both the ambulacral and interambulacral plates, more especially on the latter, these being more or less distinctly tuberculated for the support or attachment of the spines. The mouth, in echinids, is always on the under-side of the body, and either central or sub-marginal in position. The anal orifice is situated in some forms at the apex of the shell or test, but in others it is at or near the under margin.

Echinida are very numerous in existing seas. They abounded also in the seas of the Cainozoic and Mesozoic Ages, but appear to have been rare in Palæozoic times. No certain evidence of their remains in the strata of Central Canada has yet been obtained : it is unnecessary therefore in the present work to describe their leading subdivisions and genera.

------

* In the palæozoic *Bothriocidaris*, the interambulacrals form a single row. In *Tetracidaris*, a Lower Cretaceous type, there are four rows of interambulacrals at the lower part of the test, but these merge into the typical two rows at the upper part.

VIII.

HOLOTHUROIDEA.

This class comprises a small number of elongated, more or less vermiform types, protected by a thick, coriaceous integument, strengthened by calcareous wheel-like, anchor-shaped, and other spicules. The mouth is situated at one extremity of the body, and is surrounded by a circle of quinary, usually branching tentacles. The anal orifice in all the typical forms is at the opposite end of the body ; but in the genus *Rhopalodina* (Gray) it is at the same extremity as the mouth. Wheel-shaped spicules, thought to have belonged to holothurans, have been found occasionally in carboniferous and higher strata, but apart from this, fossil· forms are of very doubtful occurrence.

SUB-KINGDOM V.

VERMES.

This division, comprising the various parasitic and other worms, with some related memberless types, is of comparatively little palæontological interest. It may be subdivided into the six following classes : *Turbellaria, Platyelmintha, Nematelmintha, Rotifera, Gephyrea, Chætognatha,* and *Annelida.*

The *Turbellaria* are non-parasitic, mostly marine or fresh water forms, more or less depressed in shape, but of great length in certain genera *(Linus, Borlasia).* Fossil forms are unknown.

The *Platyelmintha* comprise the so called " flat worms " most of which are internal parasites. Fossil forms unknown.

The *Nematelmintha* comprise the so called " round worms " most of which are also permanently, or for a time, parasitic. Of late years some supposed representatives of this class have been discovered in amber and brown coal, both of Cainozoic age ; but otherwise there are no known fossil representatives.

The *Rotifera* are very minute aquatic forms, with one or more ciliated discs at the front end of the body. They are commonly known as " wheel animalculæ." There are no recognized fossil forms.

The *Gephyrea*, represented principally by the *Sipunculus*, are marine, worm-like forms with thick skin. They present in some respects connecting links between the worms and the holothurians, whence the name of the class, from γεφυρα, a bridge. Some supposed fossil forms have been cited from the Upper Jurassic Strata of Bavaria. Otherwise the class is unknown in the fossil state.

The *Chœtognatha*, comprise the single genus *Sagitta*, a small, somewhat fish-like form from the Mediterranean, with fin-representative at the posterior end of the body. Fossil forms unknown.

The *Annelida* comprise earth-worms, leeches, serpulæ, &c., and are thus classified :

Abranchiata—without visible branchiæ.
    1. *Suctoria* (leeches). Foss. Rep., unknown.
    2. *Terricola* (earth worms). Foss. Rep. unknown.

Branchiata—with distinct branchial organs.
    1. *Tubicola* or *Cephalo-branchiata :*
    2. *Errantia* or *Dorsi-branchiata:*

The *Tubicola* are marine worms with a circlet of thread-like branchiæ around the rudimentary head. Some secrete a calcareous tube or shell, and others form a protecting sheath of agglutinated grains of sand, &c, Two genera, *Serpula*, with calcareous wavy or contorted tube, and *Spirorbis*, with regularly enrolled shell (Fig. 162) are of frequent occurrence in the fossil state. They

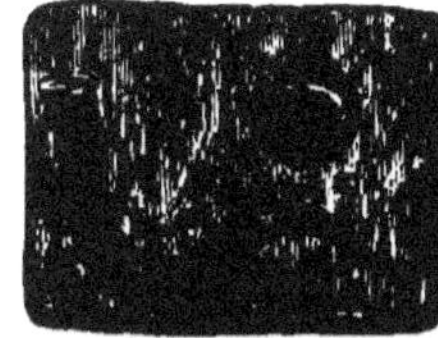

Fig. 162.
a. Serpula. *b.* Spirorbis.

date from the Cambrian or Silurian period, and are found mostly on the external or internal surface of fossil shells.

The *Errantia* are marine worms with tufts of thread-like branchiæ along the sides of the body, and without a shell or protecting sheath. They date from the Cambrian period, but many fossil examples referred to the genera *Nereites*, *Nemertites*, &c., are of somewhat doubtful nature. The narrow, cylindrical cavities found occasionally in the Potsdam sandstone of the County of Leeds, Ontario, and at some other localities, and which are commonly known as "Scolithus cavities," are thought by some observers to have been made by boring annelids, but others regard them as of fucoidal origin. Minute, horny

or chitonous bodies (Fig. 163) recognized as the
jaws of species of Errantia, were found a few
years ago in the Hudson River Strata and other
Lower Silurian rocks of Ontario by Mr. J. G.
Hinde. Two of these, greatly magnified, are
shewn in figure 163. Similar bodies from the

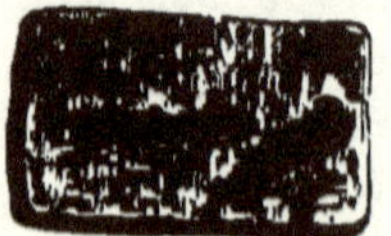

Fig. 163.
Jaws of *Errantia*.

same strata in Ohio had been previously regarded by Grinel as the
lower jaws of *Errantia*.

SUB-KINGDOM VI.

ARTHROPODA.

The Arthropoda or Articulata (as they are also called) comprise
a large series of animals, characterised typically by their jointed legs,
more or less distinctly segmented body, and bilateral symmetry.
They include both aquatic and terrestrial forms. Some are of great
palæontological interest; but many in their relations to geology are
comparatively unimportant. Four classes are universally recognized.
These comprise: 1. *Crustacea*; 2. *Arachnida*; 3. *Myriapoda*; and
4. *Insecta* or *Hexapoda*.

I.

CRUSTACEA.

The crustaceans are mostly aquatic types, with respiratory organs
(when present) in the form of branchiæ. They include: barnacles,
crabs and lobsters, wood lice, &c., among living forms; and an extinct
group, the Trilobites, with some other extinct forms, of great geological interest. By uniting some of the more closely connected orders
we may arrange the crustaceans, generally, under ten leading sub-
divisions, comprising: 1. *Cirripedia*; 2. *Ostracoda*; 3. *Phyllopoda*;
4. *Trilobita*; 5. *Merostomata*; 6. *Phyllocarida*; 7. *Amphipoda*;
8. *Isopoda*; 9. *Stomapoda*; 10 *Decapoda*.

1. *Cirripedia*:—The cirripeds form a small group of marine animals,
sedentary in their adult condition, and more resembling mollusks at
first sight than members of the articulated series. They secrete an
external, many-valved, calcareous shell; and possess a number of
delicate, plume-like cirrhi, capable of protrusion beyond the shell for
the creation of currents in the surrounding water. Some of the

more common types are pedunculated, others are sessile. In the former, to which the well known barnacles belong, the animal is attached to ships' bottoms, floating timber, &c., by a flexible, coriaceous stem ; whilst in the latter, typified by the *balanus* or " sea-acorn," the shell is fixed directly by its base to rocks and other submarine bodies, especially to those which lie between the tide marks. Fig. 164 shews the general form of a living balanus with its cirrhi protruded between the smaller opercula-like valves of its shell. Fragments of comparatively large shells, which must have averaged an inch or more in diameter, belonging to one or two species of Balanus *(B. Undevallensis, B. Hameri ?)* occur in the Post-Cainozoic " Saxicava Sand Formation " of Beauport near Quebec ; but no cirripeds are found in our lower rocks, nor have any undoubted examples been discovered in Palæozoic strata.

Fig. 164.
Living Balanus.

2. *Ostracoda :*—The Ostracods comprise a large number of generally minute aquatic forms, in which the entire body is enclosed in a bivalve shell, whence the name of "bivalve entomostracans" by which they are often known. Natatory antennæ, and several pairs of small feet (which do not serve as swimming organs), project in living forms beyond the shell. The latter is smooth in some genera, and more or less embossed or tuberculated in others. Most living forms are marine, but some *(Cypris,* &c,) are fresh-water types. The best known Palæozoic genera comprise *Leperditia* and *Beyrichia.* In *Leperditia,* the shell is comparatively thick, with straight dorsal edge ; and it commonly averages from one-fourth to three-fourths' of an inch in length. Fig 165 is an example from the Trenton formation. In *Beyrichia,* the shell is very similar in shape but much smaller, rarely exceeding the 12th of an inch in length, and its surface is tuberculated or embossed.

Fig. 165.
Leperditia Canadensis,
Nat. size : Trenton Formation.

3. *Phyllopoda :*—This sub-division may be made to include the *Copepoda, Cladocera,* and *Phyllopoda, proper*—small* aquatic types,

---

* The large, phyllopod resembling types, *Hymenocaris, Dictyocaris,* &c. of early palæozoic age, are now separated from the Phyllopods, proper, and placed in a distinct group, the *Phyllorarida* of Packard. As shewn by Packard and Claus, they appear to form a connecting link between the lower and higher crustaceans : the *Entomostraca* and *Malacostraca* of many classifications. They form the sub-division *Leptostraca* of Claus.

with flattened, natatory feet. The *Copepods* represented by the modern *Cyclops*, *Argulus*, &c., have no known fossil-representatives, and the *Cladocera* are also of very doubtful occurrence in the fossil state. The *Phyllopods*, represented by the living *Apus*, *Branchipus*, *Estheria*, &c., date apparently from the Devonian period, but no fossil examples have been found, as yet, in Ontario or Quebec.

4. *Trilobita* :—The Trilobites form an entirely extinct series of crustacea, related to the Phyllopods on one hand, and to the Merostomes on the other.   Their remains are found in Cambrian, Silurian, and Devonian strata in great numbers, and sparingly in the Lower Carboniferous beds, above which, no traces of the group have been discovered.   The trilobites present a generally oval, tri-lobed form of body, averaging about an inch-and-a-half to three inches in length ; but some examples are scarcely the fourth of an inch, whilst others occasionally shew a length of eight or nine inches.   The upper surface of the trilobite was protected by a chitonous or crustaceous shell composed of numerous pieces, in part free, and partly united by sutures.   The underside of the body seems to have been covered essentially by a soft or semi-coriaceous integument, and to have carried numerous feet, some of these being "jaw feet" as in the merostomes and copepods, and others probably branchial and natatory in their functions.

As shewn in the annexed Figure (166), the upper covering or " back " of the trilobite consists of three principal parts : (1) the *Buckler* or *Head-shield, H ;* (2) the *Body* or *Thorax, T ;* and (3) the *Pygidium* or *Caudal-shield, P.*

The head-shield is always more or less of a crescented or horse-shoe shape, with the convex side in front ; and its posterior or so-called genal angles, though rounded in some species, very commonly terminate in points or spines. Its central portion is generally in the form of a distinctly raised area known as the *glabella* (= G, Fig. 166).

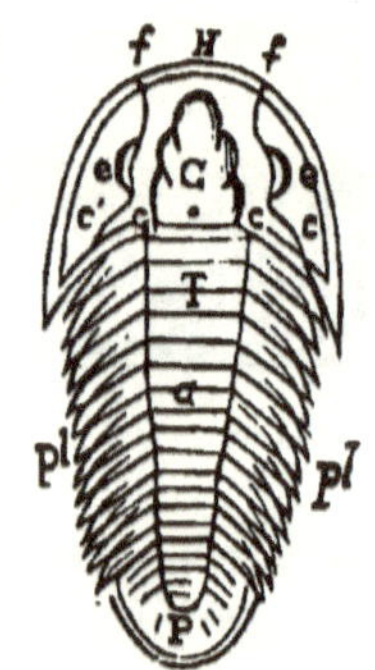

Fig. 166.

The surface of this is sometimes smooth, but is more commonly lobed, furrowed, or granulated. In certain genera *(Phacops, &c.)* the glabella is enlarged or expanded anteriorly, and in others *(Calymene,*

&c.) it is contracted in that direction. In some, again, *(Triarthrus, &c.)*, it is of nearly uniform width throughout. Very prominent, also, in some genera ; and in others, but feebly elevated and comparatively inconspicuous. The head-shield exhibits likewise, in most examples, on each side of the glabella a sutural line (=*f. f.* in Fig. 166) called the "facial suture." The hinder extremity of this sutural line terminates either at the posterior margin of the head-shield *(asaphus, &c.);* or at the angles of this *(calymene, &c.);* or at the sides, near the level of the eyes *(Phacops, Ceraurus, &c.).* In some few genera, however, the facial suture is not preceptible. The eyes, when present, occur on each side of the head-shield in the line of the facial suture, just on the outer edge of this. In all genera but Harpes they are compound, as in ordinary crustaceans, insects, &c., and in certain genera *(Dalmanites, Phacops)*, the component facets are comparatively large, forming the so-called "reticulated eye." The outer sides or "cheeks" *(gena)* of the head-shield often separate along the line of the facial suture, and are found detached. The shell is continued over the head-shield, and a so-called *labrum* or *hypostoma* is occasionally found, where the mouth was situated, on the underside. This is commonly found detached, and generally oval in form ; but in the genus *Asaphus* it presents a forked or horse-shoe shape, with the concavity at its lower or posterior margin.

The body or thorax of the trilobite consists of a series of movable rings or segments, varying in number, according to the genus or the state of development of the individual. Two, generally deeply-marked, longitudinal grooves or furrows traverse the thorax and extend into the head-shield above and the pygidium below. In the head-shield they limit the glabella. They divide the thorax into three parts—a central part or axis, and two side portions, or *pleurae* (=*a*, and *pl.* of figure 166.) These latter have their free ends rounded in some species, and pointed or even prolonged in part or wholly into spines, in others. In some genera *(e. g. ceraurus* or *cheirurus)* there is a raised band on each pleura, and in others a narrow groove. The degree of mobility with which the thoracic segments were endowed, at least in most cases, enabled the trilobite to bring the under parts of the caudal and head shield together, both for the protection of the branchial feet or more or less undefended portions of the body, and also, in all probability, to enable the creature to sink with greater rapidity into deeper water in moments of danger or alarm. Fig. 167 is an example of a trilobite in this " rolled up" condition.

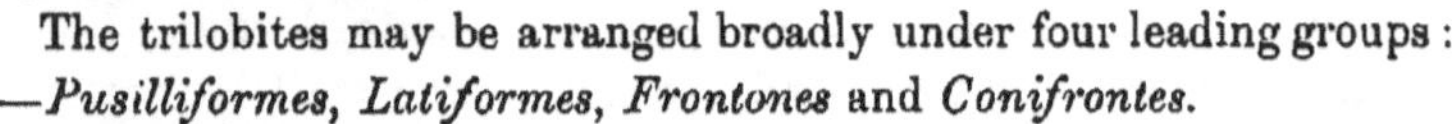

Fig. 167.
Calymene Senaria in " rolled up" condition.

The pygidium, or shelly covering of the tail or abdomen, consists of a single piece, arising probably from consolidated segments. Very commonly, it is found detached from other portions of the body. Its outline is either rounded, with smooth or digitated margin, or is more or less pointed ; and it sometimes terminates in a long spine, or in several spinous processes. It shows very frequently a distinct axis, with pleuræ-like, lateral furrows, but is quite smooth or without furrows in some species. In some genera, also, it is exceedingly small (*Paradoxides, &c.*), whilst in others (*Illænus, Asaphus, &c.*), it equals the head-shield in size.

The trilobites may be arranged broadly under four leading groups :—*Pusilliformes, Latiformes, Frontones* and *Conifrontes.*

The *Pusilliformes* constitute a group of very small and somewhat doubtful trilobites represented by the genus *Agnostus.* In this form the thorax consists of only two segments, whilst the head-shield and pygidium are quite large in comparison and of nearly equal size. The type is essentially Cambrian, but lower Silurian examples are also known. Several species have been obtained from the limestone bands intercalated with the graptolitic states of the Lévis formation (Quebec), but usually without the thorax. Figure 167 shows the head-shield and pygidium of *Agnostus Canadensis* from that formation.

Fig. 167 *bis*
Agnostus Canadensis.
Hillings.
Lévis Formation.

The group of *Latiformes* comprises a series of undoubted trilobites, mostly of considerable size and broad form, with large head-shield and pygidium—thus differing essentially from the more elongated many-ringed forms of the typical *Frontones* and *Conifrontes.* The principal genera of Canadian occurrence, comprise : *Asaphus* and *Illaenus ;* but the group includes, also, *Bathyurus, Bronteus, Lichas, Dikelocephalus* and other genera. *Asaphus* is distinguished by its eight body-segments, its large head-shield with feebly-elevated and (as a rule) unfurrowed glabella, its forked or horse-shoe-shaped hypostoma, its large and broad pygidium, and other characters. The side cheeks are often broken off, and the pygidium is frequently

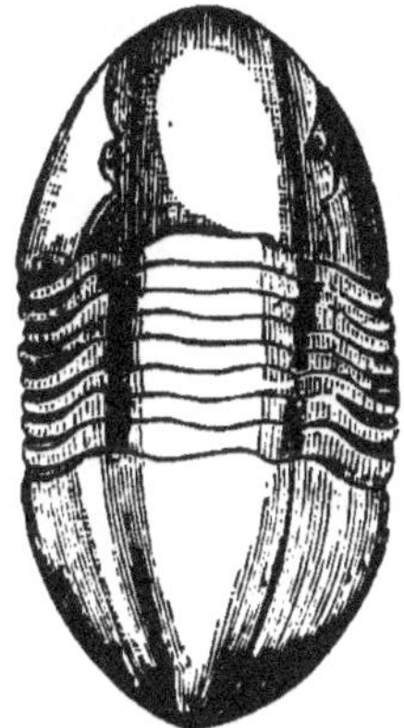

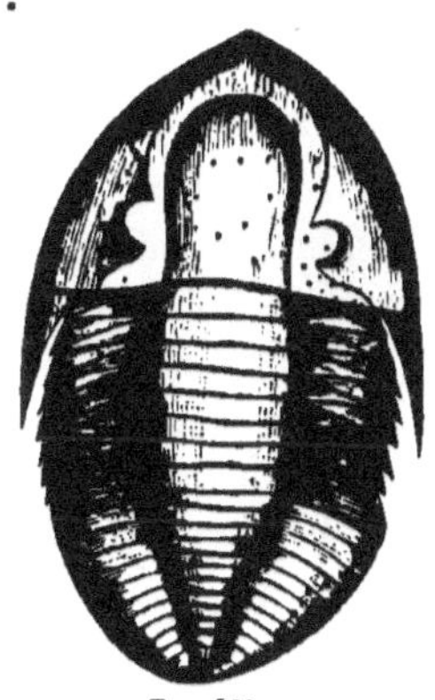

FIG. 168.

Asaphus platycephalus
(=Isotelus gigas):
Stokes. Trenton
Formation.

H.=The hypostoma.

FIG. 169.

Asaphus Canadensis : Chapman. Utica Formation.

found without the other portions of the body. The genus appears to be exclusively Lower Silurian. Our most common species comprise ; *A. platycephalus*, a more or less smooth species with rounded head angles and pleuræ, very abundant in the Trenton limestone, but occurring also in the Chazy and Hudson River Formations, and *A. Canadensis*, with horned head-shield, pointed pleuræ, and furrowed pygidium, of common occurrence in the Utica bituminous schists of Collingwood, Whitby, Ottawa and other localities. Closely related to *Asaphus* is the genus *Ogygia*, distinguished by its shield-shaped or pointed hypostoma, and its laterally-furrowed pygidium, but it is essentially a European type. The genus *Illænus* is represented by smooth, oval species, with large head-shield and pygidium, feebly-raised glabella, far-apart eyes, ten (or rarely nine) body-segments, and rounded pleuræ. It is essentially a Lower Silurian type but ranges from Upper Cambrian into Upper Silurian strata.

FIG. 170,

Illænus globosus : Billings.
Chazy Formation.

Figure 170 represents a species in a rolled up condition from the Chazy Formation. *I Americanus* (= *I. crassicauda ?*) from the Trenton limestone of the Ottawa district, is another Canadian species ; and fragmentary examples of additional species occur in the Levis formation of Quebec. *Bathyurus* is also chiefly from the same formation. Its buckler, thorax (with

nine segments) and pygidium, are nearly of equal size. The axis
usually extends to within ·a short distance of the extremity of the
pygidium, and the latter has very frequently a marginal furrow.
The genus *Bronteus* is mostly a Devonian and Upper Silurian type,
but is represented also in Lower Silurian strata. It is distinguished
by its ten body-segments with slightly-raised bands in place of furrows on the pleuræ, and its widely-expanded glabella ; and especially
by its large pygidium, with short axis, and fan-like, furrowed surface, the furrows curving upwards. The genus *Lichas* is exclusively
Silurian. It is distinguished chiefly by its nine or ten body-segments ;
its broad, backward-curving, and pointed pleuræ ; and its comparatively large pygidium with short axis, largely-serated margin and
almost longitudinal furrows. The genus *Dikelocephalus* is exclusively Cambrian. From this latter circumstance it is comm nly
placed in the family of the *Olenidæ*, in despite of its very dissimilar
aspect and structural characters. Whilst Olenus (with other
members of the Olenidæ family)has a very small pygidium, compared
with the broad head-shield, in *Dikelocephalus* the pygidium is as
large as the head-shield or even larger, and it has a very short axis
as in *Lichas* and *Bronteus* of this group. The genus, however, is
still very imperfectly defined. The number of
the body-segments is not yet known ; and the
glabella, as regards its transverse furrows,
appears to differ very considerably in different
species. Figure 171 represents the pygidium
of *D. magnificus* (after Billings) from the Levis
or Quebec formation. In some other species
referred to this genus, the outline of the pygidium is simply rounded.

Fig. 171.
*Dikelocephalus magnifi-
cus* : Pygidium, after
Billings. Levis
Formation.

The *Frontones* are distinguished by a very
large head-shield and small, or comparatively
small, pygidium. The glabella is more or less broad and prominently
developed, and the posterior angles of the head-shield are almost
always in the form of long spines or horns. The more typical
genera comprise : *Paradoxides*, with from sixteen to twenty spinous
body-segments and very small pygidium ; *Trinucleus*, with six body-
segments, large oval glabella, and head-shield surrounded by a per-

forated margin ; *Cheirurus* or *Ceraurus*, with eleven body rings, and facial-suture terminating at the sides of the head-shield ; *Encrinurus*, also with eleven body-segments and other characters as in *Cheirurus*, but with many-ringed axis to its pygidium ; and *Dalmannites* and *Phacops*, forms with eleven body-segments and facial suture also as in *Cheirurus*, with other characters described below.

Paradoxides is strictly a Cambrian type. Examples occur in New Brunswick and Newfoundland, but none appear to have been discovered elsewhere in Canada. *Trinucleus* is essentially a Silurian type, characterised by its six body-segments, its globose, strongly-pronounced glabella, and the perforated border of its head-shield. Adult forms are withont eyes. Figure 172 shews our common species *T. concentricus* from the Trenton and Hudson River (Lower Silurian) formations. *Ceraurus* is also essentially a Silurian type, although the genus ranged from the Upper Cambrian into the Devonian period. It is characterised chiefly by its eleven body-segments,

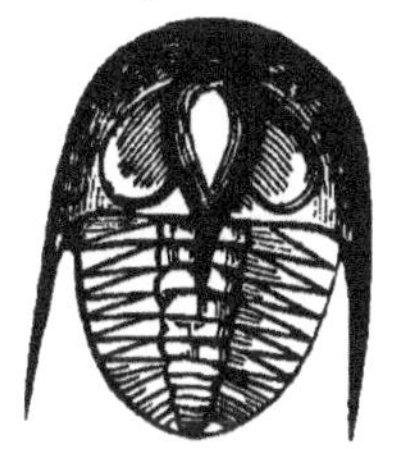

Fig. 172
Trinucleus concentricus
Trenton and Hudson
River Formations.

its pleuræ with raised bands in place of furrows, and the lateral termination of its facial sutures. By the latter character as well as by the number of its body-segments, it approaches *Phacops* and *Dalmanites ;*

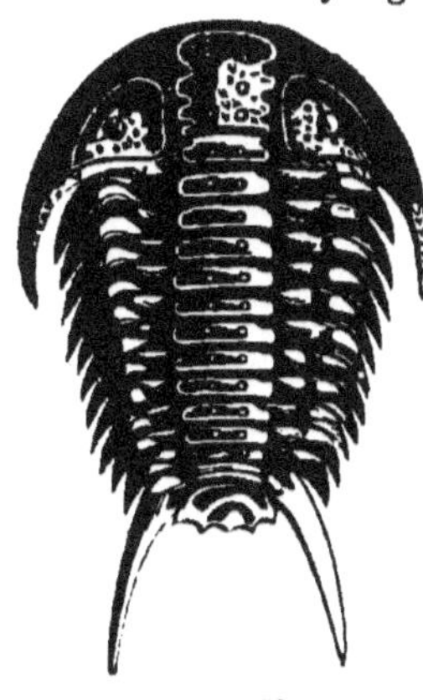

Fig. 173.
*Ceraurus (= Cheirurus) plen-
rexanthemus :* Green. Tren-]
ton Formation.

but from these it is distinguished among other characters, by its finely reticulated eyes, its glabella of almost uniform width. and the bands upon its pleuræ. Figure 173 represents a common species, *C. pleurexanthe-nus* from our Trenton limestone. Impressions of the glabella and two-horned pygidium, are especially abundant in some of these limestone beds, as at Belleville, Peterborough and elsewhere. *Phacops* is characteristic both of Silurian and Devonian strata. It is distinguished by its eleven body-segments ; its laterally-terminating facial suture; its large anteriorly expanded and usually granulated glabella ; its coarsely-facetted eyes ; and its rounded pleuræ. A common Devonian species,

17

P. Bufo, is represented in Figure 174. The genus *Dalmanites* is closely allied to *Phacops*, possessing like the latter an anteriorly expanded gla-

bella, with coarsely-reticulated eyes, and facial suture terminating at the sides of the head-shield. But it is distinguished by distinctly marked lateral fur-rows on the glabella and by its pointed pleuræ. The head-shield is also horned at its genal angles, and the pygidium in some species ends in a long spine. This is the case with our *D. limulurus*, Figure 175, from the Niagara (Upper Silurian) formation, but the caudal spine is only seen in well preserved examples. The genus is exclu-sively Silurian.

In the group of *Conifrontes*, the distinguishing character is the conical form and comparatively small size of the glabella, but in other respects these trilobites closely approach the Frontones— the head-shield as a rule exceeding the pygidium

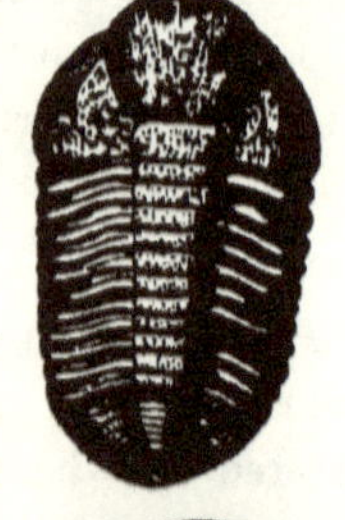

Fig. 174.

Phacops bufo : Green. Corniferous and Ha-milton Formation

in size, and the body-axis being continued into the latter, so as to ren-der the line of separation between the thorax and pygidium scarcely distinguishable. Some of the more typical genera comprise : *Olenus*, *Harpes*, *Conocephalites*, *Triarthrus*, *Calymene* and *Homalonotus*.

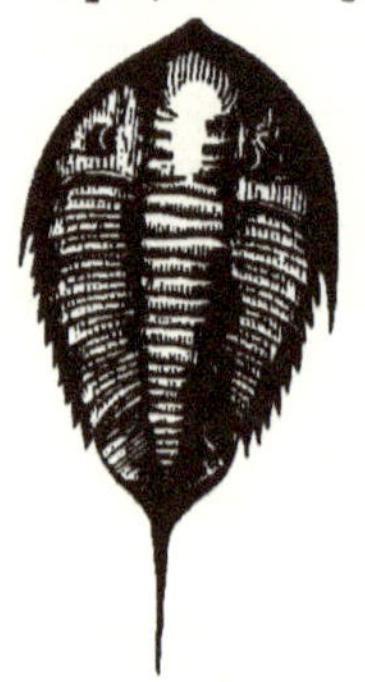

Fig. 175.

Dalmanites limulurus : Green. Niagara For-mation.

The genus *Olenus*, with from twelve to sixteen narrow, pointed body-rings, and very short pygi-dium, is of doubtful occurrence in Central Cana-dian strata, unless the "Loganellus," discovered by Mr. T. Devine in the Levis formation of Quebec, belongs to it. The type is essentially Cambrian. The genus *Harpes* is both Silurian and Devonian. It has a very large horse-shoe-shaped head-shield, with broad, perforated border, but the glabella is short and conical. The body-segments vary from twelve to twenty-five or twenty-six in number, and the pygidium is very small. Figure 176 shews the head-shield (with eyes connected by a raised band, as in *Olenus*) and part of the many-ringed thorax of *Harpes Ottwaensis* (Billings) from a specimen obtained from the Trenton Formation by Dr. Grant of Ottawa. The genus *Conocepha-*

lites is a Cambrian and Lower Silurian type, but as regards the area treated of in the present volume it occurs only in fragmentary examples in the Levis formation of Quebec. It possesses fourteen or fifteen body-segments and a glabella which narrows anteriorly and has short lateral furrows as in *Ca y mene*, a genus with which in other respects it has close affinities. The genus *Triarthrus* is peculiar to this continent, and its species are apparently confined to the Lower Silurian, Utica formation. The body-segments vary from fourteen to six-

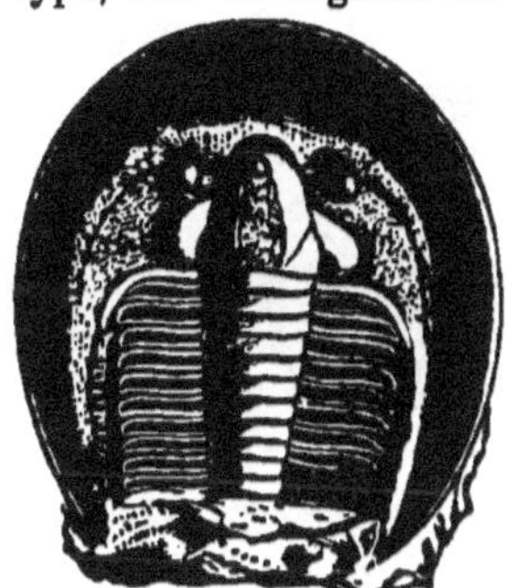

Fig. 176.
Harpes Ottawaensis : after Billings.
Trenton Formation.

teen in number, and in our most common species *T. Beckii*, each segment bears in the centre a short spine, as shown in figure 177. In another species, *T. spinosus* (Billings), a very long spine is attached to the eighth or ninth thoracic segment, and another to the neck segment.* The glabella in this genus is of nearly uniform width, not much raised, and is marked on each side by several short furrows. Impressions of the glabella of T. Beckii occur in the shale beds west of Collingvood and at Whitby and Ottawa, in great abundance. A third Canadian species, *T. glaber*, is destitute of spines. The genus *Calymene* is exclusively Silurian. its species are about equally numerous in the Lower and Upper Silurian beds, and several range from the lower into the higher series. The genus is distinguished by its thirteen body-segments ; its lobed glabella, narrowing upwards ; and by the posterior ends of its facial

Fig. 177.
Triarthrus Beckii:
Eaton. Utica
Formation.

suture terminating at the corners of the head-shield. Our common species is *C. Blumenbachii* (= *C. senaria*) with rounded head-angles and pleuræ, strongly lobed glabella, and little apparent distinction between the end of the thorax and commencement of the pygidium.

---

* See a revised description of this species by Henry M. Ami, of the Geological Survey, in a paper on the Utica Slate Formation published in the Transactions of the Ottawa Fiel Naturalists' Club : 1882.

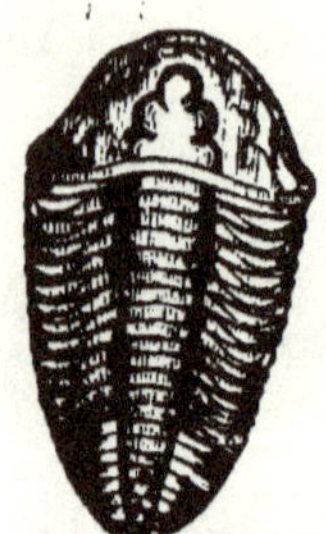

Fig. 178.
Calymene Blumen-bachii: Trenton, Hudson River, and Niagara Formations.

Examples are often found in a "rolled up" condition. The genus *Homalonotus* has also thirteen body-segments, and facial suture terminating at the posterior corners of the head-shield as in Calymene. But the glabella is unlobed and feebly pronounced, and the axis or central part of the thorax is scarcely defined —the longitudinal furrows, by which in most trilobites the axis is marked off from the pleuræ, being in this genus very indistinct. The *Homalonoti* are typically Devonian and Upper Silurian forms, but some occur in Lower Silurian strata. Figure 179 represents *H. delphinocephalus* (usually when perfect, about three inches in length) from the Niagara formation.

5. *Merostomata :*—The crustaceans of this Order comprise a single living genus, *Limulus,* and several extinct genera of related character in which the basal joints of some of the feet fulfil the office of jaws in the process of mastication. Hence the name of the group, from $\mu\eta\rho\grave{o}\varsigma$ a thigh and $\sigma\tau\acute{o}\mu\alpha$ mouth. The Merostomes may be arranged under three families: *Limulidæ, Belinuridæ* and *Eurypteridæ.*

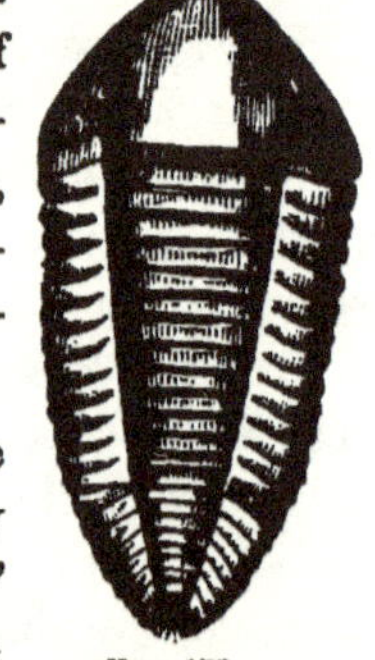

Fig. 179.
Homalonotus delphinocephalus: Green. Niagara Formation.

The *Limulidæ* are represented by the single genus, *Limulas,* species of which are popularly known as "Horseshoe crabs" or "King Crabs." They consist of three essential parts : a large crescent-shaped cephalic portion covered by a single shield, a thoracic portion covered by another shield, and an abdominal portion in the form of a long spine or "telson," the whole presenting a trilobitic aspect. In the young state, the telson is absent, and the trilobitic aspect is especially marked. The genus dates from the Triassic period, but fossil examples are unknown as regards Canadian strata.

The *Belinuridæ* much resemble the *Limulidæ* in general characters, but the thorax and abdomen in most forms are distinctly segmented. *Hemiaspis* from Silurian strata, and the Carboniferous types *Belinurus* and *Prestwichia,* are the principal genera. Examples have been discovered in the United States but none in Central Canada. The dorsal aspect in these forms is more or less distinctly tri-lobed.

The *Eurypteridae* are sometimes known as *Gigantostraca* from the large size presented by some examples. The head is covered by a single plate, and carries on the under side several pairs of masticating, and a single pair of swimming feet. The long thorax and abdomen (including the telson) consist of thirteen moveable segments.

The principal genera comprise *Eurypterus* and *Pterygotus*, in both of which (although only seen in well preserved examples) the surface of the shelly covering shews scale-like markings. In *Eurypterus*, the anterior feet are slender and antennæ-like ; whilst in *Pterygotus* the front pair are very long, and are terminated by claws or nippers. In *Eurypterus*, also, there is a long thorn-like telson, and in *Pterygotus* a comparatively short and flattened terminal-segment. Examples of *Eurypterus* in a more or less fragmentary condition are not uncommon in the Oriskany (Devonian) Formation of South-western Ontario. A restored example of *E. remipes* is represented in Figure 180. The

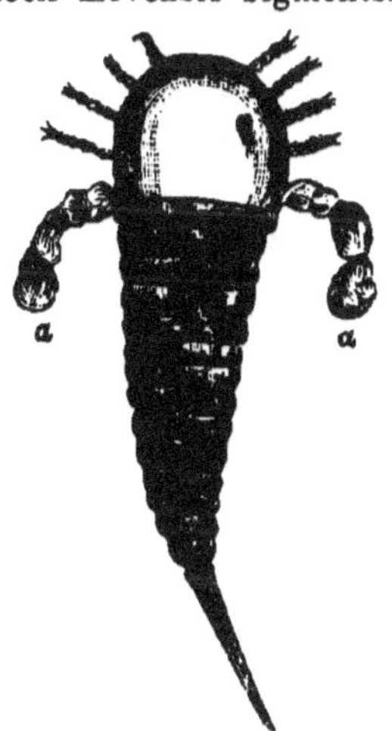

Fig. 180.

Eurypterus remipes: Devonian. *a, a,* are the swimming feet.

cephalic shield and the long thorn-like telson, are the parts generally found. The genus ranges from the Upper Silurian into Carboniferous strata. *Pterygotus* is an Upper Silurian and Devonian type.

6. *Phyllocarida :*—This sub-division is of comparatively recent adoption, principally from the researches of Packard and Claus. It includes the modern *Nebalia* (a small shrimp-like crustacean with stalked eyes) and a series of extinct forms of related character but much larger size. In these, there appears to be a blending of characters belonging to both the lower types (often grouped together under the common term of *Entomostraca*), and the higher forms (*Malacostraca*) of existing crustaceans. The Cambrian *Hymenocaris*, the Silurian and Carboniferous *Ceratiocaris*, and the Devonian *Dithyrocaris*, are the principal fossil genera. They present a large and in general laterally compressed head-shield (or cephalo-thoracic shelly covering) composed mostly of two pieces, and a jointed or ring-formed abdomen terminating in a telson of three 'or several spines. Ex-

amples have not been detected in our strata, although found in the Palæozoic rocks of the United States.

7. *Amphipoda :*—This Order (which may be made to include the *Laemodipoda* of some systems of classification) comprises a series of small shrimp-like crustaceans, represented by " water-fleas," " sand hoppers," " spectre-shrimps," &c., with sessile eyes and typically seven pairs of legs, the front pairs directed backwards, and the hind pairs forwards, whence the name of the group. Marine, fresh-water, and terrestrial types are known. Fossil forms (*Gampsonyx*, &c.) only date with certainty from the Carboniferous period, and the amphipods are unrepresented in our strata.

8. *Isopoda :*—The isopods are small crustaceans, very similar in character to the amphipods, and including, like the latter, marine, fresh-water, and terrestrial types ; but the form instead of being laterally compressed, is generally flattened from above downwards. The *Oniscus* or " wood-louse " is the typical terrestrial representative. Fossil forms of the group, date from the Devonian period, but are of comparatively little interest.

9. *Stomapoda :*—This division comprises small shrimp-like crustaceans with stalked eyes, and with branchiæ suspended from the abdomen or attached to the thoracic feet. *Squilla* and *Mysis* are typical examples. Species of the latter are familiarly known as " opossum shrimps." Fossil forms date from the Jurassic period, but are rare and of no special interest.

10. *Decapoda :*—In the Decapods—the type-forms of the Crustacea—the true feet are always in five pairs ; the eyes are stalked ; the head and thorax are united into a cephalo-thorax ; and the branchiæ are in special cavities at the sides of the latter. Three leading groups are recognized :—(1) *Macrura* or long-tailed decapods, in which the abdomen is well-devoleped, forming a powerful swimming organ. Typical representatives comprise lobsters, cray-fish, and true shrimps. Fossil forms date from the Carboniferous period. (2) *Anomura*, or defenceless-tailed decapods, in which the abdomen is unprotected by a shelly covering, and does not serve as a natatory organ. The *Paguridae* or " Hermit-crabs " are examples. These insert the abdomen into the vacant shells of whelks or other gasteropods, or keep it buried in the sand of the sea shores on which they live. Fossil forms date from the Jurassic period. (3) *Brachyura*

or short-tailed decapods, in which the abdomen is rudimentary and bent under the cephalo-thorax. They are represented by the various crabs, properly so-called. Fossil forms have been cited from Palæozoic and Jurassic rocks, but the earliest undoubted examples are Cretaceous.

II.

ARACHNIDA.

This Class is represented typically by Spiders, Harvest-Spiders, Chelifers and Scorpions, in which the respiration is aerial (tracheal or pulmo-branchial), the body and thorax united, and the legs in four pairs. In certain lower forms there are no visible organs of respiration, and the legs are variable in number.

In Spiders, the abdomen is short and unsegmented. Examples date from the Carboniferous period.

Scorpions have a long abdomen divided into segments and scarcely distinguishable from the thorax. They possess also a pair of large antennæ or palpi furnished at their extremities with nipper-claws, as in the extinct crustacean genus *Pterygotus*. From this and other characters, the latter is referred to the Scorpions by some palæontologists. The earliest known Scorpions appear in Carboniferous strata, in which, also, examples of chelifers are said to have been discovered; but no fossil arachnids belong to our rocks.

III.

MYRIAPODA.

The myriapods are represented essentially by centipedes and millipedes, in which the form is elongated and worm-like, the abdomen being divided into numerous segments each of which bears a pair or a couple of pairs of short legs. The head is distinct from the thorax, and the respiration is trachæal. Fossil examples first appear in Carboniferous strata, but are of rare occurrence. None occur in our rocks, but a species of *Xylobius* (allied to *Iulus*) was recognized by Sir J. W. Dawson, many years ago, in the hollow stem of a sigillaria from the middle or productive coal formation of Nova Scotia.

IV.

INSECTA.

This large and important division of the Arthropoda is of comparatively little geological interest.   In all insects, the head, thorax, and
abdomen are distinctly separated, the respiration is tracheal, and the
legs are always in three pairs.   Hence the name of *Hexapoda* by
which the entire division is now often known.

Some insects assume their perfect form at birth.   These belong to
the *Thysanura* (or group *Ametabola* : *e. g. Podura, Lepisma,* &c.), and
are without fossil representatives except in amber.   The recognized
forms, thus preserved, belong with one exception (*Glessaria*) to
existing genera.

Other insects undergo an incomplete metamorphosis after birth.
They belong to the section *Hemimetabola,* and comprise the *Hemip-
tera* or *Rhyncota* (insects with rostral mouth : *e. g. Aphides, Coccina,
Cicada, Nepa,* &c.), and the *Orthoptera,* including with the latter the
so-called *Pseudo-Neuroptera.*   *Orthoptera* comprise cockroaches, grass-
hoppers, locusts, &c., and the *Pseudo-Neuroptera,* the so-called " dra-
gon-flies " and related forms. .Fossil examples occur in Devonian
strata (*Platyphemera antiqua*) ; in the Carboniferous formation (*Blat-
tina, Termes, Euphemerites, Palingia*), and more abundantly in
Mesozoic and Cainozoic deposits (*Ephemera, Libellula,* &c.).

Other insects, again, undergo complete metamorphoses—passing
through the three conditions of larva, pupa and imago.   These form
the group or section *Holomltabola* and cromprise the *Neuroptera*
as now restricted (*e. g.* " scorpion flies," " caddis worms," " ant-lions,"
&c.); the *Lepidoptera* (butterflies and moths) ; *Diptera* (house-flies,
&c.) ; *Hymenoptera* (bees, wasps, ants) ; *Strepsiptera* (bee-parasites) ;
and *Coleoptera* (beetles).   The parasitic *Strepsiptera* have no known
fossil representatives.   Doubtful examples of *Coleoptera* have been
cited from Carboniferous strata, but the Order only dates with certainty from the Triassic period.   Fossil *Lepidoptera, Diptera* and
Hymenoptera are Mesozoic and Cainozoic only.   No examples of
fossil insects occur in the strata of Ontario and Quebec.   This brief
summary, therefore, is all that need be given in the present work.

SUB-KINGDOM VII.

MOLLUSCA.

The animals of this series comprise a large number of soft-bodied forms in which the blood is essentially colorless and the symmetry bilateral. In most, an external calcareous shell is present; but in some cases the shell is internal, and in others it is wanting or rudimentary. Most mollusks inhabit the sea, but a small number are fresh-water types, and a still smaller number are terrestrial. The marine oyster, whelk, nautilus and cuttle-fish, the fresh-water limnea, and the land snails, are typical examples of existing forms.

Mollusca are usually classed under two series: *Molluscoidea* and *Mollusca vera*.

A. MOLLUSCOIDEA.

In the representatives of this series there are no distinct branchiæ as in the Mollusca proper, and the heart is either absent or more or less rudimentary. Although connected by these and other organization characters, the two classes into which the Molluscoidea are subdivided, differ greatly in external configuration. These classes comprise: 1, *Bryozoa* or *Polyzoa*, and 2, *Brachiopoda*.

I.

BRYOZOA OR POLYZOA.

The bryozoons—so named from their general moss-like aspect—are minute animals of aquatic habitat. They form colonies of individuals which secrete in common a horny, calcareous or gelatinous cellular " exoskeleton " or support, and thus closely resemble Sertularians and some other compound Cœlenterates. They possess, however, a distinct oral and anal orifice, with other characters indicating a higher organization. The secreted cell-structure is either leaf-like, plumose, tubular, dendritic, heliciform, circular, or irregular, in form; and it frequently encrusts other marine bodies. The actual bryozoon may be likened to a minute, digestive sack, with oral and anal orifices, and with a wreath or circle of delicate tentacles at the oral opening. The bryozoa are exceedingly rich in genera and species; and of late years many fossil forms have been removed from the corals and referred to them, although necessarily on very uncertain evidence.

Modern bryozoa are generally divided into *Entoprocta* and *Ecto-procta*, according as to whether the anal orifice is within or without the tentacular area. The *Entoprocta* have no fossil representatives. The *Ectoprocta* are subdivided into two leading sections: 1, *Phylac-tolœmata*, in which the mouth is furnished with a skin-like "epistome" or moveable cover, and the "lophophore" or tentacle-bearing organ is of horse-shoe form, as seen in the fresh-water genus *Plumatella*, &c.; and 2, *Gymnolœmata*, in which there is no epistome, and the lophophore is circular in form. The first section or Order is supposed to have no fossil representatives. The *Gymnolœmata* are divided into four sub-Orders, two of which, *Cyclostomata* and *Cheilostomata*, include fossil examples. Living forms of these sub-Orders are exclusively marine types. In the *Cyclostomata*, the cells are tubular, and the cell-opening is terminal and as large as the diameter of the cell. The cyclostomes date from the lower Silurian period. The delicate lace-like genera, *Fenestella* and *Ptilo-dictya*, are common Canadian examples; and certain other forms *(Monticulipora, &c,)* gen-erally referred to the HYDRO-CORALLA, are also placed under this Order by some palœontolo-zists. In the *Cheilostoma* the cells are oval or egg-shaped, and the cell-opening is narrower than the general width of the cell. It is also furnished in many forms with an articulated lid

FIG. 181.
Fenstella Etegans.
Upper Silurian. Hall.

or operculum. This sub-order is only known with certainty to date from the Jurassic period: we have consequently no examples in Central Canada.

II.

BRACHIOPODA.

The brachiopods are separately-living, marine, molluscoid animals, provided with a bivalve shell. This latter character, and their gen-eral aspect, causes them on superficial observation to seem very nearly allied to the lamellibranchiates or ordinary bivalve mollusca, but, rightly considered, they are far more closely connected with the bryozoons. Their chief characteristic is the possession of a pair of long, ciliated " arms," coiled up within the shell, and supported in some forms by calcified structures. These so-called " arms " undoubtedly represei t to some extent the tentacular organs of the bryozoons. Living brachiopods are for the greater part attached to the sea-floor by a flexible pedicel, passing through a foramen in the shell or between the bivalves ; but in some cases the attachment is made by the shell itself, without a pedicel.

In the seas of the Palæozoic Age, brachipods abounded and far exceeded in numbers the lamellibranchiate mollusca. During the Mesozoic periods, they were still numerous, but they diminished greatly in the succeeding Cainozoic Age, and at present about 100 living species only are known. Whilst most of the Palæozoic genera are extinct, some few—as *lingula, rhynconella, terebratula,* &c.—still offer living representatives.

Fio. 182.
Diagram of a typical brachiopod in normal position.

The two valves of the brachiopod are of unequal size, but álways equilateral. As regards the latter character, therefore, a straight line drawn vertically through the middle of each valve will divide the shell into two symmetrically equal parts. This serves to distinguish at a glance a brachiopod shell from the shells of ordinary bivalve mollusca, or, at least, frum the great majority of these, as some few, the Pectens for example, have nearly equilateral shells. The larger valve in brachiopods is almost always the ventral valve ; its "beak " or "umbo " is sometimes very prominent, sometimes depressed or inconspicuous ; and in some genera it is perforated for the passage of the pedicel. The smaller or dorsal valve is in some cases articulated by short projecting processes or " teeth " to the larger valve, whilst in other cases the articulation is very indistinct ; and in one group of forms, including the *lingulidæ*, &c., the valves are without articulation. The two valves along the so-called " hinge-line " (the upper part of the shell as conventionally Jrawn in figures) are either in close contact or connected by an intervening " area." The latter is usually striated. In its centre, or under the beak or umbo of the larger shell, there is sometimes a smaller triangular area or "deltidium." This is sometimes perforated for the passage of a pedicel, and is sometimes covered by a "pseudo-deltidium.' The hinge-line in some genera is straight or horizontal (*orthis, stro phomena, spirifer, &c.*), and in others curved (*rhynconella, terebratula &c.*). The outer surface of the shell may be smooth, or (and mor commonly) it may be marked by fine or coarse radiating lines. I many genera the centre of the ventral valve (more rarely the dorsal shows a depression or "sinus," and there is a corresponding elevatio or "mesial fold" on the other valve. The substance of the shel may be "punctate" or "impunctate"—*i.e.* traversed, or not, b minute pores; but this character often becomes inconspicuous throug fossilization.

The arms or tentacles in the interior of the shell are alway attached to the smaller valve. In some genera (*orthis, &c.*), thes are without supporting calcareous structures. In other genera *(rhyn conella, &c.*) they are attached to two very short processes or " crura.' In others (*spirifer, &c.*), they are wound around two calcareou spires. In others, again, *(terabratela, waldheimia, &c.)* they ar supported by a short or long calcareous loop. Unfortunately, thes internal structures in fossil examples are often not directly observ able. As a rule, the shell must be carefully ground down with fin emery on an iron plate, or artificial sections must be prepared fo their recognition. In like manner, the muscular impressions o the inner surface of the valves are only occasionally observabl These impressions differ in shape and mode of arrangement, but ar usually in pairs; four to six in number.

The brachiopods are classed under two principal divisions. In on (*Articulata* of Huxley, *Arthropomata* of Owen, *Testicardines* c Brown, *Clistenterata* of King, *Apygia* of Zittel) the valves have more or less distinct hinge or interlocking teeth, and (as seen i the living forms) the intestine is without any anal opening. In th other (*Inarticulata* of Huxley, *Lyopomata* of Owen, *Ecardines* c Brown, *Tretenterata* of King, *Pleuropygia* of Zittel), there is n proper hinge, and the intestine terminates in an anal orifice at th side of the body.

Division 1—*Articulata* or *Clistenterata* :—Brachiopods with hinge shell. In living forms, the intestine closed :

This division includes (principally) the following families :—1

Spiriferidæ ; 2, *Atrypidæ ;* 3, *Terebratulidæ ;* 4, *Rhynconellidæ ;* 5, *Strophomenidæ ;* 6, *Produetidæ.*

Family 1. *Spiriferidæ* : Shell with both valves convex. Smaller valve with internal calcareous spires, tapering laterally.

The principal genera of Canadian occurrence comprise :—*Spirifer,* with straight hinge-line, and beaks comparatively near together (Silurian to Lias) ; figures 183, 184, 185. *Cyrtina,* with straight hinge-line and large area, the beaks consequently far apart (Silurian to Trias) ; fig. 186. And *Spirigera* (or *Athyris*), with curved hinge-line ; Silurian to Trias) ; fig. 187.

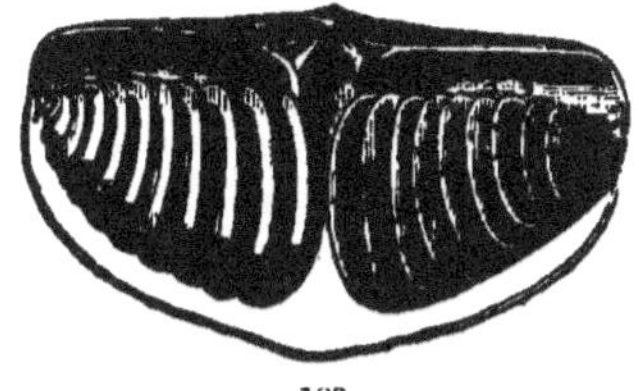

183.

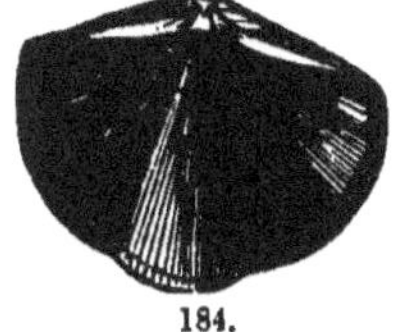

184.

184 *bis.*

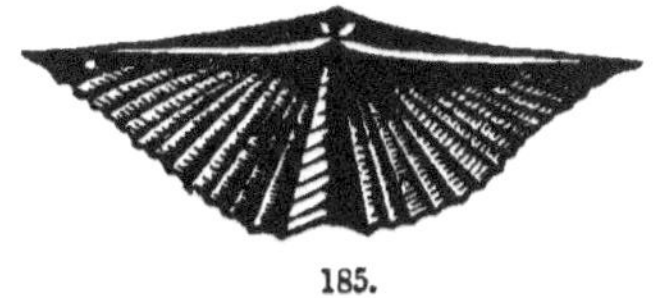

185.

Fig. 183. Interior of dorsal valve of a spirifer shewing position of spiral arm-supports.

Fig. 184. *S. radiatus* : Niagara Formation.

Fig. 184 *bis.* *S. gregarius* : Corniferous (Devonian) Formation.

Fig. 185. *S. mucronatus* : Devonian.

186.

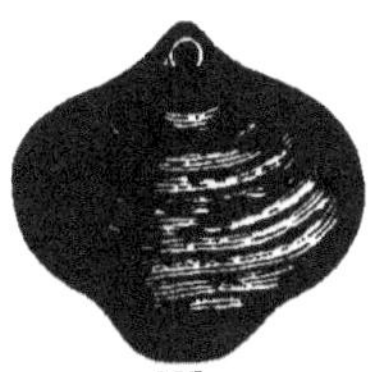

187.

Fig. 186. *Cyrtina.* Devonian.

Fig. 187. *Spirigera (Athyris) concentrica* : Devonian.

Note :—The *Spiriferidæ,* in accordance with their more prominent characters, may be grouped in two sections and four sub-sections, as follows :

§ 1. *Librati* : Hinge-line straight. A distinct area :

    A. *Vicinales :* Beaks near together ; Type-forms : *Spirifer, Spiriferina.*

    B. *Disjuncti* : Beaks widely apart ; Type-forms : *Cyrtia, Cyrtina.*

§ 2. *Arcuati* : Hinge-line curved. True area, absent or indistinct.

    A. *Vicinales* : Beaks near together ; Type-forms : *Spirigera (= Athyris) Retzia.*

    B. *Disjuncti* : Beaks widely apart ; Type-form : *Uncites.*

Family 2. *Atrypidæ* : Shell with internal spires attached to th[e] hinge-line of the dorsal valve. No area. Shell bi-convex.

This family differs from that of the Spiriferidæ by the direction of the spiral supports. The only typical genus *Atrypa* (= Spirigerina) much resembles the spirifers with curved hinge-line. Our most common species is *A. reticularis,* fig. 188, an Upper Silurian and Devonian type of wide geographical range.

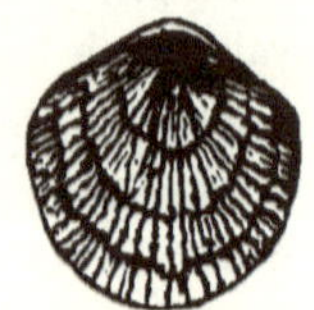

FIG. 188.
Atrypa reticulari[s]
Upper Silurian.
Devonian.

Family 3. *Terebratulidæ* : Shell with both valves convex. Internal arm-supports in the form of a short or long calcareous loop.

This family has several living representatives, notably, *Terebratula* dating from the Devonian period, and Waldheimia, from the Triassic In the former, the internal loop is short, and in the latter compara tively long and deeply recurved. The genus *Centronella,* an extinct Devonian type is the principal if not the only Canadian representa tive of the family. Fig. 189 represents *C. glans-fagea* of Hall.

Family 4. *Rhynconellidæ* : Shell with both valves convex; usually strongly ribbed. Internal arm-supports in the form of very short, more or less inconspicuous points.

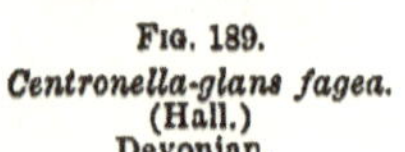

FIG. 189.
Centronella-glans fagea.
(Hall.)
Devonian.

This family includes the typical genus *Rhynconella,* dating from the Silurian period and still surviving, and the extinct genera *Pentamerus* (Silurian to Carb.), *Stricklandia,* and *Camerella* : the two latter, Silurian only. Several species of *Rhynconella* and a common *Pentamerus* are represented in figures 190 to 193.

190.   191.   192

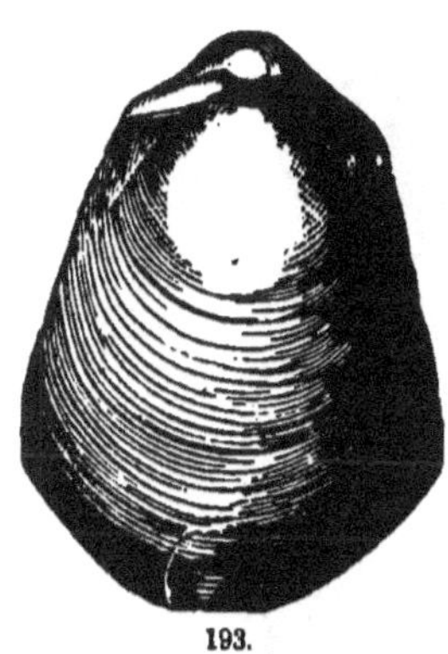

193.

Fig. 190. *Rhynconella plena*, Chazy Formation.

Fig. 191. *R. increbescens*, Trenton Formation.

Fig. 192. *R. (Camerella) varians*, Chazy Formation.

Fig. 193. *Pentamerus oblongus*, Clinton and Niagara Formations.

Family 5. *Strophomenidœ*: — Shell, mostly, with one valve convex and the other flat or concave, but in some forms biconvex. No internal arm-supports. Hinge-line straight.

This family is entirely extinct and almost exclusively palæozoic. Its more characteristic genera comprize: (1) *Strophomena*, with concavo-convex or plano-convex shell and greatest width at hinge-line: the four most distinct muscular impressions in a single row: Silurian to Carboniferous period: (2) *Leptœna*, much like strophomena but muscular impressions very long: Silurian to Liassic period. (3) *Orthis*, shell flatly bi-convex or plano-convex, with greatest width below the hinge-line ; the four most distinct muscular impressions not in a single horizontal row: Cambrian to Carboniferous. (4) *Platystrophia*, with strongly bi-convex shell : one valve depressed in the centre and the other with mesial fold, thus greatly resembling a spirifer shell. Surface of shell strongly ribbed : Silurian to Carboniferous. Representatives of these genera are shewn in the following figures.

194.

196.

197.

Fig. 194. *Strophomena alternata* : Trenton and Hudson River Formations.

Fig. 196. *Leptœna sericea*: Trenton and Hudson River Formations.

Fig. 197. *Orthis testudinaria* : Trenton and Hudson River Forms. *O. elegantula*, an Upper Silurian form, much resemble this species.

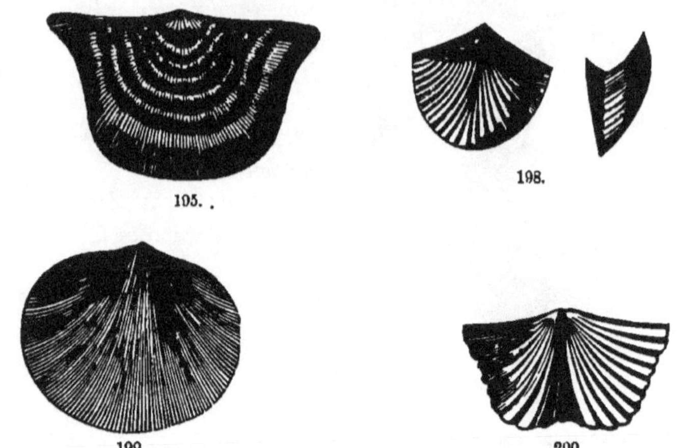

Fig. 195. *S. rhomboidalis* (= S. *depressa*) : Upper Silurian and Devonian.

Fig. 198. *O. tricenaria* : Trenton Formation.

Fig. 199. *O. vanuxemi* : Devonian.

Fig. 200. *Platystrophia lynx*, the *Delthyris lynx*, afterwards *Spirifer lynx*, and *Orthis lynx*, of old publications : Trenton formation. The plications on the mesial fold and sinus serve as a distinctive character.

Family 6. *Productidæ* : Shell concavo-convex, free or attached by surface of larger valve. Area absent or very narrow. Hinge-teeth often rudimentary. No internal arm-supports.

This family is entirely extinct. Its chief representatives comprise : *Productus*, with very convex ventral valve and spiny surface, a characteristic Carboniferous genus, but occurring also in Devonian and Pernian strata ; and *Chonetes*, with flat, transversely elongated shell, bearing spines along] the hinge-margin. The latter genus ranges from Silurian into Carboniferous strata, but the family is unrepresented in the rocks of Central Canada.

Division II. *Inarticulata* or *Tretenterata* :—Brachiopods with hingeless shell. In living forms, the intestine terminates in an anal opening on the right side of the body. There are no calcified arm-supports.

This Division comprises the following families : 1, *Trimerellidæ* ; 2, *Craniadæ* ; 3, *Discinidæ* ; 4, *Lingulidæ*.

Family 1. *Trimerellidæ* :—Shell, thick, calcareous, without fora-men. Rudimentary hinge-teeth sometimes present. A broad, bipartite plate, with supporting septum, in the interior of each valve.

This family is entirely of Silurian age. It is represented chiefly by the genera, *Monomerella*, with flat, central plate ; and *Linobolus* and *Trimerella*, with the edges of the central plate curved outwards. In all, the ventral valve has a large area and deltidium, and the smaller valve is nearly circular. Both valves are smoo.h or with-out external ribs, and moderately convex in form. Several species of these genera occur in our Silurian strata, more especially in the Upper Silurian Guelph Formation. Examples are commonly in the form of casts. These shew two or three deep sinusses at or near the beak of the shell.

Fig. 201.
Cast of *Trimerella
acuminata*
(Billings.)
Guelph Formation

Family 2. *Craniadæ* :—Shell thick, calcareous, finely-perforated, more or less circular or square-shaped, without foramen. Beak of both valves, sub-central.

This family dates from the Silurian period, but its typical genus *Crania* is especially characteristic of Cretaceous strata. The internal muscular impressions and presence of a nose-shaped septum produce the appearance of a face or skull—whence the name of the genus. Our strata are without recognized representatives of the family.

Family 3. *Discinidæ* :—Shell composed of horny and calcareous layers, more or less circular in form, with central or sub-central beak, and foramen (usually in the form of a narrow slit) in the larger valve.

This family is represented chiefly by *Discina*, *Orbiculoidea*, and *Trematis*. The horny part of the shell in these genera (as in the *lingulidæ*) is composed essentially of calcium phosphate. Examples, mostly of small size occur in our Silurian strata. The genus, *Discina* has living representatives. The other genera are extinct.

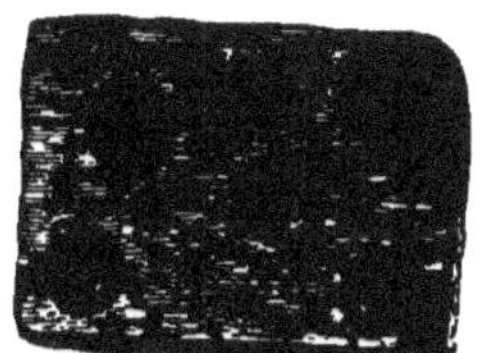

Fig. 202.
*Discina (or Orbiculoidea)
Circe ?*
Trenton Formation.

Family 4. *Lingulidæ* : Shell composed of horny and calcareous layers, the horny por-tion consisting of calcium phosphate. Com-monly oblong or oval in shape, more rarely

18

circular.   Usually smooth, or marked simply by concentric lines of growth.

This family, of the highest antiquity, still offers living representatives.   The principal genera comprise *Lingula*, *Lingulella*, and *Obolus*.   The shell in *lingula* is without a foramen, the pedicel, as seen in living species, passing out between the valves.   In *lingulella* and in *obolus*, there is a narrow marginal foramen in one valve. These genera date from the Cambrian period.   In the strata of Ontario and Quebec, examples of lingulæ are of frequent occurrence. Usually the shell is dark and lustrous; but in examples from the Medina formation of Hamilton, Ontario, it preserves its normal sub-pearly and iridescent aspect.   Some of our more characteristic species are shewn in the annexed figures.

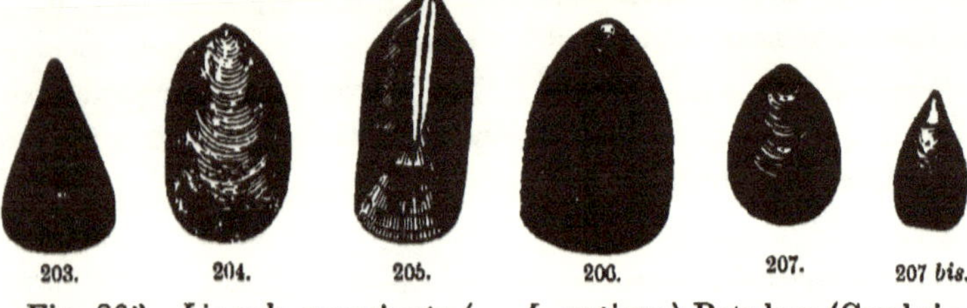

203.   204.   205.   206.   207.   207 bis.

Fig. 203.   *Lingula acuminata* (= *L. antiqua*) Potsdam (Cambrian) formation.

Fig. 204.   *L. Quebecensis* : Levis formation.

Fig. 205.   *L. Lyelli* : Chazy formation.

Fig. 206.   *L. quadrata* : Trenton formation.

Fig. 207.   *L. obtusa* : Utica and Hudson River formations.

Fig. 207 *bi*.   *L. oblonga* (Conrad) : Medina and Clinton formations.*

---

* This small species of lingula is abundant in some of the red, Medina strata of Hamilton, Ontario.   In many examples the substance of the shell is preserved.

B. MOLLUSCA VERA.

The mollusca proper, as distinguished from the molluscoidea, possess distinct branchiæ, or, in default of these, a pulmonary sac; and they have also, typically, a systematic heart. In most, the body is enclosed in a calcareous shell, as in the oyster, whelk, &c.; but in some forms the shell is internal, and in others it is absent or rudimentary. They fall primarily under two sections: *Aglossata*, in which the mollusk has no "lingual ribbon" or odontophore; and *Glossophora* or *Odontophora*, in which the animal is provided with a long, narrow "tongue" or "radula" thickly set with rows of minute, rasping teeth, of variable form and mode of arrangement in different genera. The *Aglossata* comprise a single class, *Lamellibranchiata*. The *Glossophora* are sub-divided into the following classes: *Scapho-poda; Pteropoda; Gasteropoda; Heteropoda;* and *Cephalopoda.*

CLASS 1. LAMMELLIBRANCHIATA. (= CONCHIFERA, PELECYPODA, &c.)

The lamellibranchs are entirely aquatic, and mostly marine types. They secrete an external, calcareous shell composed of two separate valves, united, in most forms, by interlocking points or teeth, and by a narrow elastic ligament. All are headless. They are commonly classed under two leading sections: *Asiphonida* and *Siphonida*. In the latter the animal possesses a pair of long or short siphonal tubes, capable, more or less, of extension beyond the shell. One of these tubes serves for the ingress of water, and the other for the expulsion of this after the air has been extracted from it, as well as for the removal of effete matters generally. The Asiphonida are without these siphonal tubes, although there is an approach towards them in one of their families, the *Unionidæ*. These sections are again sub-divided into two groups or Orders, as in the following arrangement:

Asphonida $\begin{cases} 1.\ \text{Pleuroconcha.} \\ 2.\ \text{Orthoconcha.} \end{cases}$

Siphonida $\begin{cases} 3.\ \text{Integripalliata.} \\ 4.\ \text{Sinupalliata.} \end{cases}$

1. *Pleuroconcha*: In this subdivision, the valves of the shell are mostly of unequal size; and, in normal position, one valve is the *under*, and the other the *upper* valve. In the interior of each valve there is one large muscular impression, or one large and one

small impression—whence the terms *Monomyaria* and *Heteromyaria*, often applied to the members of this group. Some of the more common genera comprise: *Ostræa* (dating from the Carboniferous period); *Spondylus* (dating from the Triassic period); *Pecten* (dating from the Devonian period); *Lima* (dating from the Carboniferous period); *Avicula* (dating from the Silurian period); *Mytilus* (dating from the Triassic period); and *Pinna* (dating from the Devonian period). Among the most frequent forms of Palæozoic age, occurring in our strata, *Ambonychia radiata* (Family *Aviculidæ*), and *Modiolopsis modiolaris* (Family *Mytilidæ*) may be especially cited.

Fig. 208.
Ambonychia radiata.
Trenton and Hudson
River Formations.

Fig. 209.
Modiolopsis modiolaris.
Hudson River Formation.

2. *Orthoconcha* :—This group, so named by the French palæontologist Alcide d'Orbigny, is the *Homomyaria* of some authors. The valves of the shell in the normal position of the animal are *right* and *left*,* and the muscular impressions are two in number and of practically equal size. In these respects, therefore, the *Orthoconcha* are more closely allied to the *Siphonida* than to the *Pleuroconcha*, although they are without siphonal tubes, properly so-called. The marine forms *Arca*, *Nucula* and *Leda* (all dating from the Silurian period), *Trigonia* (dating from the Liassic period), and the freshwater *Unio*, are some of the more representative genera. The fossil genus *Megalomus* or *Cyrtodonta* probably belongs to the *Arcaceæ*. One species, *M. Canadensis*, found mostly in the form of casts of comparatively large size, is very characteristic of the (Upper Silurian) Guelph formation. Other species (*Cyrtodonta* of Billings) are of common occurrence in the Hudson River formation. A species of *Leda* (*L. truncata*) is equally characteristic of the Post-Glacial

* The conventional position adopted in drawings does not represent the normal position of the living lamellibranchiates.

deposits—known, after Dawson, as the Leda clay formation—of the Province of Quebec.

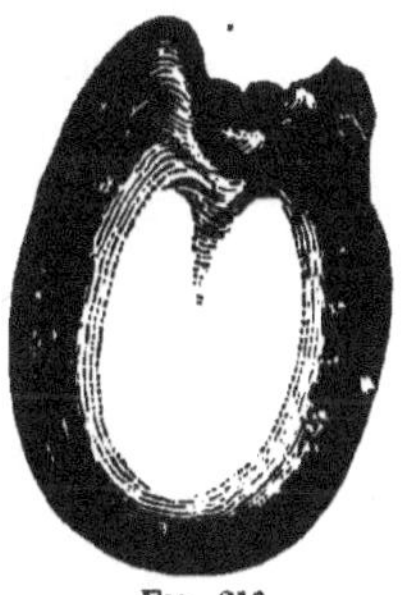

FIG. 210.
Cast of *Megalomus (Cyrtodonta) Canadensis:*
(Hall). Guelph Formation.

FIG. 211.
Leda truncata.
Post Glacial.

3. *Integripalliata* :—The shell and normal position of the animal in this group are the same as in the *Orthoconcha*, but the animal possesses a pair of short siphonal tubes. The muscular impressions are connected by a narrow linear depression, marking the edge of the mantel, and this is continuous or without any bend or sinus— whence the term " Integripalleata." *Tridacna* (dating from the Miocene period), *Cardium* and *Lucina* (dating from the Silurian period), *Cyclas* and *Cyrena* (dating from the Jurassic period), and *Crassatella* (from the Cretaceous), are some of the more typical genera. The annexed figures represent two of our Palæozoic forms.

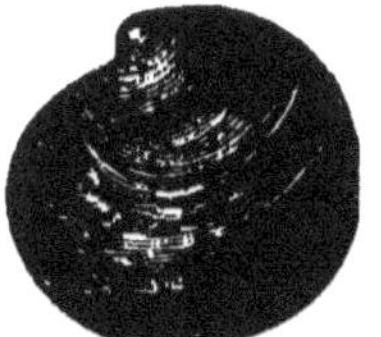

FIG. 212.
Lucina proavia : (Hall).
Devonian.

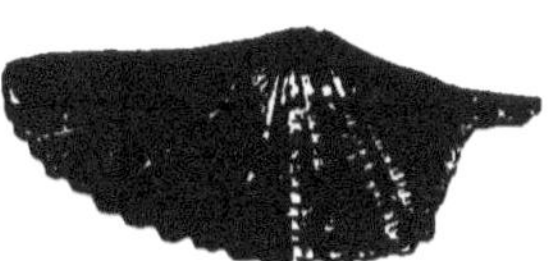

FIG. 213.
Conocardium trigonale : (Conrad).
Devonian.

4. *Sinupalliata* :—In this group the animal possesses a pair of long, more or less extensible, siphonal tubes ; and the muscular impressions are connected by a parallel groove with a fold or sinus where the tubes occur. Representatives of the group are hardly known in Palæozoic strata, but are abundant in Mesozoic and Caino- zoic formations and in existing seas. Some of the more characteristic

genera comprise : *Venus, Cytherea, Tellina,* and *Mactra* (all dating from the Jurassic period), *Donax* and *Solen* (both dating from the Eocene) and *Mya, Saxicava,* &c. (dating from the Miocene period). Species of *Mya, Tellina* and *Saxicava* are very characteristic of the Saxicava-Sand formation (Post-Glacial) of the Province of Quebec.

FIG. 214.

Mya truncata.
Post-Glacial.

FIG. 215.

Tellina grœnlandica.
Post-Glacial.

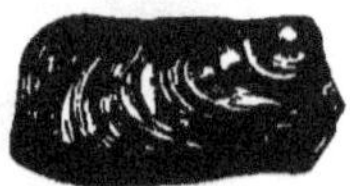

FIG. 216.

Saxicava rugosa.
Post-Glacial.

CLASS II. SCAPHOPODA.

This class comprisés but one family, that of the *Dentalidæ,* in which the head is rudimentary (but furnished with a lingual ribbon or radula) and the foot in the form of a hollow, conical organ, adapted for boring—whence the name of the class, from ςχαφος, a spade. The typical genus Dentalium, in which the shell is in the form of a hollow cone, straight or curved, and open at both ends, dates from the Silurian or Devonian periods, but is without representatives in our strata.

CLASS III. PTEROPODA.

The Pteropods form a small series of pelagic, free-swimming mollusks, frequenters of the open ocean. They possess a rudimentary head, with a fin-like, natatory organ on each side. Some are naked types, others secrete a light shell of variable form. They are commonly classed in two orders : 1. *Thecosomata,* with, typically, a thin external shell ; and 2. *Gymnosomata,* including the naked forms, with more or less distinct head. The modern genera *Hyalaea* and *Limacina* (both dating from the Miocene period) are types of the first Order ; and the recent genus *Clio* is the principal representative of the second Order.

Two essentially palæozoic forms, *Tentaculites* (Silurian, Devonian), and *Conularia* (Silurian to Lower Jurassic), are commonly regarded as thecosomatous pteropods, but their true position is still uncertain.

Fig. 217.
Tentaculites Ornatus
Upper Silurian.

Fig. 218.
Conularia Trentonensis:
Lower Silurian.

The shell, in *Tentaculites*, is tubular, of narrow diameter, tapering to a rounded point, and transversely ringed. In *Conularia*, it is more or less conical and four-angled, but usually flattened by com-pression. Its surface is marked by a few longitudinal furrows, and by numerous fine lines, resembling rows of punctures, arrange transversely in zigzag form as shewn in figure 216. Some examples are several inches in length.

CLASS IV. GASTEROPODA.

In the mollusks of this class there is a more or less distinct head and all are furnished with a radula or lingual ribbon, thickly set with minute teeth. The number, form and arrangement of the latter constitute valuable classification-characters as regards living species, but are without, or of only indirect, palæontological application, Typical gasteropods possess also a fleshy expansion or so-called foot on the under side of the body, by which locomotion is effected : hence the name of the class. The greater number secrete an external uni-valve shell, mostly spiral in form, but some few, as the slugs, are naked, or possess merely a rudimentary shell ; and in the *chitons*, an exceptional group, the shell is composed of several pieces. In many of the spiral forms, the mouth or aperture of the shell can be closed when the animal has retired within it, by a shelly or horny plate, known as an "operculum." Some gasteropods, as the common snails, are terrestrial; others, as the *limnea*, and *planorbis*, species of which are of common occurrence in our lakes and streams, inhabit fresh-water; but the greater number are marine types.

The class may be subdivided into four sub-classes : *Pulmonata, Opisthobranchiata, Placophora,* and *Prosobranchiala*—the great majority of species belonging to the latter.

Pulmonata :—The animals of this sub-class possess, in place of branchiæ, a simple lung-sac by which they breathe air directly from the atmosphere. They comprise land and fresh-water types, and are chiefly represented by the genera *Helix, Planorbis, Limnea,* &c., fossilized examples of which, belonging to recent species, occur in some of our Post-Glacial deposits ; but the sub-class dates from at least the Carboniferous period. All the Pulmonata are hemaphrodites.

Opisthobranchiata :—This sub-class comprises marine types in which the branchiæ are situated behind the heart. Like the *Pulmonata* (and the preceding class of *Pteropods*) all are hemaphrodites. They are commonly divided into *Nudibranchiata* (*e. g. Eolis, Doris*) in which there is no shell ; and *Tectibranchiata* (*e. g. Bulla, Actæonella,* &c.) mostly with external shell. The latter date from the Triassic period, but are of no special interest.

Placophora :—in this sub-class there is only one family, the *Chitonidæ*, dating from the Silurian period but without known representatives in Canadian strata. The *Chitons* are abnormal gasteropods of flat, oval or elongated form, in which the upper surface of the body is protected by a series of eight more or less tile-like plates. They are marine, sexually-distinct types.

Prosobranchiata : This sub-class includes all the typical branchiferous gasterpods in which the branchiæ are placed in front of the heart. The species are marine forms with distinct sexes. The shell is commonly more or less spiral ; but whilst in some forms the spire is very long, in others it is quite short or almost obliterated ; and in some few genera the shell is patelliform. The sub-class admits of a natural division into two leading groups or sections : *Holostomata* and *Siphonostomata*.

Holostomata : In this section, the aperture or so-called mouth of the shell presents a continuous or entire margin : hence holostomatous types are often called " round-mouthed gasteropods," but the same character is presented by the *Helicidæ* (snails) and other *Pulmonata*. It indicates, as a rule, vegetable-feeding types. Examples of *Holostomata* are of common occurrence in all fossiliferous rocks, and they abound on existing sea-coasts ; but whilst they are associated at present and in Cainozoic and Mesozoic strata, with siphonostomatous types, in the rocks of the various Palæozoic periods they appear to form the sole representatives of the prosobranchiate gasteropods. Carnivorous forms of mollusca were then mainly represented by predatory, coast-frequenting cephalopods, which afterwards became extinct. The holostomous gasteropods of our Silurian and Devonian strata belong chiefly to the following genera: *Pleurotomaria*: Shell with few whorls ; the lower whorl large, with slit in outer lip. The slit becomes closed with the growth of the shell, but its position is indicated by a more or less distinct band (fig. 219). *Murchisonia* : Shell spiral or sub-spiral, with many whorls ; a slit or band present, as in *pleurotomaria* (fig. 220). *Bellerophon* (formerly referred to the *Heteropoda*) : Shell boat-shaped, somewhat globular, equilaterally curved ; a slit in outer lip often present (fig. 221). *Cyrtolites*, like *Bellerophon*, but mostly compressed ; with distinctly-seen whorls and keeled back. *Euomphalus* (including *Ophileta**) : Shell discoidal, more or less regularly enrolled ; with, typically, sunk spire, large umbilicus, and sub-angular whorls and aperture. *Maclurea* (formerly referred to the *Heteropoda*): Shell with flat spire and rapidly enlarging outer whorl, furnished with a thick operculum, shewing two projections on its inner side (fig. 222). *Platyceras* (= *Acroculia*) : Shell bonnet-shaped or patelliform, with very large aperture (fig. 223). *Platyostoma* : Shell turbinate, few-whorled, the lower whorl very large, often projected laterally, with aperture wider than high. *Holopea* : Shell turbinate, few-whorled ; aperture round, or higher than wide, with continuous margin (fig. 224). *Cyclonema* : Shell turbinate, few-whorled, much like *Holopea*, but with somewhat interrupted margin. *Subulites* : Shell elongated, cylindrical, with comparatively narrow aperture (fig 225).

---

* *Ophileta* is properly a *Euomphalus* with slightly raised spire ; or a low-spired *Straparollus*, if *Straparollus* and *Euomphalus* be considered distinct.

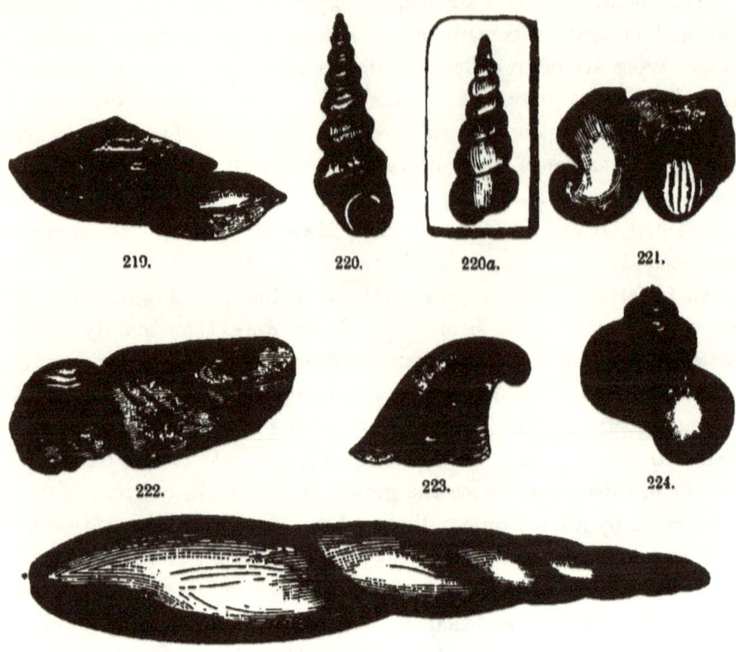

Fig. 219. *Pleurotomaria Progne* (Billings). Lower Trenton formation.

Fig. 220. *Murchisonia gracilis.* Trenton and Hudson River formations.

Fig. 220a. Cast of *M. gracilis.*

Fig. 221. *Bellerophon bilobatus.* Trenton and Hudson River formations.

Fig. 222. *Maclurea Logani* (shell and operculum). Lower Trenton formation.

Fig. 223. *Platyceras angulatum.* Trenton formation. Other species are without the strong ribs on the shell; and in some, the shell is spine-bearing.

Fig. 224. *Holopea Guelphensis.* Upper Silurian.

Fig. 225. *Subulites elongatus* (Internal cast): Lower Silurian.

These and other species of gasteropods occur very commonly in the form of internal casts, and in that condition their specific distinctions are not, in all cases, easily determined. Many so-called species, moreover, are founded on very slight differences, and should rank as varieties only.

Siphonostomata :—The representatives of this section are essentially carnivorous types. They are distinguished by having the aperture of the shell knotched at both of its extremities, or otherwise by the aperture being extended into a slit tube or so-called canal. Their remains are apparently unknown in Palæozoic strata, but become abundant in Mesozoic and Cainozoic beds, and their representatives are still more abundant in the seas of the existing period. The more typical families comprise : the *Cerithiadæ, Purpuridæ*, and *Fusidæ*, dating from the Triassic period ; the *Srombidæ, Buccinidæ* and *Nerineidæ*, dating from the Jurassic period ; the *Muricidæ, Volutidæ, Cancellaridæ, Conidæ*, and *Cypraeidæ*,'dating from Cretaceous times. The annexed figure represents a species of *Buccinum* from the Post-Glacial, Saxicava-sand formation of Beauport, near Qubec.

Fig. 226.
Buccinum undatum
Post Cainozoic and living.

CLASS V.—HETEROPODA.

The mollusks of this class are regarded by many zoologists as simply an Order *(=Nucleobranchiata)* of the Gasteropods. They are free-swimming, pelagic types, and thus resemble in habit the Pteropods, but are of much higher organization. They possess a well-developed head, with perfectly formed eyes, &c., and the sexes are distinct. Some are without a shell ; others have a very small, thin shell, protecting only a portion of the body ; and in those of the Family *Atlantidæ*, there is a light shell of sufficient size to contain the entire animal, and this in some cases is furnished with an operculum.

Two families are commonly recognized : 1, *Firolidæ* (or *Pterotrachœidæ*), in which natation is partly performed by a so-called " tail fin," including the naked genus *Firola*, and also the *Carinaria* with small, delicate, patelliform shell ; and 2, *Atlantidæ*, in which there is a light, hyaline shell into which the animal can withdraw its body. .The only known fossils are some small shells of two species of *Carinaria*, from European strata of Miocene age. The extinct genera, *Bellerophon*,

Cytolites, Maclurea, &c., formerly regarded as belonging to this class, are now placed under the holostomatous gasteropods.

CLASS VI. CEPHALOPODA.

This class is represented by highly-organized mollusca, furnished with a distinct head, large eyes, and a central mouth (with horny, beak-like jaws) surrounded by a series of tentacles, or so-called "arms," which serve as organs of prehension and locomotion. The Nautilus, Argonaut, Octopus, Sepia or Cuttle-Fish, and Loligo or Squid, are its principal living representatives. These are commonly classed under two leading sections or orders, *Tetrabranchiata* and *Dibranchiata,* in accordance with the number of their branchiæ. The Nautilus is the only living representative of the Tetrabranchiate Cephalopods, but many extinct genera in which the character of the shell is more or less akin to that of the Nautilus, are usually regarded as belonging to the same order. Of late years, however, doubts have arisen on this point with respect to some of these extinct forms, which are now supposed to have been more nearly related (as regards the animal, apart from the shell) to living Dibranchiate types. They are chiefly represented by the extinct ammonites and related genera. Hence, the recent adoption of three sections or orders in the classification of the Cephalopoda—namely : (1) *Tetra-branchiata* ; (2) *Ammonoidia,* and (3) *Dibranchiata.* The *Tetra-branchiata* are essentially ancient types in a state of decadence, the Nautilus (with greatly diminished species) being the only living representative. The *Ammonoidea* are extinct, typically Mesozoic, forms. The *Dibranchiata,* on the other hand, although including some extinct genera (as the Belemnite of Mesozoic age,) are essentially recent types. These relations are shown in the following diagram :

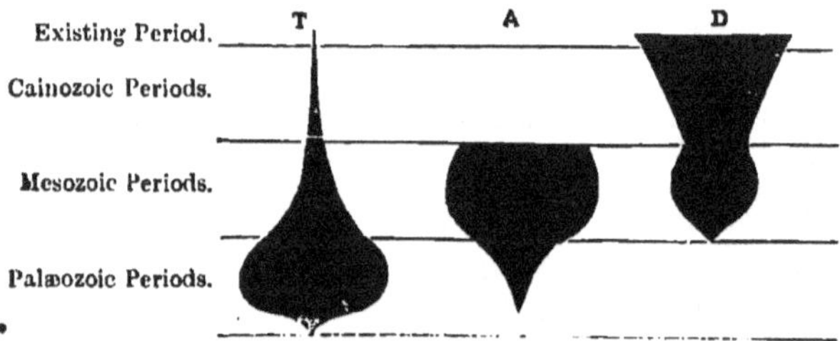

Fig. 227.

Diagram, shewing geological relations of the Tetrabranchiate, Ammonitoidal, and Dibranchiate Cephalopods, respectively.

Tetrabranchiata : In this nearly extinct Order, the shell—judging from the single surviving genus, the *Nautilus*—is external, and is divided into numerous chambers by concave or slightly sinuated partitions, known as "septa." These are formed successively during the growth of the cephalopod—the animal inhabiting the last-formed outer chamber. The septa are traversed by a long tube or "siphuncle," passing through all the chambers, and serving to keep the earlier-formed portions of the shell in connection with the animal's body. This tube or siphuncle varies in size, shape and position. In some forms it is of narrow diameter ; in others, comparatively wide. In some, again, of uniform, gradually tapering shape ; in others contracted at regular intervals, in which case it is known as a "beaded "

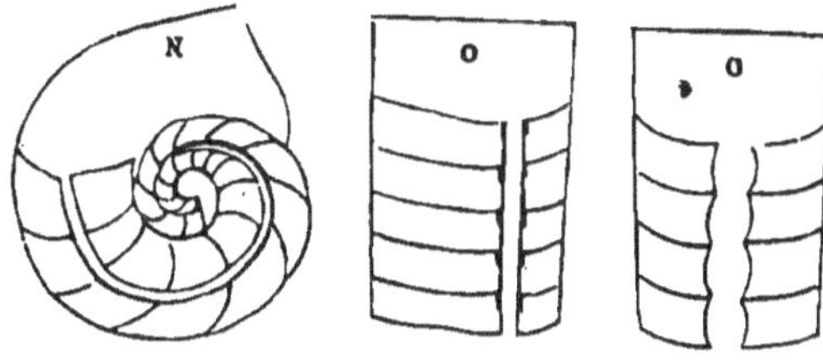

Fig. 227.
Section of shell of *Nautilus* (N), and of *Orthoceras* (O and O'), shewing form and position of septa and siphuncle.

siphuncle ; and as regards position it may be central or more or less excentric. Worn sections of the orthoceras shell, especially when the siphuncle is beaded and the outer portion of the shell is destroyed, much resemble the vertebral column of a fish, and are often taken for this by quarrymen and others. With regard to external form, the shell may either be convolute (as in the nautilus), or simply curved or more or less conical. The concave edges of the septa generally shew on the external surface of the shell ; but these must not be confounded with the lines of growth, ribs, striæ, and other surface markings and ornamentation. The mouth or aperture of the shell is commonly open or expanded, but in certain forms it is more or less contracted, and in some few almost closed—a narrow (somewhat key-hole shaped) opening only remaining. Merely the arms or tentacles of the animal could have been protruded through this : always supposing the shell to have been external. Occasionally, in the larger orthoceratites especially, the posterior portion of the siphuncle is filled with a solid conical secretion of calcic carbonate, which probably served as ballast to counteract the extreme buoyancy of the shell, (see figure 232).

The Tetrabranchiate Cephalopods (thus limited) may be arranged as first indicated by Barrande in two leading groups, named by Fischer, *Retrosiphonata* and *Prosiphonata*.   In the first, the funnel-shaped sheaths which surround the siphuncle at each septum, point backwards, and the latter forwards (as in most of the Ammonitoidal Cephalopods); but in Canada we have no known examples of the Prosiphonata group.   Canadian representatives of the *Retrosiphonata* may be defined as follows :

A. Septa extending completely across the shell :

§ 1. *Aperture of shell contracted to a narrow orifice of definite form.**
   Shell, straight—*Gomphoceras*.   Silurian to Carboniferous.
   Shell, curved—*Phragmoceras*.   Silurian.
   Shell, enrolled—*Lituites*.   Silurian.

§ . *Aperture of shell essentially open :*
   Shell, straight—*Orthoceras*.   Cambrian to Trias.
   Shell, curved—*Cyrtoceras*.   Cambrian to Permian.
   Shell, enrolled—*Nautilus*.   Silurian to Recent.

B. Septa incomplete, confined to one side of the shell.

   Representatives very imperfectly known : they form probably a single genus, *Ascoceras* confined to Silurian strata.

---

* A subdivision of this kind has been objected to, first, because an approach towards a contracted aperture is seen occasionally in other genera of the tetrabranchiate series ; and, secondly, because it is exhibited by both straight and curved forms. But in *gomphoceras*, &c., the contracted aperture assumes a regular, definite shape ; and as it can hardly have been present in the earlier-formed chambers of the shell, but only in the last, abruptly enlarged body-chamber, it evidently indicates a marked change in the condition of the animal. As regards the second objection, surely the definite shape of the aperture is quite as good a classification-character, used subordinately, as the external configuration of the shell. In the recent classification of Zittel *(Handbuch der Pal.)* gomphoceras, because its shell is straight, is placed in the family of the *Orthoceratidæ*, whilst the slightly-curved *Cyrtoceras* is placed in a distinct family *(Cyrtoceratidæ)* with *Phragmoceras*. And, in like manner, *Trochoceras, Nautilus, &c.*, are formed into other distinct families, based on the shape of the shell. There can be little doubt, however, that *gomphoceras* and *phragmoceras* are more nearly related than *gomphoceras* and *orthoceras ;* and, on the other hand, that *cyrtoceras* is more closely related to *orthoceras* than to *phragmoceras*. Whilst adhering to the outer shape of the shell as a leading-classification character in these tetrabranchiate forms, Zittel departs from it entirely in the more or less closely related *Ammonitoidea* (still represented by Owen and other authorities as tetrabranchiate), and he thus places the straight *Baculites*, the hooked *Hamites*, the spiral *Turrilites, &c.*, in one family with *Lytoceras* and other completely enrolled nautiloidal forms.

The genus *Orthoceras* is our most commonly occurring type of this division. It presents a long, more or less conical shell, with simple concave septa. The exterior surface of the shell is smooth or delicately striated in most species; but in some it is transversely ringed, and in others marked with longitudinal ridges and furrows. The Siphuncle is either central or excentric in position, and either of small or large diameter. In the latter case it sometimes contains a solid cone of calcareous matter, as shown in figure 232. Very frequently it presents a so-called beaded form, expanded and contracted at regular intervals (fig. 233), and this becomes especially marked in certain forms from Lake Huron and elsewhere, regarded by some geologists as forming distinct sub-genera under the names of *Actinoceras* and *Huronia*. Figures of several of our characteristic species, with examples also of other genera, are shown in the annexed engravings.

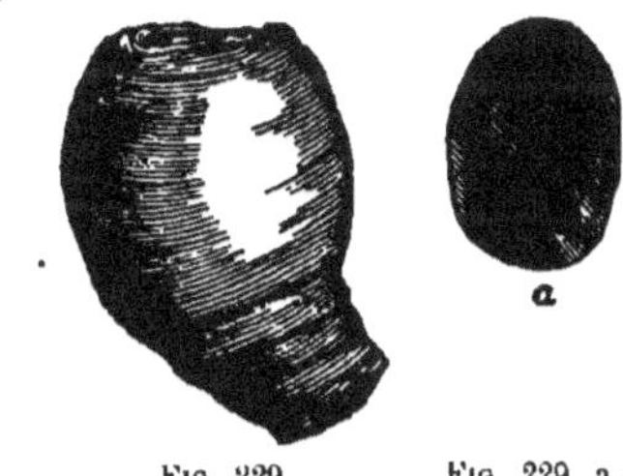

FIG. 229.   FIG. 229. a.

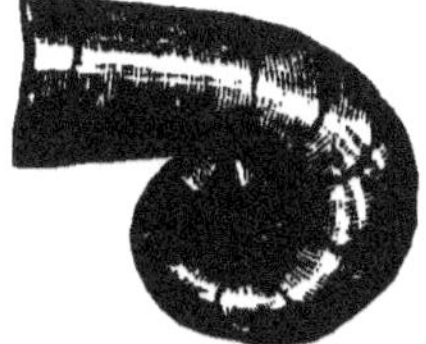

FIG. 230.

FIG. 235.

Explanation of figures :

Fig. 229. *Phragmoceras Hector.* (Billings.) Guelph Formation. 229 a. Aperture of *P. Hector.*

Fig. 230. *Lituites undatus* (Hall). Lower Silurian.

Fig. 235. *Ascoceras Townsendi* (Whiteaves). Guelph Formation, Ontario.

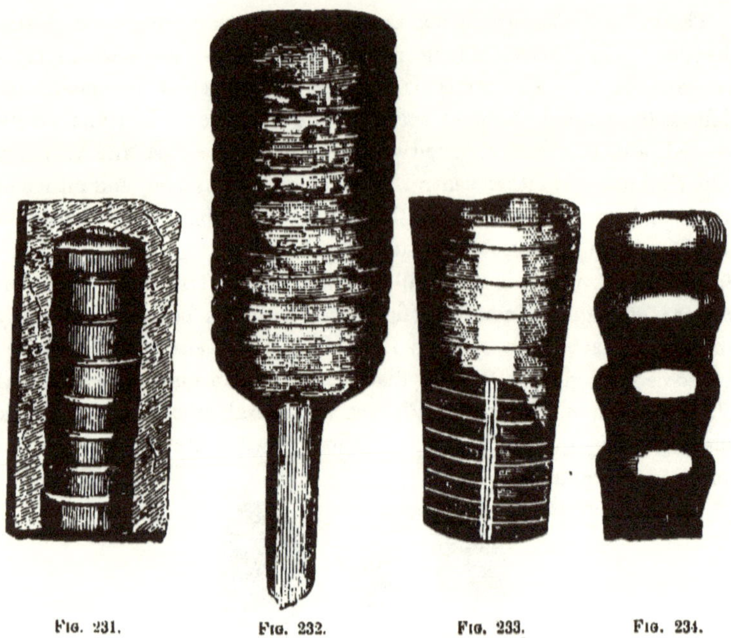

Fig. 231.   Fig. 232.   Fig. 233.   Fig. 234.

Explanation of figures :

Fig. 231. *Orthoceras Lamarcki* (Billings).   Calciferous Formation.

Fig. 232. *O. (Endoceras) proteiforme* (Hall).   Trenton Formation.

Fig. 233. *O. (Ormoceras) crebriseptum* (Hall).   Hudson River Formation.

Fig. 234. Part of siphuncle, natural size, of *O. (Huronia) vertebrale*. Niagara Formation, Lake Huron.

Ammonoidea : This Order, separated of late years, only, from the Tetrebranchiata, comprises a number of extinct forms, more especially characteristic of Mesozoic formations, although including two Palæozoic genera, *Clymenia* and *Goniatites*.   In most forms the shell is convolute or closely enrolled ; but straight, curved, spiral and other shapes are occasionally presented, especially by Cretaceous genera. In all, the shell is divided into chambers by septa, which are slightly undulated in *Clymenia ;* more prominently undulated or indented in *Goniatites ;* and, as a rule, highly complicated with numerous lobes or foliations in the *Ammonites proper.*   The siphuncle is small and simple. In *Clymenia* it runs along the inner margin of the shell ; and along the outer margin in *Goniatites, Ceratites* and *Ammonites proper.* In *Clymenia, Goniatites, Ceratites* and most *Ammonites* the shell is enrolled or more or less nautiloidal in aspect, but is straight in *Baculites*, hooked or only partially enrolled in *Hamites*, spiral in *Turrilites*, and variously modified in other forms.

A few examples of *Goniatites* have been cited from the Devonian Rocks of Western Ontario, but these appear to be of rare occurrence, and are not well characerised. The Ammonites and other representatives of the Order, although abundant in the cretaceous strata of the North-West and British Columbia, are wholly unknown in Palæozoic formations, and are, therefore, never met with in the rocks of Central Canada.

Dibranchiata :—This Order—distinguished by the presence of two branchiæ, and the possession of eight or ten powerful arms or prehensile organs furnished (typically) on their inner side with hooked suckers—is rich in living representatives. The Argonaut, Octopus, Sepia or Cuttle-fish, Squid, Spirula, and the extinct Belemnites, are its principal types. The female Argonaut—the so-called " paper Nautilus"—secretes an external boat-shaped shell, consisting of a single chamber. In all other forms of Dibranchiate Cephalopods, there is only an internal, often horny, shell; or the animal is practically without a shell. The Order is commonly sub-divided into two sub-Orders :—*Octopoda*, with eight arms, and *Decapoda* with eight arms and two longer tentacles. In the living *spirula* and the extinct *belemnites* (both belonging to the latter sub-division), the internal shell is furnished with a " phragmocone" or set of air-chambers divided by simply concave septa, and thus resembling to some extent the external shell of the Tetrabranchiate and Ammonitoidal Cephalopods.

Remains of this sub-Order are unknown in Palæozoic rocks ; and with the exception of the *belemnites*, confined to Mesozoic strata, they are rare in higher formations.

19

SUB-KINGDOM VIII.

TUNICATA.

The forms of this sub-Kingdom—Tunicates or Ascidians—were classed until recently among the *Molluscoidea*, and are still placed there by many authorities. They comprise a series of marine animals of small size, in which the body is enclosed in a thick, sack-like integument. Some are simple, others compound organisms, at least in the adult state. They have been raised to their present position from certain very striking resemblances between the young *ascidian* and the *amphioxus* or lancelet, first detected by Kowalewsky in 1866. As the *amphioxus* is now definitely regarded as a low form of the fish class, the tunicates are thus looked upon as constituting a connecting link between the invertebrata and higher types. Apart from this, they have no palæontological interest, fossil forms being entirely unknown.

SUB-KINGDOM IX.

VERTEBRATA.

This sub-Kingdom includes all animals of bilateral symmetry, possessing a vertebrated backbone, or a permanent chorda dorsalis, or both. Limbs may be present or absent. Vertebrates thus comprise, typically, all animals with internal cartilaginous or bony skeleton. The only exception is the gelatinous *Amphioxus* or Lancelet; but a chorda dorsalis is present in this peculiar type, as in all embryonic and some adult vertebrates.* The sub-Kingdom is separated into the following classes :—1, Fishes ; 2, Amphibians ; 3, Reptiles ; 4, Birds ; 5, Mammals. The annexed diagram shews the life-range and relative importance at different epochs, so far as revealed by fossil evidence, of these various Classes of Vertebrata.

---

* The chorda dorsalis or notochord is the forerunner or embryonic representative of the vertebral column. In some fishes it is retained permanently, but in the great majority of vertebrates it is merely a transitory organ.

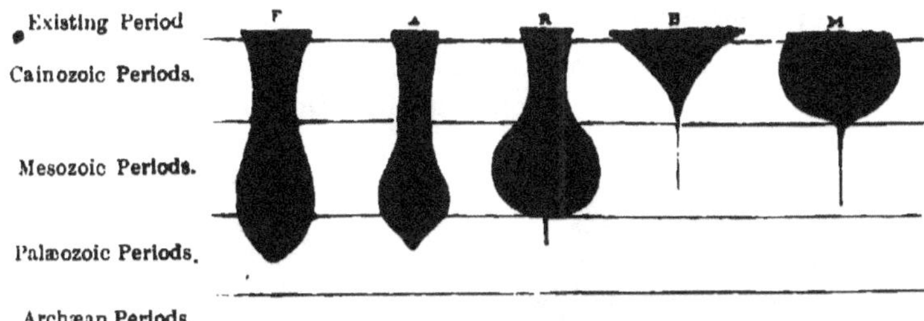

As the remains of vertebrates in the rock formations of the Provinces to which this book refers are practically of little import- ance, it will not be necessary to enter into classification details.

A few spines belonging apparently to an extinct cestraciont shark (*Machœracanthus sulcatus*) have been obtained from the Corniferous

Fɪɢ. 237.   Spine of Castraciont *(Machœracanthus sulcatus.)*   Corniferous Formation.

and Hamilton formations of South Western Ontario ; and in the succeeding Portage-Chemung formation ganoidal scales are occasion- ally found.   The spines are several inches in length, curved, and more or less distinctly furrowed.   The remains of several species of ganoid fishes, in a more or less fragmentary condition, have also been found of late years in the Upper Devonian rocks of Scaumenac Bay in Eastern Quebec.   These are referred to *Pterichthys (* or *Bothrio- lepis), Acanthodes* and *Phaneropleuron.*   See descriptions and figures by Mr. Whiteaves in the Transactions of the Royal Society of Canada for 1886.

In the comparatively modern Post-Glacial deposits, entire skeletons of some existing teleostean species—*mallotus villosus* (the capelin), and *cyclopterus lumpus* (the lump-sucker)—occur in clay modules at Green's Creek on the Ottawa, and at other places in that district, at considerable elevations above the present sea-level.   Bones of a living seal (*phoca grœnlandica*) occur in the same deposits.   The Post- Glacial sands and clays around Hamilton and elsewhere have yielded occasionally a few bones and teeth of our existing black bear (*Ursus Americanus*), wapiti, beaver and other existing forms; together with teeth and portions of the tusks of two extinct proboscidians : the mammoth, an extinct elephant of the Asiatic type (probably *Elephas primigenius**), and the common mastodon (*M. Ohioticus*).

Fig. 238.

Molar tooth of *Elephas primigenius.*

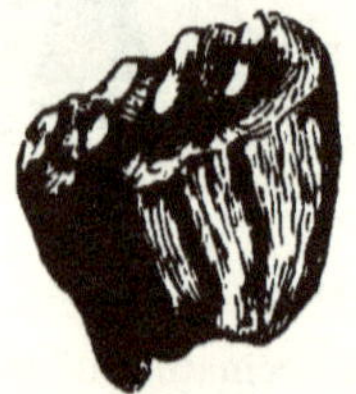

Fig. 239.

Molar tooth of *Mastodon Ohioticus.*

---

* It has been proposed to subdivide the genus *Elephas* into two distinct genera : *Euelephas*, for the Asiatic type, with narrow, linear tooth-bands, short ears, &c.; and *Loxodon*, for the African type, with tooth-bands of lozenge-like form, large ears and other more or less divergent characters ; but this subdivision is not generally adopted. *E. primigenius*, again, is subdivided by some authorities into several species or sub-species, as *E. Jacksoni*, *E. Americanus*, &c.

PART V.

SYSTEMATIC OUTLINE OF THE GEOLOGY OF CENTRAL CANADA, COMPRISING THE PROVINCES OF ONTARIO AND QUEBEC.

As explained in preceding sections of this work, the various rock-formations which make up the outer portion or so-called crust of the earth do not form a single unbroken series, but belong to various epochs of formation. These epochs comprise five primary divisions or ages: determined partly by the superposition of their rocks, and partly by the fossilized organic bodies which many of these rocks contain. The ages thus recognized are as follows:

5. The Present or Existing Age.
4. The Cainozoic Age.
3. The Mesozoic Age.
2. The Palæozoic Age.
1. The Archæan Age.

These ages, although distinct in the main, offer, as in all historic periods, a gradual passage from their older into their newer epochs. The rocks by which they are made known to us are of three principal kinds: *Sedimentary* or *Stratified rocks*, represented by ordinary sandstones and limestones, clay-slates, clays, sands, &c., many of which contain the fossilized remains of plants and animals living on the earth when the rocks in question were formed; *Foliated* or *Stratiform-crystalline rocks*, represented by gneiss, mica-schist, crystalline limestone, &c.; *Eruptive rocks*—partly crystalline, as granites and syenites; partly compact or aphanitic in texture, as ordinary traps and basalts; partly scoriaceous, as ordinary lavas; and partly vitreous, as obsidian.

In the Dominion of Canada, viewed in its entirety, examples occur of all these rocks, and each of the five great geological ages is represented by rock-formations; but in the separate Provinces of the Dominion, the rock-representatives of some of the geological ages are unknown. Thus within the boundaries of Ontario and Quebec there are no known rock-formations of Mesozoic or Cainozoic age. Viewed broadly, the older rock-formations lie in the northern and eastern portions of the Dominion ; the newer, in the west.*

PROVINCE OF ONTARIO.

INTRODUCTORY NOTICE.

The southern limit of this Province extends westward along the St. Lawrence River, from a few miles above the junction of the St Lawrence and the Ottawa, through Lake Ontario, and the greater part of Lake Erie to the River Detroit. Its south-western and western limit runs through Lake St. Clair, Lake Huron and Lake Superior as far as the mouth of Pigeon River, and from thence to the north-west angle of the Lake of the Woods. Its north-west and northern boundary runs from that point to Winnipeg River, and then north-easterly by Lac Seul and St. Joseph's Lake, along the Albany River to James' Bay, the southern point of Hudson's Bay ; and its eastern boundary passes from the latter directly south to Lake Temiscamingue, and from thence down the Ottawa to near the junction of that river with the St. Lawrence. The area of the Province (exclusive of the portions of the great lakes within its boundary) is computed to equal 181,800 square miles.

On passing south-westerly, parallel with the River St. Lawrence from the County of Glengarry, the eastern extremity of Ontario, we traverse a gently undulating district, rising from about 100 to 250 feet above the sea-level and extending from the Province boundary to the vicinity of Brockville. This district is underlaid by limestones and sandstones of Silurian and Cambrian age, and is of good fertility. The Nation River flows through it in a north-easterly direction and falls into the Ottawa. A little west of Brockville, a gneissoid, crystalline district, comparatively wild and rocky, is traversed for about 40 miles to the neighborhood of Kingston. This extends northwards into the great Archæan region, which it connects, south of the St. Lawrence, with the mountainous district of the Adirondacks. West of Kingston, a gently-undulating, agricultural district, underlaid mostly by Lower Silurian limestones, is again traversed. This extends, with an avergage elevation of about 250 feet near the shore of

---

* The rock-formations of Ontario and Quebec alone come under review in the present work. A synopsis of the geology of the other Provinces will be found in the author's " Outline of the Geology of Canada.

Lake Ontario, to Hamilton at the head of the Lake. Northwards
the ground rises in a succession of terraces to about 1,000 feet, and
then descends towards Lake Simcoe to about 700 feet above the level
of the sea. The northern edge of this Lower Silurian area, from
Kingston to Georgian Bay, abuts against the Archæan, gneissoid
region, and the junction of the Silurian limestones with the gneissoid
Laurentian rocks of the latter is marked by an almost continuous
line of small lakes. At Hamilton the ground rises abruptly into the
great escarpment known as the Niagara escarpment, which extends
from the Niagara River to Cabot's Head on Georgian Bay. Westward, the escarpment forms the edge of a broad table-land, which,
from an average elevation of about 1,100 feet, stretches on one side
to Georgian Bay (578 feet above the sea-level), and on others to Lake
Erie (565 feet) and the southern extremity of Lake Huron (578
feet). This high land is underlaid by Upper Silurian and Devonian
strata, and is of great fertility over the larger portion of its area.
The Manitoulin Islands in Georgian Bay are made up mostly of
Lower and Upper Silurian strata. All the other portion of the
Province except the remote and still imperfectly known region
around James' Bay, in which the strata are chiefly or wholly of
Upper Silurian and Devonian age, belongs essentially to the great
Archæan region of Canada, and is underlaid by Huronian and Lau-
rentian gneissoid rocks, with a comparatively small development of
Lower Cambrian strata about Thunder Bay and a few other points on
Lake Superior. This vast Archæan region, although densely wooded,
is of a more or less wild and inhospitable character, and as a rule ill
adapted for agricultural occupation : but it abounds in mineral
wealth, and especially in ores of iron, copper and silver.

From this rapid outline of the topography and general features of
Ontario, it will be seen that three great geological areas may be
recognized within the limits of the Province. A line drawn from
James' Bay southwards to Lake Ontario will traverse these three
leading areas, as shewn in the annexed sketch-section :

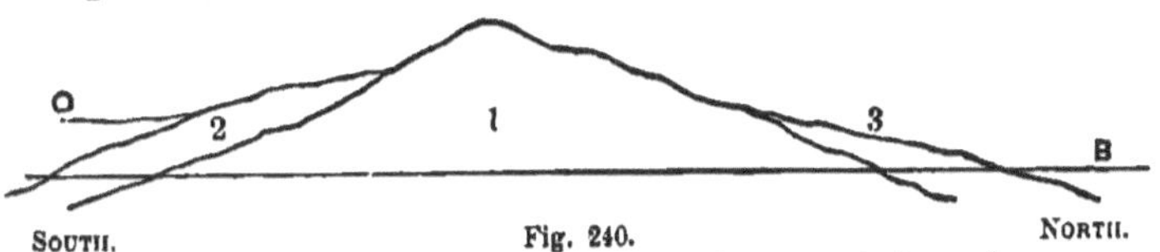

With the exception of the remote and unsettled region around

James' Bay, these large areas may be conveniently divided, in order to facilitate their description, into several smaller areas or districts, as in the following distribution :

1. THE ARCHÆAN AREA.
    1ᵃ. The District of the Upper Lakes.
    1ᵇ. The Eastern Archæan District.

2. THE SOUTHERN PALÆOZOIC AREA.
    2ᵃ. The Lower Ottawa District.
    2ᵇ. The Lake Ontario District.
    2ᶜ. The Erie and Huron District.
    2ᵈ. The Manitoulin District.

3. THE NORTHERN PALÆOZOIC AREA.

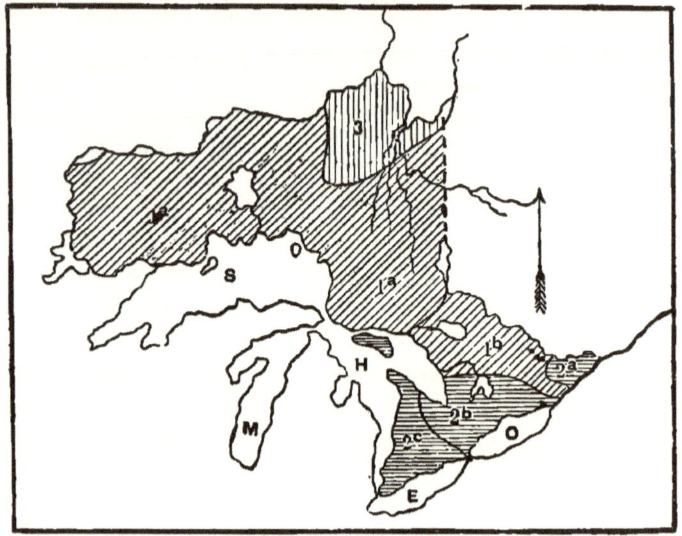

FIG. 241.

Sketch Map of Province of Ontario, shewing geological areas.

Explanation : The diagonally shaded space shews the great Archæan area : 1ᵃ indicating the District of the Upper Lakes, and 1ᵇ the Eastern Archæan District.

The horizontally shaded space shows the Southern Palæozoic area : 2ᵃ denoting the Lower Ottawa District ; 2ᵇ, the Lake Ontario District ; 2ᶜ, the Erie and Huron District ; and 2ᵈ, the Manitoulin District.

The vertically-shaded space (3) indicates the portion of the Northern Palæozoic area lying within the Province boundary.

S=Lake Superior ; M=Lake Michigan ; H=Lake Huron ; E=Lake Erie ; and O=Lake Ontario. The unshaded district on the right is part of the Province of Quebec.

DISTRICT OF THE UPPER LAKES.

This district may be described in general terms as extending west of Lake Temiscamingue, over the almost entire north-west portion of Ontario, exclusive of the still little known Palæozoic region around James' Bay. It forms for the greater part a densely-wooded, but rugged and mountainous region, broken up by numerous bodies of water, and underlaid essentially by crystalline and semi-crystalline rocks of the Laurentian and Huronian series. The surface of Lake Temiscamingue is 612 feet, that of Lake Huron 578 feet, and that of Lake Superior 600 feet above the sea. From these levels the ground rises more or less abruptly to an average height of 1,000 to 1,500 feet, with here and there a few points of still greater elevation. The rocks within its area comprise : (1), representatives of the Laurentian and Huronian series, occupying, as remarked above, the greater part of its surface ; (2), some succeeding strata of supposed Lower Cambrian age ; (3), many eruptive granites and trappean masses ; and (4), overlying Glacial and Post-Glacial superficial deposits.

Laurentian and Huronian rocks :—The Laurentian rocks of this district consist for the most part of ordinary red and gray gneiss, in more or less inclined and often nearly vertical beds, or presenting a highly contorted lamination. These alternate with darker beds containing hornblende, or in some cases augite, and with occasional bands of crystalline limestone. Black tourmaline or schorl and opaque dark red garnets are among the more commonly-occurring accidental minerals of these gneissic rocks. The Huronian representatives, although distinct enough in their entirety, closely resemble in many cases the Laurentian rocks of the district, and cannot always be readily separated from these. As a rule, however, the texture is less crystalline or less granitoidal, and slaty or semi-crystalline conglomerates appear among them. Quartzites, varying in tint from colorless or pale green to dark gray and black, are especially abundant ; and many of these hold pebbles or fragments of jasper or of gneiss, or pass into siliceous slates or slaty conglomerates. In some cases also they are more or less feldspathic, and consist of an intimate mixture of quartz and orthoclase or other feldspar. In addition to these quartzites and slate conglomerates, dark green chloritic and hornblendic rocks form the more characteristic representatives of the Huronian series.

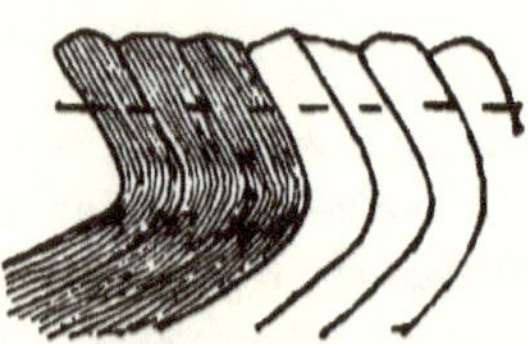

FIG. 242.

The dotted line in this diagram indicates the level below which the rocks are concealed.

The stratigraphical relations of the two series, Laurentian and Huronian, in this district, have not yet been clearly made out. The mineral characteristics, and especially the presence of conglomerates holding gneissoid and other fragments, lead undoubtedly to the conclusion that the Huronian beds are of later formation than the Laurentian; but, as pointed out by Dr. Selwyn, the Huronian appear in many places to pass under the latter. This can only be explained by the assumption of great overturned or reversed dips, as shewn roughly in fig. 242.

As regards distribution, whilst the Laurentian rocks cover perhaps the greater portion of the district now under description, large areas within it are occupied by Huronian beds. The latter form essentially a series of wide bands ranging in a north-east and south-west direction. The largest perhaps of these Huronian areas lies immediately west and south-west of Lake Temiscamingue, extending in the latter direction to near Killarney on Georgian Bay, and along the north shore of Lake Huron and the back country to beyond Goulais Bay in the south-east angle of Lake Superior. This area is traversed for about   miles by the Canadian Pacific Railway, from the Wahnahpitae River, by Sudbury, to Spanish Forks, where Laurentian rocks come up ; and fine sections may be seen in the railway cuttings, especially in the vicinity of Sudbury Junction, and westward, where large deposits of copper ore occur. The copper pyrites of Eagle Lake (near Lake Huron) and that of the Bruce Mines, now apparently exhausted, occur in quartz veins traversing the same Huronian formation. About Sudbury and along the branch line to the Algoma Mills on Georgian Bay the Huronian quartzites and conglomerates are broken through by many dykes and erruptive masses of diorite and syenite. Another Huronian area of considerable size extends around Michipicoten Harbour, and along Michipicoten, Magpie and Dog Rivers, for some distance inland. Here the rocks are more or less ferruginous, and are broken through by some large granitic masses. Huronian beds occur also further west on Lake Superior along the course of Pic River. And again, with granitic intrusions, in the back country between Black Bay and the International boundary on Pigeon River. Long belts of Huronian rocks range also towards the north-east from the eastern shore of Lake Nepigon. Still further west, Huronian beds, mostly in the form of hornblendic, chloritic and nacreous schists, with some clay-slates and quartzites, extend around the shores and through the numerous rocky islets of the Lake of the Woods. On the north-east shore and adjacent islands of this lake, especially round Big Stone Bay, gold-bearing quartz veins have been opened at several localities. Eruptive granites are of common occurrence also throughout this portion of the Huronian area.

Lower Cambrian Strata: Animikie and Keweenian Forma'ions: —The actual age of these formations is still somewhat uncertain, but they belong most probably to an early Cambrian period. But although of post-Archæan age, they form part of the great Archæan region of North-western Ontario, and are thus legitimately described with the latter. They were designated originally by Sir William Logan as the "Upper Copper-bearing Rocks of Lake Superior." Two series or separate formations are recognized. The lower formation, now known as the Animikie formation,* is made up principally of black slates with subordinate stratifications of white, gray and black chert (in places anthracitic), dark gray ferruginous dolomite, occasional layers of altered sandstone, and bands of trappean matter, mostly composed of dark hornblende and greenish white feldspar, and frequently porphyritic. An enormous trappean overflow, with well-marked sub-columnar structure, caps the entire formation, as seen in the bold promontary of Thunder Cape, as well as at McKay's Mountain, and on Pie Island and elsewhere around Thunder Bay. The higher formation, known as the Keweenian (or Keweenian and Nepigon), consists of white and red calcareous sandstones and marls, beds of conglomerate, and numerous interstratified trappean bands, the whole overlaid as in the lower series by a great trappean overflow· These traps or greenstones are more or less compact or fine-granular in texture as regards those which occur west of Thunder Cape and which are thus associated with the Animikie series ; whilst the more eastern displays, or those connected more especially with the higher

---

* The earliest name bestowed on this series was that of the *Kaministiquia Formation*, in the first edition of this work, published in 1864. The name was derived from the Kaministiquia River of the Thunder Bay country, along the course of which these rocks are principally developed.

Keweenian series are very generally amygdaloidal (see page 182). Both series are traversed by numerous trappean and dioritic dykes (often distinctly porphyritic) in which a transverse columnar structure, as first pointed out by Sir William Logan, is very conspicuous. Both series also are interpenetrated by mineral veins carrying native silver, silver-glance, galena, zinc blende, copper-pyrites, and other ores, some of which are more or less auriferous. These are referred to below.

The Animikie formation extends from Pigeon River eastward across the Kaministiquia and around the shore of Thunder Bay to a little beyond Thunder Cape. The Keweenian formation stretches from this point around Black Bay (on the west shore of which it abuts against a large mass of granite), and across Nepigon Bay, St. Ignace and adjacent islands ; and it occupies also a broad area on the south, west and north sides of Lake Nepigon. Michipicoten Island with its cupreous greenstone dykes, and one or two headlands on the east shore of Lake Superior, likewise belong to this series.

Superficial Deposits :—Over the floor of crystalline rocks by which this vast region is essentially underlaid, Drift clays and boulders, and Post-Glacial clays and sands, with other recent accumulations, are spread in many places. Glacial furrows and striæ also are seen in numerous localities. The prevalent direction of the striæ is decidedly towards the south-west, although some run nearly south, others east of south, and others, again, almost east and west. Where commonly seen two or more sets of striæ occur together and thus intersect each other. Drift clays are seen in many of the river channels, and boulders are of very general distribution. The latter are accumulated in some places in long ridges or morains at the opening of valleys or along the lower slopes of hills. Whilst many of these boulders are of essentially local origin—and thus consist on the north shore of Lake Huron of jasper-conglomerate, quartzite, and other Huronian rocks, and of Laurentian and Copper-Bearing rocks about Lake Superior—examples of limestone boulders, containing Middle or Upper Silurian and Devonian fossils, occur in places throughout the region. These latter must have come generally from the northern country which lies beyond the height of land and extends north-easterly to Hudson's Bay. The Post-Glacial deposits of the district consist in their lower beds of stratified clays and sands, referred by the

Survey to the " Saugeen " division ; and in their higher beds, of ridges and widely-spread accumulations of fine sand only, referred to the " Algoma " series. The stratified clays (of gray, red, and buff colors) are seen in places on the north shore of Lake Huron and St. Mary's River, and prominently in the high terraces around the Sault, as well as on the Pic River, and in the ancient banks of the Kaministiquia and other rivers of Lake Superior. The higher sands are also displayed in the Kaministiquia banks, and more or less throughout the Nepigon country, where they form in places hills and ridges of considerable elevation. Also on the Pic River, and largely around Michipicoten Harbour, and on the Goulais River ; and along nearly all the river valleys, and over many intervening tracts of country north of Lake Huron.

Economic Minerals :—At many localities within this vast region, and notably in the Huronian, Animikie and Keweenian areas, the presence of great mineral wealth has been fully proved. But very little has been done in the way of actual mining. Many of the so-called mines of the district are merely mineral locations on which only a few test pits have been sunk. And where mining has been attempted, it has in several instances been prematurely abandoned owing to want of capital or other causes. Somewhat extensive operations, however, are now being carried on in the vicinity of Sudbury, and at the Rabbit Hill and Beaver Mines north of Thunder Bay. The more important metalliferous localities are given in the following list :

Gold :—This metal has been found both in the free state, and in small quantities varying from a few pennyweights to over an ounce per ton, in the copper-ore deposits of Sudbury and surrounding country, in Huronian rocks. Also in the same geological formation at the Lake of the Woods, chiefly around Big Stone Bay, where it occurs in quartz veins with iron and copper pyrites, zinc blende, &c. The principal " mines " are known as the George Heenan, Winnipeg Consolidated, Keewatin, Gold Hill, Pine Portage, and Sultana Mines, but of late little or no work has been done upon them. Gold-bearing pyrito-quartzose veins occur also in Huronian rocks at Victoria Cape, Lake Superior ; and in the dark slates of the Animikie formation north and west of Thunder Bay, but in that district the ores are chiefly argentiferous. In a vein carrying galena, zinc blende

and copper pyrites in the Keweenian formation of Black Bay (Enter-
prise mine), the author found a small percentage of gold, averaging
about 18 dwts. in the ton of ore.

Silver:—Nearly all the pyritous veins of this district hold small quantities of silver, more especially when the vein-matter carries intermixed
copper pyrites, iron pyrites, zinc blende and galena; but paying
amounts are confined almost wholly to the calc-spar (or calc-spar and
quartz) veins around or in the vicinity of Thunder Bay. These veins
traverse, as a rule, the black slates of the Animikie formation, but
some occur in the Keweenian series. The most remarkable of these
veins is that which runs through Silver Islet, a small rock lying a
few miles east of Thunder Cape, where the slate is in contact with a
comparatively broad and persistent dyke of sub-crystalline diorite.
The vein cuts this transversely and carries native silver, silver glance
and galena, with subordinate shews of pyrites, blende, plumbago, &c.,
in a gangue of calcite mixed in places with quartz and a little fluor
spar. Although now abandoned, the vein has yielded an amount of
ore (mostly native silver) valued at the lowest estimate at more than
three millions of dollars. The workings extended far under the level
of the lake; and when the mine was closed in 1884 its shaft had
been carried down to a depth of 1,230 feet. Other silver-bearing
veins traverse the Animikie formation immediately north and west
of Thunder Bay. Many of these have shewn very rich bunches of
ore, and several, comprising more especially the Rabbit Mountain,
Beaver Mountain, Badger, and Silver Mountain veins, are now being
successfully worked. A good deal of work has also been done on
the Shuniah or Duncan vein, but the work on this vein is now
stopped for a time. Other mining locations in this neighborhood, on
most of which, however, merely trial pits have been sunk, comprise
the Silver Creek, Singleton, Porcupine, Crown Point, Spar Island, and
other properties. Silver-bearing veins are also known on Pie Island
in Thunder Bay. The gold-bearing veins of the Shebandowan district
and the Lake of the Woods are likewise more or less argentiferous.*
Many of the Huronian copper ores of Sudbury also hold small
amounts of silver; and a vein of argentiferous galena (some portions
of which hold over 100 oz. of silver in the ton) has been worked, off

----

* In some of the Shebandowan ores the presence of tellurium was first detected by
Dr. W. H. Ellis of the School of Practical Science, Toronto.

and on for some years with fluctuating results, near Garden River, east of the Sault. Other argentiferous galena veins have been recognized in the vicinity of Lake Temiscamingue.

Copper :—The Keweenian formation on Michipicoton Island, Lake Superior, has yielded native copper in disseminated masses and small strings, and somewhat extensive mining operations have been carried on at the north-west of the island, although with not very encouraging results. The same formation at Black Bay is traversed by veins containing intermixed copper pyrites and galena, carrying small amounts of gold and silver; and at Prince's Location, west of Thunder Bay, both copper pyrites and copper glance occur in the dark Animikie slates which there form the country rock. Veins containing promising amounts of copper pyrites have been discovered also in Neebing and other townships in the Thunder Bay district; but the most important deposits of copper ore lie in the Huronian rocks of the region now under review. The veins at the Bruce and Wallace mines, Lake Huron, after yielding large supplies of copper (in the form of the yellow and horseflesh ores) for many years, are now thought to be exhausted, and mining operations at that immediate locality are abandoned. Some promising veins, however, are now being opened at Echo Lake, a short distance inland. Vast deposits of copper ore, probably in the form of both veins and stocks, occur to the north-east, in the Sudbury district, and these are now beginning to be largely worked.

Lead and Zinc Ores and Other Economic Minerals of the District : —Promising amounts of galena occur in many of the veins which traverse the Animikie, Keweenian and Huronian rocks around Thunder Bay and adjacent country, as in Neebing Township, and at Silver Lake and Black Bay. Also at Garden River, east of the Sault Ste. Marie and elsewhere. In most of these veins the galena is intermixed with blende and copper pyrites, and carries more or less silver. Some comparatively broad veins carrying workable quantities of zinc blende have been discovered at Blende Lake and Silver Lake immediately adjacent to Thunder Bay, and also a few miles inland on the line of the Canadian Pacific Railway. Large iron deposits, at present only partially explored, occur at various localities within this region, as in the vicinity of the Pic River, and at Michipicoten, Eagle Lake, and elsewhere; but some of these deposits hold a good deal of siliceous rock-matter, and are rather ferruginous schists than work-able ore. Many however are rich in metal, and are especially free from titanium and phosphorus. Finally, it may be mentioned, the granite masses which occur so abundantly within the district, yield excellent building stone, and have been largely quarried near the Lake of the Woods and at other spots for use in the construction of the Canadian Pacific Railway.

THE EASTERN ARCHÆAN DISTRICT.

This district is a south-eastern extension of the great archæan district of the northern lakes, described above. It is separated only conventionally from the latter, and chiefly for facility of description. At the same time, it presents certain points of difference: more especially in the absence of any clearly recognised Huronian strata, and in the apparently total absence of the Animikie and Koweenian Formations. Its bands of crystalline limestone differ also very generally from those of the more northern and western region by containing various crystalline silicates and other minerals: pyroxene, zircon, garnets, brown tourmaline, phlogopite mica, apatite, and graphite, being especially prevalent.

Its north-western boundary is a conventional line running from Lake Temiscamingue to a short distance beyond French River on the north shore of Georgian Bay. Its northern and north-eastern boundary is the river Ottawa to the vicinity of Arnprior; and from this point its eastern limit runs south by Pakenham, Carleton Place, Perth, and Charleston Lake, to the St. Lawrence. The rocks of the district form a narrow-belt along the St. Lawrence, roughly, between Brockville and a few miles west of Gananoque. From the latter point, its southern boundary passes through the back townships of Frontenac, Addington, Hastings, Peterborough, and Simcoe, and strikes Georgian Bay near the mouth of the Severn. The average elevation is about 800 feet above the sea. Lake Nipissing lies at an elevation of 640 feet, but the ground to the south and east of the lake is considerably higher. Around Haliburton, for instance, the elevation above the sea-level exceeds 1000 feet. Lake Nipissing is its largest body of water. Numerous smaller lakes, as Lake Opiongo and the **Muskoka** lakes, lie within its area; and an almost continuous chain of these extends along its southern border. As a rule, the country is of a broken, hilly character: vast masses of gneissoid rock standing in many places high above the ground. The entire district, however, is more or less densely wooded.

Gneissoid, Laurentian rocks underlie the district generally. These consist for the greater part of ordinary and hornblendic gneisses, traversed by numerous granite veins, and interstratified with subordinate beds of dark-green pyroxenic rock, crystalline limestone, quartzite, and siliceous slate,—some beds of conglomerate appearing in places near the upper part of the series. The green pyroxenic bands are mostly associated with deposits of iron ore. In some parts of the district, long rugged belts of red syenitic granite (as seen in the Huckleberry Hills near Marmora, and elsewhere) traverse the country in a general N.E. and S.W. direction, and produce a synclinal structure in the intervening tracts. At many spots, as around Stoney Lake and elsewhere, the gneiss is almost free from mica; and where it is traversed by granite veins, the latter are usually very feldspathic and more coarsely granular than the gneiss. Good exposures occur more or less all over the district, especially on the lake shores and islands, and along the high ridges which traverse the district generally. The beds at most spots are tilted at high angles, and are frequently much corrugated and contorted, as shewn in the annexed sketch.

Outliers :—The Laurentian rocks of this district are overlaid here and there by outlying patches of Lower Silurian strata. On the Bonnechere River in Renfrew, and in the vicinity of Pembroke, outliers of this kind, composed of light-coloured Chazy sandstones and Black River

FIG. 243.

Portion of a granite vein traversing contorted micaceous-gneiss. Vicinity of Haliburton, Ontario.

limestones, rest directly on the Laurentian gneiss. Several outliers of Black River and Trenton limestone (see under the Lake Ontario District) cover detached areas of considerable extent in the gneissoid country immediately north of Madoc in the county of Hastings.

Drift deposits :—Over many portions of the district, unstratified clays with boulders of gneissoid rock, and higher or post-glacial

20

stratified clays and sands, are largely distributed; and it is chiefly where these occur that the district admits of agricultural occupation.

Glacial striæ of this period are seen on the exposed surfaces of many of the harder rocks. In some localities, as in most parts of Renfrew and other eastern sites, the striæ have a very general south-easterly direction; whilst in the west, as about Lake Nipissing and Georgian Bay, and in the northern parts of Hastings, Peterborough, and Victoria, the prevalent direction is towards the south-west.

Economic Minerals :—This district is especially rich in deposits of iron ore; and, in addition, it contains auriferous mispickel, galena, apatite, marble, and mica, in workable quantities The iron ore con-sists chiefly of magnetite (see page 83), but valuable deposits of hematite replace this at some localities. As a rule, these oxidized ores form large irregular masses or "stocks," and are very generally associated with the green pyroxenic rocks and the crystalline lime-stones of the district. Many are exceedingly rich and pure, holding from 65 to 70 per cent. metallic iron,* with consequently very little intermixed rock-matter ; and although pyrites is occasionally present, the amount of sulphur and phosphorus is in general quite low. But some of these magnetites are rendered unmarketable in consequence of the presence of titanium in comparative excess. An enormous deposit of titaniferous magnetite occurs on lot 35 of the 4th con-cession of Glamorgan ; and another in the Township of Tudor in North Hastings.† The deposits of workable ore, however, far exceed in number those which are unavailable from the presence of titanium. These workable deposits occur all over the district, and their presence in new localities is constantly being discovered. Some of the best known occur in the townships of McNabb, Bedford, Crosby, Sherbrook, Wollaston, Faraday, Glamorgan, Tudor, Madoc, Marmora, Belmont, Limerick, Minden, and Snowden.‡

---

* The maximum percentage of metallic iron in perfectly pure magnetite is 72.41. and in hematite, 70. See descriptions of these ores in Part II.

† See descriptions and analyses by the author, in Transactions of the Royal Society of Canada for 1884. Analyses and brief descriptions of many of the iron ores of Central Canada, by the author, will also be found in the Transactions of the Royal Society for 1885.

‡ Other townships will no doubt be rapidly added to those of the present list. It should be pointed out, however, that as these iron ore deposits are essentially in the form of "stocks," great care should be exercised in determining the probable dimensions of a deposit, before purchasing, or putting up expensive works in connection with it. Happily, in the diamond drill, we have the means of readily testing the size and purity of ore-masses of this character.

The auriferous mispickel of the district occurs in the form of veins, running mostly parallel with the stratification, in the township of Marmora. The mispickel (see page 77) is essentially in a quartz veinstone in which "free gold" is often visible. Bitter spar or dolomite usually accompanies the quartz; and iron pyrites, brown and blackish-green mica and other minerals are also commonly present. The veins have a general N.N.W. and S.S.E. direction, and dip westerly at an average angle of 30° or 35°. A layer of talcose slate forms in most cases one or both of the walls. The annexed sketch-section, fig. 244, shews the general character of the ground in the

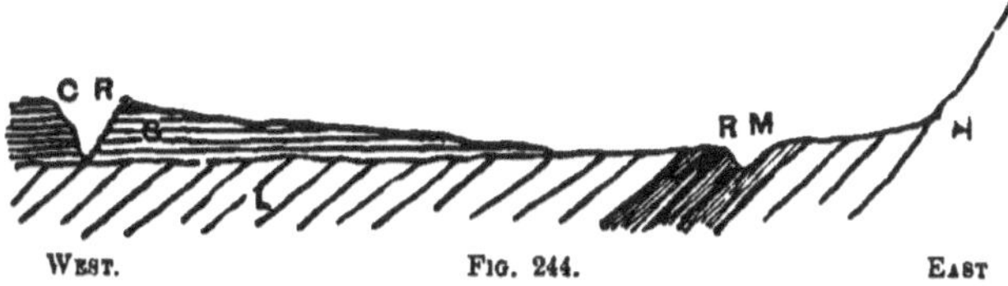

more southern portions of the eighth, ninth, tenth, and eleventh concessions of the township. H, indicates the syenitic Huckleberry Hills ; R M, the River Moira, here reduced to a comparatively small stream ; C R, Crow River, with the site of Marmora village adjoining; L, the inclined Laurentian beds ; and S, the overlying and nearly horizontal Lower Silurian strata. The shaded portion of the Laurentian rocks indicates the principal site of the gold-bearing veins. The amount of gold carried by the mispickel of these veins, varies from one or two to seven or eight ounces in the ton ; and the average yield of the undressed veinstuff, generally, is rarely less, and commonly higher, than from fifteen to eighteen dwts. per ton of ore. The gold averages in fineness about 23 carats, very little silver being combined with it. The large amount of arsenic in the mispickel also adds to the value of the ore.

Galena veins occur within the district in the townships of Loughborough, Bedford, Lansdowne, Ramsey, Galway, Somerville, Tudor, Marmora, Lake, Limerick, and elsewhere ; but hitherto these veins have not been successfully worked.

Apatite—the "phosphate" of commerce—occurs in North Burgess and North Elmsley, and at other localities in the country between

A few borings put down just beyond the supposed limits of the deposit, and in the central part of the latter, will prove its extent and depth; and the cores brought up by the drill will show the character of the deposit throughout its mass.

Kingston and Perth, mostly in veins associated with calcite, phlogopite, and pyroxene. At present however, these deposits are
practically idle, the phosphate now shipped from Canada coming
essentially from the left bank of the Ottawa in the Province of
Quebec. The township of Burgess has also yielded some good mica;
and marble quarries have been opened, among other places, in the
townships of McNabb, Fitzroy, Barrie, and Elzevir. The fine grey
marble of Arnprior in the township of McNabb, has been largely
used in the interior of the Parliamentary Buildings at Ottawa.

THE LOWER OTTAWA DISTRICT.

This is essentially an agricultural area, underlaid by Palæozoic
formations in comparatively undisturbed stratification. It occupies
the country between the right bank of the Ottawa and the left bank
of the St. Lawrence, extending to the Province boundary near the
junction of these rivers. On the west, it is bounded by a line extending roughly from Brockville to the vicinity of Perth, and from
the latter point to the Ottawa a little north of the mouth of the
Madawaska. It presents a generally level surface, although some
bold escarpments, in part connected with faults, occur within its
limits, especially around Ottawa city, and in the townships of Cumberland, Alfred, L'Orignal and Hawkesbury. The height above the sea
at the junction of the Ottawa and St. Lawrence rivers, is about 60
feet; and at the foot of the Chaudière Falls, about 118 feet. From
these levels, the district rises near its north-west boundary to about
400 feet, and its average elevation may be placed at from 250 to 300
feet above the sea. In some parts of Cumberland and other townships, extensive swamps occur, but viewed generally, the district is
well timbered and of good fertility. The Rideau Canal passes
through its central portion, and a large part of its more southern
area is drained by the South Nation River, which rises near the St.
Lawrence in Edwardsburg, and flowing north-east, falls into the
Ottawa, in the township of Plantagenet.

The strata of the district belong to the Cambrian and Lower
Silurian series, but these are overlaid in many places by Drift

deposits and still more recent accumulations. The Cambrian strata comprise representatives of the Potsdam and Calciferous formations; and the Lower Silurian series—or "Cambro-Silurian" of the present Survey—is made up of Chazy, Trenton, Utica, and Hudson River beds. These strata, as a rule, are practically horizontal, or dip at an angle only of two or three degrees.

The Potsdam strata are mostly brown, red, and white sandstones, with some quartzose conglomerates and a few interstratified beds of dolomitic limestone. They form a more or less continuous belt of no great width, curving from the St. Lawrence, below Brockville, north-westerly to the township of Bedford, and from thence in a north-easterly direction to Pakenham on the Ottawa. Instructive exposures may be seen on the river bank between Brockville and Prescott,[*] at Charleston Lake in the township of Escott; in the vicinity of Perth (where *Protichnites* and *Climactichnites* impressions, figures 116, 117, occur); at Otty Lake in Drummond; and at various places in the township of Nepean, where the formation is represented by a white, fine-grained sandstone. Fossils are rare, but at some spots, the small, narrow-beaked brachiopod, *lingula acuminata*, and the still problematical "scolithus cavities," are present in the sand-stone beds. The *Protichnites* and *Climactichnites* impressions occur more especially on ripple-marked sandstones in the immediate neigh-bourhood of Perth.

The Calciferous strata, which consist chiefly of dolomitic and sandy limestones—the "bastard limestones" of the country parlance—ex-tend over a considerable area along the inner edge of the Potsdam belt more especially in the Counties of Leeds, Grenville, Lanark, Carleton, and Russell. The principal fossils in these strata, comprise, a small narrow lingula, *L Mantelli*; a low-spired gasteropod *Euomphalus (or Ophileta) compacta*; some *Pleurotomariæ*, not unlike fig. 219; and the orthoceras *O. Lamarki*. Exposures occur on the river-bank near Prescott; at Smith's Falls and elsewhere on the Rideau canal; and at various points in the townships of Oxford, Young, and Edwardsburg; but throughout its extent it is much obscured by overlying Drift deposits.

The other portions of this district, extending to the Ottawa river, are occupied by Lower Silurian formations. The first of these in ascending order, the Chazy formation, is made up principally of light-coloured, thin-bedded sandstones, followed in places by grey or dark-brown limestones largely employed in the manufacture of hydraulic cement or "water-lime." In these limestones, examples of the ostracod *Leperditia Canadensis* (fig. 165), or a closely allied species, are especially numerous. The brachiopod *Rhynconella plena* (fig. 190) is also very characteristic, and examples occur in great profusion throughout most of the Chazy strata. *Camerella varians* (fig. 192) is another characteristic Chazy type. Many other brachiopods, with *Illenus globosus* (fig. 170) and other trilobites, and numerous gasteropods (*Murchisonia, Pleurotomaria, etc.*) are of common occurrence in these strata. In outlying patches on the gneissoid rocks near Pembroke, etc., in Renfrew, some of the lower Chazy beds contain small nodular masses of a dark-brown colour, regarded as fish-coprolites. They consist of impure phosphate of lime mixed with shells of lingulæ in a fragmentary condition. In these beds also examples of *Lingula Lyelli* (fig. 205) and pleurotomariæ are abundant. Good exposures of the Chazy formation, generally, occur elsewhere at L'Orignal and Hawkesbury on the lower Ottawa, and at various points in Lochiel, Cornwall, Nepean, Huntly, and adjacent townships.

The Chazy beds are succeeded by Black River and Trenton strata, mostly represented by very fossiliferous limestones of a dark grey colour. These occur in force about Ottawa City, as well as in Cumberland township, and throughout the Counties of Russell, Stormont, and Carleton, generally. Among the more typical fossils, the following may be especially cited: *Glyptocrinus decadactylus* (fig. 154), *Lecanocrinus elegans*, and other crinoids, mostly however in the form of stem fragments; many cystideans as *Glyptocystites Logani, etc.*, (fig. 158); *Hemicystites Billingsii* (fig. 161); numerous brachiopods, as *Strophomena alternata* (fig. 194), *Orthis testudinaria* (fig. 191); various species of *modiolopsis* and other lamellibranchiates; numerous gasteropods, as *Maclurea Logani* (fig. 222), *Pleurotomaria Progne* (fig. 219), *Subulites elongatus* (fig. 225), and others; various orthocertites; and several trilobites, more especially *Asaphus platycephalus* (fig. 168), *Cheirurus pleurexanthemus* (fig. 173), and *Trinucleus con-*

centricus (fig. 172). Characteristic exposures of these strata occur at Barrack Hill (Ottawa City); Green's Creek in East Gloucester; near Dunning's Mills in Cumberland; near McCaul's Mills in Clarence; at the High Falls of the South Nation River in Cambridge; and at various sites in L'Orignal, West Hawkesbury, Lochiel, Kenyon, Cornwall and adjacent townships.

The Utica formation, consisting essentially of dark-brown or black bituminous shales, overlies the Trenton, but occurs only in small areas around Ottawa City, and in parts of Cumberland, Clarence, and Plantagenet. Its more characteristic fossils, as occuring in the shales of the Rideau and elsewhere in the immediate vicinity of Ottawa,* include several graptolites, as *Diplograptus pristis* (fig. 132) and other species; various crinoids, brachiopods, lamellibranchiates and gasteropods; and the trilobites, *Asaphus Canadensis* (fig. 169) and *Triarthrus Beckii* (fig. 177). The Hudson River Formation, which follows the Utica Formation in ascending order, is also but slightly developed in this district. It is represented essentially by thin-bedded calcareous sandstones, seen here and there in the townships of Osgoode and Russell, and at points near Ottawa City. The more characteristic fossils of the Formation (*Leptœna sericea*, fig. 196; *Strophomena alternata*, fig. 194; *Ambonychia radiata*, fig. 208; *Calymene Blumenbachii*, fig. 178; *etc., etc.*), are referred to more fully under this formation as occurring in the LAKE ONTARIO DISTRICT.

These various strata, as stated above, are overlaid very generally by Drift deposits and other superficial accumulations. The Drift formation is represented by boulder clays in some places, and by gravel heaps and scattered boulders in others. These latter are arranged here and there in long ridges which occasionally cause obstructions and rapids in the river courses, as at Green's Creek and L'Orignal on the Ottawa. At Barrack Hill and other places around Ottawa, and also in other parts of the district, the limestones beneath the boulder clay often shew ice-grooves and other signs of glacial action. The grooves and striæ have mostly a south-east direction, but some are only a few degrees east of south. To these Drift or Glacial deposits proper, succeed in many places beds of stratified clay, sand and gravel, as a rule free, or practically free, from boulders. They contain shells

---

* See a complete list in papers by Mr. Ami, contributed to the Field Naturalists' Association of Ottawa.

of marine or estuary species of mollusca now living in the St. Lawrence Gulf. At Green's Creek, and at spots on the Madawaska, 200 feet above the present sea-level, some of these later deposits contain also calcareous nodules inclosing well-preserved examples of the capelin *(Mallotus villosus)*, lumpsucker *(Cyclopterus lumpus)* and other small fishes of existing species, with occasional impressions of modern leaves, as those of the *populus balsamnifera, etc.* To still more recent formations belong the deposits of shell-marl and peat, indicating the sites of ancient ponds and swamps, which overlie the surface rocks of the district in Cumberland, Plantagenet, Clarence, Gloucester and other townships. Several of these peat deposits are of wide extent.

In addition to good building-stones, the more important economic products comprise : The white sandstone of the Potsdam formation of Nepean, available as a glass material ; the dolomitic limestone of the Chazy formation of the same township, which furnishes the celebrated " Hull Cement ;" the " shell limestone," also belonging to the Chazy strata, which is largely dressed at L'Orignal for manteltops, tombstones, etc.; and the great peat beds referred to above.*

LAKE ONTARIO DISTRICT.

This district, separated from the palæozoic district of the Lower Ottawa by the intervening gneissoid belt which crosses the St. Lawrence between Brockville and Kingston and extends south-wards into the wild region of the Adirondacks, is underlaid essentially by Lower Silurian formations in nearly horizontal beds. Viewed generally, the strata present a dip of about half-a-degree, or less, towards the south-west ; and, thus, in proceeding westward from Kingston to Hamilton we pass from older to newer formations, as shewn (but with necessarily exaggerated dip, as in all ordinary sections) in the annexed diagram. The latter, however, extending still further westward, shews the outcropping strata of the Erie and Huron district, as well.

The southern limit of the present district forms the entire Canadian shore of lake Ontario. Its eastern and northern limits are

---

* Dr Robert Bell, of the Geological Survey of Canada, has recently expressed the opinion (Ottawa Evening Journal, Febuary 4th, 1888) that a supply of "natural gas" would probably be obtained by boring at suitable sites around Ottawa City.

bounded by the crystalline, Archæan regions already described. Its western boundary is the high escarpment which runs from the Niagara river, near Queenston, in a general westerly direction beyond Grimsby, to the back of Hamilton, then northwards by Dundas, Georgetown, Bellefontaine and Orangeville, to the northern part of Nottawasaga, and from thence north-westerly by the " Blue Mountains," etc., to Cabot's Head on Georgian Bay. From the latter point, eastward, the district includes the shore of the Bay to a short distance beyond the mouth of the River Severn, where the crystalline rocks of the Archæan region come up from beneath its strata. The Lake Ontario District thus includes portions of the Counties of Frontenac, Addington, Hastings, Peterborough, Victoria, Simcoe, Peel, Halton, Wentworth and Lincoln, with the whole of York, Ontario, Durham, Northumberland and Lennox. Numerous lakes, of which Lake Simcoe is the largest, lie within its area, or form a more or less continuous band along its northern edge where the Archæan country commences. The River Trent, which rises in the latter, and flows through a series of small lakes into the Bay of Quinté and Lake Ontario, is its most important river. Among other streams flowing southwards are the Salmon or Shannon, the Moira, the Don, Humber, and Credit. Northward-flowing rivers include the Scugog, which flows from the small lake of that name, 800 feet above the sea-level, into Sturgeon Lake; the Holland river which flows into Lake Simcoe, the northern waters of this lake communicating by the Severn with Georgian Bay; and the Nottawasaga, on the extreme west of the district, flowing into Nottawasaga Bay, a broad inlet of the Georgian Bay waters. In its surface features the district is generally of an undulating character, with but few abrupt inequalities of level. The ground rises gradually from Lake Ontario (232 feet above the sea) in a series of ridges or terraces running in a general east and west direction. These ridges are composed of Drift materials, mostly sand and gravels filled with boulders of various kinds, brought down from northern sources during the Glacial and Post-Glacial epochs, by glaciers or by floating icebergs, when the land was necessarily beneath the sea (*vide* page 207). The highest ridge in Albion and King townships has an elevation of from 700 to 750 feet above Lake Ontario, but becomes gradually lower in its eastern extension. Near the village of Stirling, in Hastings County, it averages 515 feet above the ordinary level of the lake; and a few miles east of this spot it merges into the general levels of the country. Lake Simcoe to the north is 704 feet above the sea, and Balsam Lake (the northern part of which lies within the crystalline area already described) is still higher, its elevation being 820 feet above the sea. Belmont Lake and Rice Lake are each nearly 600 feet, and Scugog Lake (in the midst of the Drift ridges) is nearly 800 feet above the sea level.

The strata of the District consist essentially of Lower Silurian formations, overlaid extensively by Glacial and Post-Glacial deposits ; but on its extreme eastern border a few indications of the (Upper Cambrian) Potsdam formation have been recognised; and along its western limits, the (Upper Silurian) Medina and Clinton formations outcrop beneath or upon the great Niagara escarpment, and connect the Lake Ontario District with the Erie and Huron region of the Province. These strata, apart from a few subordinate anticlinals, exhibit a slight inclination only towards the west or south-west; and they are altogether free from intrusions of trappean or other eruptive rocks. In ascending order, and succeeding each other in successive outcrops from east to west, they comprise representatives of the Potsdam, Black River and Trenton, Utica, Hudson River, Medina, and Clinton formations.

The *Potsdam formation* is very sparingly represented. It occurs in the form of a thin band (or in broken patches) of sandstone and conglomerate lying along the south-western edge of the crystalline gneissoid region in the townships of Loughborough and Storrington. Exposures occur on Loughborough Lake and on Knowlton and Eel Lakes Some of the beds in Storrington are readily friable, and yield a refractory sand employed as a lining for iron furnaces. The best display is at the north end of Knowlton Lake, where the Potsdam strata form a high cliff composed of thin-bedded, more or less ferruginous sandstones of a red and brownish-green colour. A sandstone bed of the same formation, available as a material for furnace hearths, occurs at the Frontenac mining location in Loughborough township, about 12 miles north of Kingston.

The Calciferous and Chazy Formations which intervene, in the Palæozoic area of the lower Ottawa, between the Potsdam and the Trenton formations, have not been recognized within the present district.

The *Trenton (or Black River and Trenton) Formation* is represented chiefly by dark-grey thick-bedded limestones overlaid by limestone shales. The upper beds are in places exceedingly fossiliferous, and most of the lower beds yield excellent building stones. One of these latter beds is capable of employment, for the greater part, as a lithographic stone. It forms a continuous band, running north-west from the vicinity of Kingston, through Marmora, etc., to Georgian Bay. The thickness of the formation in this district averages about 700 or 750 feet. The northern boundary runs from the St. Lawrence a little east of Kingston, through North Hastings, Peterborough, South Victoria, Ontario, and Simcoe, to near the mouth of the River Severn on Georgian Bay—a chain of small lakes indicating its course throughout the greater part of the distance. North of these lakes, gneissoid Laurentian rocks occur in highly tilted beds, whilst the Trenton (or Black River) strata, on the south, occupy a nearly horizontal position. The southern boundary of the formation constitutes the coast of Lake Ontario from Kingston to the neighbourhood of Newcastle a few miles west of Port Hope. From that point the formation turns towards the north-west, and passing across Durham, Ontario, and Simcoe, comes out on Nottawasaga Bay, west of Collingwood. The intermediate country, however is very thickly overlaid in most places by Drift and superficial deposits. Some of the more frequently occurring fossils comprise : *Lithophycus Ottawaensis* (fig. 111); *Stromatopora rugosa* (fig. 127) ; *Monticulipora (or Stenopora) fibrosa* (fig. 136); *Columnaria alveolata* (fig. 142); *Petraia corniculum* (fig. 148) ; *Glyptocrinus decadactylus* (fig. 154) ; *Leperditia Canadensis* (fig. 165) ; *Asaphus platycephalus* (fig. 168) ; *Trinucleus concentricus* (fig. 172) ; *Cheirurus pleurexanthemus* (fig. 173) ; *Calymene Blumenbachii* (fig. 178) ; *Strophomena alternata* (fig. 194); *Leptæna sericea* (fig. 196); *Orthis testudinaria* (fig. 197); *Orthis tricenaria* (fig. 198); *Platystrophia lynx* (fig. 200); *Orbiculoidea Circe* (fig. 202); *Lingula quadrata* (fig. 206); *Ambonychia radiata* (fig. 208); *Conularia Trentonensis* (fig. 218); *Pleurotomaria Progne* (fig. 219); *Murchisonia gracilis* (fig. 220); *Maclurea Logani* (fig. 222); *Platyceras angulatum* (fig. 223); *Subulites elongatus* (fig. 225); *Orthoceras proteiforme* (fig. 232).

Characteristic exposures of the Black River and Trenton strata of this district may be seen more especially at the following spots: Kingston city and environs, where the limestone beds contain in places crystallized examples of celestine (page 131), gypsum (page 131), and other minerals; Clare River in Sheffield township; Crow River near Marmora; Shannonville; River Moira at Belleville and elsewhere; Ox Point, Bay of Quinté; Rednersville on the south shore of the Bay; Trenton, Healy's Falls, and elsewhere, on the Trent River; environs of Peterborough; river banks at Lakefield; Quarry near Burleigh Falls on Stony Lake; Bobcaygeon; Balsam Lake, south shore; Fenelon Falls; River Scugog near Lindsay; north east shore of Lake Couchiching; and Lake St. John in Rama township, where the strata are seen in unconformable contact with the underlying Laurentian gneiss.

The *Utica Formation* is made up almost entirely of dark-brown or black bituminous shales, interstratified here and there with beds of dark limestone, the total thickness of the Formation in this district being under 100 feet. The shales weather light grey, and yield by atmospheric disintegration a soil of much fertility. In some localities, as in the townships of Collingwood and Whitby, they are sufficiently bituminous to yield a considerable quantity of mineral oil by distillation. The Collingwood shales, for example, have yielded as much as 20 gallons of oil to the ton of shale; but the distilleries have long been closed, in consequence of the large and cheap supply of mineral oil furnished by the petroleum wells of Western Ontario and the United States. The more characteristic Utica fossils comprise: *Diplograptus pristis* (fig. 132); *Asaphus Canadensis*, the pygidium especially, (fig. 169); *Triarthrus Beckii* (fig. 177); *Lingula obtusa* (fig. 207); *Rhynconella increbescens* (fig. 191), and some other species which occur in both the underlying Trenton and overlying Hudson River Strata—the formations constituting strictly a single group. The trilobites *Asaphus Canadensis* and *Triarthrus Beckii* appear however to be confined to the Utica strata. These range along the shore of Lake Ontario from a little west of Port Hope to a short distance west of Whitby; sweeping from these points, in a comparatively narrow band, towards the north-west, and coming out on Nottawasaga Bay a mile or two west of Collingwood; but within the intervening space the formation is entirely concealed by over-lying Drift and superficial deposits. Good exposures may be seen, however, in the vicinity of Whitby, and on the lake shore under the " Blue Mountains " a few miles west of Collingwood harbour.

The *Hudson River Formation* consists essentially of shaly thin-bedded sandstones, mostly of a greenish-grey or drab colour but weathering rusty-brown. These shaly sandstones, formerly known as Lorraine shales, are interstratified here and there with a few cal-careous beds. The total thickness of the Formation in this district is about 750 feet. Its fossils are in great part identical with those of the underlying Trenton and Utica beds, these three so called formations forming properly a single subdivision of the Lower Silurian series: dolomitic and other limestones characterising the lower part; bituminous shales (with more or less distinct fossils) the central portion; and arenaceous shales, holding many of the lower limestone fossils, the upper part. Some of the more common fossils of the Hudson River strata comprise : *Climacograptus bicornis* (fig. 132); *Diplograptus pristis* (fig. 132*); *Monticulipora fibrosa* (fig. 136); *Glyptocrinus decadactylus* (fig. 154); *Calymene Blumen-bachii=C. senaria* (fig. 178); *Asaphus platycephalus* (fig. 168); *Trinucleus concentricus* (fig. 172); *Strophomena alternata* (fig. 194) ; *Leptœna sericea* (fig. 196); *Ambonychia radiata* (fig. 208) ; *Modiola modiolopsis* (fig. 209); *Murchisonia gracilis* (fig. 220); *Bellerophon bilobatus* (fig. 221); *Cyrtolites (Bellerophon) ornatus* ; and *Orthoceras crebriseptum* (fig. 233). The formation constitutes the shore-line of Lake Ontario from the River Rouge, in the township of Pickering, to the River Credit, west of Toronto. From these points, although much obscured by overlying Drift and more recent deposits, it extends to the north and north-west, and forms the shore of Nottawasaga Bay in the townships of Collingwood, St. Vincent, Keppel and Albemarle. Good exposures, at most of which fossils may be obtained, occur on the banks of the Humber, Mimico, Etobicoke, and Credit ; also at Point Boucher on Nottawasaga Bay, and at Cape Rich, Point William, Cape Crocker, and Point Montresor, farther west.

The *Medina Formation* is regarded as forming the base of the Upper Silurian series It is chiefly made up of red marls and soft red sandstones, interstratified with red and green arenaceous shales, and capped by a bed of fine-grained sandstone—commonly known as the "grey-band"—averaging about ten or twelve feet in thickness, but presenting in places a thickness of over thirty feet. This grey sandstone is largely quarried for building purposes in the township of Nottawasaga, and at Dundas, Hamilton, and other places along the great Niagara escarpment. The Medina formation, presenting collectively a thickness of about 600 feet, extends along the shore of Lake Ontario westward from the vicinity of the River Credit to the escarpment (or so called "mountain") at Hamilton; and it occupies the strip of land lying between the escarpment and the lake as far as the Dominion boundary on the Niagara River. Northwards, it extends through the townships of East and West Flamborough, Nelson, Caledon, Mono, Mulmur, and Nottawasaga : from· whence, turning westward, it continues along the course of the escarpment, but with greatly diminished thickness, and loss of the "grey-band," to the base of the bold promontory known as Cabot's Head on Georgian Bay. Fossils are of comparatively rare occurrence in this formation. The principal comprise : the fucoid *Arthrophycus Harlani* (fig. 113) and a small lingula, *L. oblonga* (fig. 127). The best exposures occur about Wellington Square, Hamilton, Dundas, Queenston, and Georgetown, and at the Dennis quarries in Nottawasaga.

The *Clinton Formation* which immediately succeeds the Medina, consists at its lower part essentially of green, red, and greyish shales, in places more or less ferruginous; and at its upper part mostly of dolomitic limestones. The lower portion should properly be referred to the Medina Formation, and the upper part to the Niagara. The lower shales hold casts, often in great abundance, of *Arthrophycus Harlani* (fig. 113), and remains of a small coral (or bryozoon) *Hælipora fragilis*. The formation, generally, follows the course of the Niagara escarpment between the Niagara River and Georgian Bay— gradually increasing in thickness, from a few feet to about 100 feet, between these points. Conventionally, all the beds lying above the

" grey-band " and beneath the limestone containing shells or casts of the brachiopod *Pentamerus oblongus* (fig. 193) are given to the Clinton Formation. Exposures occur, chiefly, on the Welland Canal and elsewhere in the vicinity of Thorold; also, with greatly increased thickness, immediately around Hamilton; in road-cuttings at Dundas; in the gorges of the Noisy and Mad Rivers in Nottawasaga; at spots on Beaver River; and at Cape Chin, Cape Commodore, and Cabot's Head on Georgian Bay. The celebrated "Thorold cement " is manufactured from a limestone of this Formation.

Glacial and Post-Glacial Deposits :—The greater portion of the Lake Ontario District is overlaid by clays, sands and gravels of the Glacial and Post-Glacial periods. Beneath these deposits, many strata, limestones especially, are seen when first uncovered, to be striated and polished by glacial action. The striæ run in some places towards the south-west, and in others towards the south-east; and the same rock-surface occasionally exhibits both S.W. and S.E. striations. These so called superficial deposits admit in ascending order of the following subdivisions : (i.) Boulder Clay formation; (ii.) Erie Clay formation ; (iii.) Saugeen Clay and Sand formation ; (iv.) Artemisia Gravel formation ; (v.) Algoma Sand formation ; and (vi.) Recent deposits, proper, as shell-marls, bog iron ore, and peat.

The Boulder Clay or "Till " consists of thick beds of boulder-holding clay, without subordinate stratification. The boulders vary in size from small pebbles to masses of considerable dimensions ; and they consist for the greater part of gneissoid and other crystalline rocks brought down from the Archæan northern region by glaciers or icebergs during the long period of cold which more or less immediately preceded the existing epoch. See *ante*, page 207. Many of the included boulders, even those of small size, shew marks of polishing and striation. In some places, especially on ridges and high lands within the district, the Drift is represented by accumulations of boulders alone, without accompanying clay. Stratified clays, also holding boulders in some localities, overlie the unstratified drift clays very generally, and are known as " Erie clays." They are more or less calcareous, and yield white or light-coloured bricks. Deposits occur at North Toronto, Cobourg, Belleville, Dundas, and numerous other places within the district. The so called Saugeen clays (which yield red bricks) and their accompanying sands succeed the Erie clay

deposit, and occur more or less all over the Province.   Beds of coarse
gravel mixed with boulders, known as the Artemisia Gravel formation, occupying a somewhat higher horizon than that of the stratified
Saugeen clays and sands, occur especially (as regards this district)
along the base of the Niagara escarpment in the townships of Arte-
mesia, Osprey, Mulmur, Mono, and Albion, where they form the
chief mass of the " Oak Ridges."   These latter extend from the escarpment eastwards, and rise in the township of King and adjacent
townships to elevations of from 700 to 760 feet above Lake Ontario.
In many places the gravels are seen to be distinctly stratified, and
here and there they hold gneissoid and other boulders of large size.
The Algoma sands consist of still higher accumulations of white or
light-coloured lacustrine sands, practically free from stones or boulders,
and often shewing signs of oblique stratification.   These lacustrine
sands occur extensively throughout the district, as seen along the
line of the Canadian Pacific Railway between Toronto and Peterborough, in cuttings on the Grand Trunk between Scarborough and
Hamilton, around Barrie on Lake Simcoe, and elsewhere.   Many
fresh-water shells *(Unio complanatus, Cyclas similis, Planorbis
trivolvis, Limnæa palustris, etc.,)* identical with those now living in
our lakes and streams, occur at various levels in these and other related beds—their presence in these deposits apparently indicating the
former union of our lake-waters into one vast fresh-water sea.   In
this case, the water must have been held up in the east by a greater
elevation of the gneissoid belt of rock which crosses the St. Lawrence
between Brockville and Kingston, and expands into the wild district
of the Adirondack Mountains in the State of New York ; or perhaps
by an enormous glacier, descending from the latter region and extending northwards into Canada.   The shells of recent species of
mollusca found in the Post-Glacial deposits east of this gneissoid
belt, belong to marine or brackish-water species ; whilst those within
the Lake Ontario District, and country to the west, are all fresh-
water types.   The most abundant of the recent deposits, proper, of
this region, are the beds of shell marl which occur at numerous localities, forming the floor and margin of small lakes and swamps.
This substance is a white or light-coloured calcareous deposit, containing minute shells of species of *cyclas, planorbis,* and other fresh-
water mollusca.

THE ERIE AND HURON DISTRICT.

This district—occupied throughout by comparatively undisturbed limestones and other strata of the Upper Silurian and Devonian series, with overlying Drift clays and sands, and more recent superficial deposits—is essentially an agricultural area of great fertility. It lies immediately west of the Lake Ontario District, and is separated from the latter by the great Niagara escarpment which runs, as stated above, from the Niagara River—by Queenston, Thorold, Grimsby, Hamilton, Dundas, Georgetown, etc.,—to Cabot's Head on Georgian Bay. On the south, the district is bounded by Lake Erie; and on the west by Lake Huron. The greater portion of its area forms an elevated table-land of from 1,000 to 1,200 feet above the sea. Along its north-eastern edge the ground rises in places to an altitude of nearly 1,600 feet; but it slopes gradually towards Lake Erie on the south (565 feet above the sea), and towards Lake Huron (578 feet above the sea-level) in the west. Its surface, except where cut by river-valleys, is generally even; and the district presents a marked contrast to the lower region of Lake Ontario by the almost total absence of inland bodies of water. It is traversed, however, by many important rivers, and especially by the Grand River, flowing into Lake Erie; the Thames, flowing into Lake St. Clair; and the Maitland and Saugeen, flowing into Lake Huron. The eastern and north-eastern boundary, along the great escarpment, is also cut through by numerous smaller streams. These enter the Lake Ontario District, consequently, through deep ravines, many of which are of a very wild and picturesque character.

The strata of the district consist, in its more eastern portions, of Upper Silurian representatives, with various Devonian formations in the west. They follow each other (in ascending order) from north-east to south-west, generally, and comprise: The Niagara formation (with some Upper Clinton beds), the Guelph formation, the Onondaga or Gypsiferous, and the Eurypterus or Lower Helderberg formations, of the Upper Silurian series; and the Oriskany, Corniferous, Hamilton or Lambton, and Chemung-Portage formations, of Devonian age. These strata, although practically undisturbed, are affected by several moderate anticlinals running across the more central part of the district in a general east and west or south-west direction; and

21

it is thought that the petroleum of this part of the region has been brought towards the surface by fissures resulting from these anti-clinals. A transverse or nearly north and south fold, forming a trough or synclinal filled with higher Devonian strata (of the Hamilton or Lambton formation), also occurs in the south-western portion of the district between Lake Erie and the south point of Lake Huron.

The *Niagara Formation* in this district, is made up of dark-gray calcareous shales and thick-bedded limestones, both of which are more or less magnesian and bituminous. Its lower limit is regarded, con-ventionally, as indicated by a magnesian limestone holding shells of the brachiopod *Pentamerus oblongus* (fig. 19); but this "Pent-amerus bed" is referred by the New York geologists to the upper part of the underlying Clinton formation. At Niagara Falls, the dark shales present a thickness of about 80 feet, and form the lower portion of the escarped face over which the cataract breaks, whilst the upper portion of the cliff is composed of thick-bedded limestones. Along the gorge, the shales are mostly concealed by the slope or talus of detrital matter which rests against the cliff face ; but they may be seen on the side of the steep road which leads from the old ferry to the Clifton House, and at several other spots. Some of the beds of this formation yield excellent hydraulic lime, much of the "Thorold cement" being manufactured from the lime obtained from them. Many of the Niagara beds are rich in fossils. The more common species comprise : The corals: *Favosites Gothlandica (=F. Niagarensis* fig. 137*);* and *Halysites catenulatus* (the so-called "chain coral" fig. 139); the Bryozoon, *Fenestella elegans* (fig. 181) ; the Brachiopods : *Pentamerus oblongus* (fig. 193) ; *Orthis elegantula* (fig. 197) ; *Spirifer Niagarensis* and *S. radiatus* (fig. 184) ; and the Trilobites, *Calymene Blumenbachii* (ranging upwards from earlier strata, fig. 167, 178) ; *Homalonotus delphinocephalus* (fig. 179) and *Dalmannites limulurus* (fig. 175). The formation extends a few miles westward from the edge of the great line of escarpment— already described as running from the Niagara River, by Hamilton, Georgetown, etc., to Cabot's Head on Georgian Bay—and then passes under the succeeding Guelph formation. It thus marks the eastern and northern limits of the table-land of which the Erie and Huron district largely consists. Good exposures occur more especially at the Niagara Falls and on the adjacent banks of the river (where the limestones are overlaid by terraces of fresh-water clay and sand of Post-Glacial age); also on the Welland Canal near Thorold; along the upper part of the escarpment or "mountain" by Grimsby, Hamilton, Dundas, Ancaster, Rockwood, etc.; on the River Credit in Caledon, as at Bellefontaine and elsewhere, and on the Nottawa and Beaver Rivers, where it forms high and precipitous cliffs; at various other points in Mulmur, Nottawasaga, Artemisia, and Euphrasia; about Owen's Sound; and at Cape Paulet, Cape Chin, and the upper part of Cabot's Head. In the immediate vicinity of Rockwood, some large caverns occur in a dolomitic limestone (thickly interspersed with crinoid stems) belonging to this formation; and deceptive veins and strings of galena have been noticed in the same township (Eramosa), as well as in the Niagara strata of Mulmur and Clinton.

The *Guelph Formation* is represented for the greater part by white or light-coloured dolomites of a peculiar semi-crystalline or granular texture, containing, among other fossils, large casts of a lamellibranchiate mollusk, the *Megalomus Canadensis* (fig. 210). The coral, *Amplexus laxatus* (fig. 143); the brachiopod, *Trimerella acuminata* (fig. 201); several species of *Murchisonia, Holopea Guelphensis* (fig. 224); and *Phragmocerus Hector* (fig. 229), are also characteristic. The formation extends over a considerable area, chiefly in the counties of Waterloo, Wellington, and Grey, but it is greatly concealed by overlying Drift and other superficial deposits. The principal exposures occur on the River Speed in the vicinity of Guelph; at Elora, near the junction of the Grand and Irvine Rivers, where the strata form high cliffs; also on other parts of the Grand River, as at Fergus, Preston, and Galt; near Hespeler, again, on the railway between Guelph and Brantford; and on the rocky Saugeen River in Bentinck. At most of these localities excellent building-stones are obtained.

The *Onondaga or Gypsiferous Formation* consists, in Canada, of thin-bedded dolomites of a yellowish or pale grey colour, associated with greenish calcareo-argillaceous shales and with large masses or irregular beds of gypsum. These deposits appear to have been largely formed from precipitates thrown down in ancient salt-lakes or bays in which an active evaporation was going on. They contain only a few obscure traces of organic remains; but hopper-shaped and prismatic casts, derived from crystals of ordinary salt, soluble sulphates, etc., are not uncommon in some of their beds. The gypsum is mostly of an earthy or granular texture, and is always more or less mixed with carbonates. The formation enters Canada a short distance above the Falls on the Niagara River, and includes the whole of Grand Island in this portion of its area. From thence, it follows the general outcrop of the Guelph formation to the vicinity of the Saugeen River on Lake Huron. It thus includes portions of Welland, Haldimand, Brant, Oxford (north-east corner), Waterloo, Perth and Bruce; but throughout much of this area it is covered by Glacial and other superficial deposits. Exposures may be seen near Waterloo village, Bertie township, on the Niagara River; along the Grand River between Cayuga and Paris, and near the Don Mills; on the Upper Saugeen, near Ayton and Neustadt, in Normanby township; around Walkerton on the Saugeen, in Brant, and at various points down the river: as at the elbow in the south-west corner of Elderslie, a little below Paisley. The gypsum or "plaster" deposits are chiefly quarried at Cayuga, Indiana, and York, in the township of Seneca; at Mount Healy and elsewhere in Oneida; largely around Paris; and at various places in Brantford township. Some of the dolomitic and argillaceous shales of the formation, as those which outcrop near Walkerton, etc., furnish valuable material for the manufacture of hydraulic cement; and it is apparently from this formation that the brine—obtained in the vicinities of Goderich, Seaforth and Clinton, by deep borings through overlying deposits— is essentially derived.

The *Lower Helderberg or Eurypterus Formation*, as occurring in this district, represents merely a small portion of the "Lower Helderberg group" of New York. It is made up of a few thin-bedded dolomites, with some interstratified shales and a brecciated bed (composed chiefly of dolomite fragments) at its base. These strata, which collectively do not exceed fifty feet in thickness, appear to represent the lower portion—the "Water-lime" or "Tentaculite limestone"—of the Helderberg series proper. They are chiefly characterized by the presence of fragmentary examples of a peculiar crustacean, *Eurypterus remipes* (fig. 180), belonging to the Mero-stomata, an almost extinct order, but represented in existing nature by the Limulus or Xiphosure. The formation probably extends as a thin band along the western border of the Onondaga formation between Lake Erie and Lake Huron, but the only known exposures are in the Lake Erie townships of Bertie and Cayuga.

The *Oriskany Formation*, the first in ascending order of the Devonian series, is also but sparingly present in the district. It is represented chiefly by a layer of chert or hornstone (containing much iron pyrites) at the base, with a succeeding brecciated béd (made up in part of chert fragments), and some quartzose grits or sandstones : the entire thickness varying from about six to ten feet. Its fossils are chiefly identical with those of the overlying Corniferous strata, but are mixed in places with Upper Silurian types. Some of the more common comprise : *Favosites Gothlandica* (fig. 137); *Zaphrentis prolifica* (fig. 144) ; *Strophomena rhomboidalis* (fig. 195); *Atrypa reticularis* (fig. 188) and *Calymene Blumenbachii* (fig. 178 . The formation enters Canada in the township of Bertie in the north-east corner of Lake Erie, and appears to run as a thin band along the southern edge of the Eurypterus or Onondaga formation at least as far as the country of Norfolk ; but the only known exposures occur at Bertie, Dunn, North Cayuga, Oneida, and Windham. From the exposure in North Cayuga, a little north of the Talbot Road, good millstones have been quarried.

The *Corniferous Formation*, as recognized in Western Canada, includes the " Onondaga limestone " and " Corniferous limestone " of New York geologists. It is made up essentially of more or less bituminous limestones, containing, in places, nodular masses of chert, or interstratified with bands of that substance, and associated here and there with beds of calcareous sandstone and bituminous shale. The thickness of these strata, collectively, is estimated at about 200 feet. The limestones contain, as a rule, a great abundance of silicified fossils, mostly brachiopods, corals, and crinoidal stems. The common forms comprise : The corals, *Michelinea convexa* (fig. 138) ; *Syringopora Maclurei* (with coarse cell-tubes, fig. 140) and *S. Hisingeri* (with narrow-cells, fig. 141) ; and the simple horn-shaped types, *Zaphrentis prolifica* (fig. 144) and *Z. gigantea*, the latter often a couple of inches in diameter and five or six inches in length. The brachiopods, *Strophomena rhomboidalis* (fig. 195) ; *Atrypa reticularis* (fig. 188) ; *Spirifer mucronatus* (fig. 185) ; *S. gregarius* (fig. 184 *bis*) ; *Spirigera concentrica* (fig. 187) and *Stricklandia elongata*—most of which occur also in higher Devonian strata. The trilobite *Phacops bufo* (fig. 174) is also a common Corniferous type. In the district under review, the formation occupies two large areas, separated by a broad intervening belt of the succeeding Hamilton or Lambton formation. The more eastern of these areas extends over portions of Welland, Haldimand, Norfolk, Brant, Oxford, Perth, Huron and Bruce ; and the western area occupies parts of Lambton, Kent and Essex. Exposures occur more particularly on or near the shore of Lake Erie in the following townships : Bertie, Humberstone (as at Rama's farm, near Port Colborne, a noted fossil locality), Dunn, Rainham, Walpole and Woodhouse. Also in North and South Cayuga ; near Woodstock village ; and at St. Mary's (another locality especially rich in fossils). The formation outcrops likewise at various places in Carrick, Brant, Bruce and Kincardine ; and again, further south, as near Port Albert and Goderich, and in the vicinity of Amherstburg in Malden. At many of these localities, and especially at the large exposure in Malden, building-stones of very superior quality are obtained.

The *Hamilton or Lambton Formation,*[*] as defined in Canada, represents merely the middle portion of the " Hamilton formation " of the New York geologists. It consists mainly of soft calcareous shales associated with some beds of encrinal limestone. The fossils are identical for the greater part with those of the Corniferous formation ; but the brachiopods, *Spirifer mucronatus* (fig. 185) ; *Spirigera concentrica* (fig. 187) ; *Atrypa reticularis* (fig. 188) ; and *Orthis Vanuxemi* (fig. 199) are especially abundant in these higher beds. The formation in this district is estimated at about 250 feet in thickness. It extends across the counties of Norfolk, Elgin, Kent, Middlesex and Lambton, and also the south part of Huron, but is much obscured throughout this area by overlying clays, sands, and other Drift and superficial deposits. The best exposures occur in the township of Bosanquet, in the north west corner of Lambton. The formation is chiefly of interest as constituting the essential petroleum

---

[*] The name by which this formation is commonly known, is derived from the village of Hamilton, in Madison county, New York. It is often supposed, in Canada, to refer to our city of Hamilton, on the western extremity of Lake Ontario, where the strata belong to a much lower horizon—that of the Medina formation, lying at the base of the Upper Silurian series. In consequence of this very prevalent misconception (of which some curious instances might be given), the writer proposed some years ago to call this group of strata, as occurring in Canada, the *Lambton formation*, after the county in which it is principally developed in Western Ontario.

area or *oil district* of Western Canada, although the deeper borings, from which the petroleum is chiefly obtained, appear to pass through its strata into the underlying Corniferous formation. Natural springs have been noticed in various parts of the district, as in Mosa, Enniskillen, Zone, Orford, etc. In the township of Enniskillen, overflows from springs of this kind have formed deposits of solid bitumen or "mineral tar," varying in thickness from an inch or two to nearly a couple of feet, and extending over an acre or more of ground. One of these deposits or "gum beds," in the northern part of the township, is covered by several feet of Drift clay ; and in places it presents a leafy or shaly texture, and contains impressions of leaves and insects. As proved by the very different results obtained in many instances from closely contiguous borings, the petroleum is evidently confined to comparatively narrow and tortuous channels, within limited belts of country. These belts are characterized, both in the United States and Canada, by the presence of anticlinals, by which a more or less fissured condition of the strata has been produced. The petroleum in these fissures is almost always accompanied by salt or brackish water, and inflammable gas is usually emitted on the first tapping of the fissure. As a rule, the wells become gradually impoverished, and frequently end by yielding water only. The petroleum, as first obtained, is of a dark colour and more or less viscid consistency. When decolourized and purified it loses about forty per cent—five barrels of crude oil yielding about three barrels of refined oil.

The *Portage, or Portage-Chemung Formation*, as seen in Canada, is made up of dark bituminous shales, holding in places large calcareous concretions, and also much iron pyrites, with occasional fish-scales and spines, and impressions of long-flattened stems of a calamite (fig. 119). Here and there these shales become coated, by weathering, with a yellow crust of oxalate of iron. The formation extends probably over a considerable area around Lake St. Clair and the adjoining country ; but it is thickly overlaid by Drift and superficial deposits throughout the greater part of this area, and the only well-recognized exposures occur at Kettle-point, or Cape Ipperwash, in the township of Bosanquet, on Lake Huron, and at one or two places in the townships of Warwick and Brooke. As seen at these spots, the thickness of the formation does not exceed twelve or fifteen feet.

Deposits of Glacial, Post-Glacial and more recent age are spread very generally over the Silurian and Devonian strata of this district—a wide break in the geological succession occurring here as in other parts of the Province. These deposits comprise in ascending order : (i.) The Boulder-clay or Lower Drift formation ; (ii.) the stratified "Erie Clay" formation ; (iii.) The Saugeen Clay and Sand formation ; (iv.) The Artemisia Gravel and (v.) the Algoma sand deposits ; with (vi.) a series of recent accumulations, proper.

The *Lower or true Drift Formation* consists of a thick deposit of unstratified clay, holding Laurentian and other boulders. It appears to be essentially a Glacier formation, derived in chief part from the grinding action of ice on the surface of the subjacent rocks. As a rule, it is greatly concealed from observation ; and much of its original substance has undoubtedly been removed, or otherwise re-arranged in the form of succeeding deposits, by the subsequent action of water. The Upper Drift deposits consist principally of dark blue or gray calcareous clays, arranged in distinct layers, and containing, as a rule, numerous stones and boulders, but no shells or other fossils. These "Erie clays" yield white or light-yellow bricks. In places, as near Brantford, etc., their planes of stratification are greatly contorted. Good displays occur along the north shore of Lake Erie generally ; also near Clark Point, etc., and adjacent coast of Lake Huron ; and at Woodstock, St. Mary's, London, and elsewhere throughout the district.

The Lower Fresh-water deposits, or "Saugeen clays and sands," where in contact with the "Erie" or underlying boulder-holding clays, commonly rest on the denuded surface of the latter. The clays present a very general brown colour, and although more or less calcareous, they yield, as a rule, red bricks. All are distinctly strati-fied, and, in most cases, boulders are but sparingly present in them. Good exposures, showing the blue or Erie clay below, occur more especially near Port Talbot and Port Stanley on Lake Erie. Others may be seen around Woodstock ; and also between Clark Point and Port Frank on Lake Huron, as well as at various spots on the Saugeen, around Walkerton, and throughout the township of Brant generally. Layers of sand and gravel are commonly associated with these clays ; and a deposit of coarse gravel with boulders—the "Artemisia gravel" of the Geological Survey—extends largely throughout the country between the Grand River and Georgian Bay, and is seen near Brantford and elsewhere to overlie the blue or Erie clay. In many parts, again, of the Erie and Huron district (as in the more western portion of the Lake Ontario region, already described) the brown clays, or in their absence the underlying deposits, are capped by lacustrine sands and gravels, some of which contain shells of fresh-water mollusca—species of unio, cyclas, amnicola, valvata, planorbis, physa, limnæa, melania, &c.—still inhabiting our lakes and streams; and similar shells are occasionally found in the clay beds. Terraced deposits containing fresh-water shells of this kind occur especially around Niagara Falls. Other examples have been seen in the vicinities of Paris, Brantford, Walkerton, &c. In addition to these Glacial and Post-Glacial accumulations, various deposits of still more recent origin occur within the district. The principal comprise : the sandy flats of the Grand River and other streams; the beds of shell-marl which underlie and margin most of the swampy areas ; the ochre deposits of the counties of Middlesex and Norfolk ; and the peat beds of Humberstone and Wainfleet on Lake Erie.

THE MANITOULIN DISTRICT.

This district may be regarded as an outlying portion of the Ontario and Erie districts combined, the Silurian strata of these extending into it, and ranging continuously throughout its area. It comprises the Great Manitoulin Island, eighty miles in length, lying along the north shore of Lake Huron, with the La Cloche and other smaller islands between the Manitoulin and the Lake Huron coast, and Cockburn Island, St. Joseph's Island, Campement d'Ours, &c., farther west. Drummond Island belongs also, geologically, to the district, but lies beyond the Canadian boundary. The more northern portion of the Great Manitoulin contains numerous lakes, and its north shore is indented by comparatively deep bays. These appear to lie in synclinal folds, formed by a series of undulations (with north and south axes) which traverse the Island throughout its length.* The strata of the district succeed each other from north to south, the

---

* See "Reports on the Manitoulin Islands," by Dr. Robert Bell, by whom these anticlinals were first pointed out, in Geological Survey Reports for 1866 and 1867.

general dip being in the latter direction. They comprise a slight development of Huronian quartzites, with representatives of the Chazy, Black River and Trenton, Utica, Hudson River, Clinton, Niagara, and Guelph formations. The Huronian outcrops occur principally in the form of bare, rocky ledges, but are only seen at one or two places, principally at Shequenandod village, near the eastern extremity of the Great Manitoulin, on La Cloche Island, and on the Island of Campement d'Ours, near the entrance to St. Mary's River. Exposures of reddish marls and light-coloured sandstones on the north side of La Cloche Island are commonly referred to the Chazy division ; and the thin bedded sandstones (provisionally known as the Ste. Marie sandstones), which occur in small outlying patches on Campement d'Ours and St. Joseph's Island, are thought, as regards their geological positon to represent the same formation. The southern portions of La Cloche Island, and the smaller islets immediately west, are occupied entirely, or essentially, by dark gray dolomitic limestones of the lower part of the Trenton (or so-called Black River) series—the higher beds merging into the Trenton division proper, and supporting at one or two points small strips or patches of Utica shale. The Bigsby, Thessalon, and other rocky islands farther west, and a large part of Campement d'Ours and St. Joseph's Island, belong to the same series. From the La Cloche group of islands these Trenton strata extend across the intervening channel, and crop out in several places as a fringe along the north coast of the Great Manitoulin. They show principally in the Manitowaning headland, and also between Little Current and West Bay. Southwards, the dark bituminous shales of the Utica formation come up, and range entirely through the islands. Good exposures occur at Shequenandod village (where the shales incline against an outcrop of Huronian quartzite), and at Cape Smyth. At the latter spot the formation is capped by a considerable thickness cf arenaceous shales and sandstones, very rich in fossils, belonging to the Hudson River series. This latter formation ranges also entirely through the Great Manitoulin, and extends over Barrie Island on the north. It is followed along its southern border by a series of strata holding Clinton fossils. These strata consist mostly of light-coloured dolomites capped by a bed of red marl—the best soil on the island, according to Dr. Bell, resulting from the disintegration of the latter. South of these

Clinton beds, a steep escarpment of Niagara limestone runs through the central part of the island, facing the north, following in its strike the same east and west direction as the other formations of the district. From the top of the escarpment, the limestones extend in a series of steps to the south shore, where they become covered in places by patches of semi-crystalline dolomites belonging to the Guelph formation. The southern portion of the Great Manitoulin is thus occupied entirely by these Upper Silurian strata, and broad shelves of bare limestone-rock form large portions of its surface. A continuous outcrop occurs along this south shore ; but the best exposures are seen in the lower beds along the line of escarpment, more especially about Lake Manitou and Lake Wolsey (and intervening country), and around the south shore of Bayfield Sound. Cockburn Island, immediately west of the Great Manitoulin, is also underlaid throughout its whole extent by Niagara limestone.

The other formations recognized within the district consist of Drift clays and higher sand deposits—the " Algoma Sand " of the Geological Survey. The clays appear to belong essentially to the lower or unstratified Drift. They occur in great thickness upon St. Joseph's Island, and are overlaid very generally on Cockburn Island by the higher Algoma sands. Both divisions are also seen on the Great Manitoulin and elsewhere throughout the district. Detached boulders, mostly of Huronian rock, occur likewise in many places ; and glacial striæ and furrows are seen on almost all the exposed rock-surfaces. The striæ have a general south-westerly direction, but vary from a few degrees west of south to about S. 50° W.

Petroleum springs occur both on the Great Manitoulin and on some of the other islands of the Manitoulin group. The petroleum appears to come from the Utica shales ; but although the formation has been penetrated and even traversed by several wells or borings, no permanent supply has hitherto been obtained.

THE NORTHERN PALÆOZOIC AREA.

This area comprises a large extent of comparatively flat country, ranging around the south shore of James' Bay to beyond the Province boundary on Albany River, and stretching from these points in

a general south-westerly direction towards the " Height of Land."
Moose River on the east, and Albany River on the west, flow mainly
within its limits.

The area still remains practically unexplored, but it is known to
be underlaid around its more southern and western portion by cal -
careous strata of Upper Silurian age (resting immediately on the
Huronian and Laurentian rocks of the great Archæan region around
the Height of Land); and northerly and easterly, by Devonian
formations extending to the shores of the Bay.   These palæozoic
strata, however, are almost entirely concealed by a thick surface
covering of Glacial and Post-Glacial clays.   Where visible on Moose
River, the underlying rock, according to Dr. Bell (Report of 1875)
consists of bituminous grey and drab limestone, holding Devonian
fossils* ; and on the banks of this river Dr. Bell observed a deposit
of spathic-iron-ore weathering into brown hematite, and also grey and
white gypsiferous deposits.   In. the post-glacial clays (Report for
1877) the same observer discovered a considerable bed of lignite,
varying in thickness from eighteen inches to six feet.   These post-
glacial deposits contain also in many places detached crystals of
gypsum (see page 131), and shells of *mya truncata, tellina græn-
landica, saxicava rugosa* (figs. 214, 215, 216, page 278), so characteristic of the post-glacial deposits of Eastern Canada.

Our knowledge of the country around James' Bay is derived
principally from the valuable Reports on Hudson's Bay by Dr. Bell
of the Geological Survey of Canada, but the region referre1 to in
these Reports lies mostly beyond the area now under consideration,
and, indeed, beyond the Province of Ontario.†   The following ex-
tract, however, from the Report for 1879 applies in chief part to the
present district.   " To the south and south-west of James' Bay, in
the latitude of Devonshire and Cornwall, there is a large tract in
which much of the land is well-wooded, but although little or no rock
comes to the surface over an immense area, neither the soil nor the

---

* In addition to the Devonian species cited in Dr. Bell's report, the author has found in col-
lections made by the stipendiary magistrate, Mr. E. B. Borron—*Zaphrentis prolifica, Z. gigantea, Cystiphyllum Senacaense, Amplexus laxatus, Phragmoceras (Hector?)*, and the pygidium
of *Dalmanites limulurus* or a closely allied species.

† Useful information may also be derived from the Report presented to the Ontario Govern-
ment by Mr. Borron, in 1882, on the portion of the Hudson's Bay basin lying within the province boundary.

climate is suitable for carrying on agriculture as a principal occupation until we have passed over more than half the distance to Lake Winnipeg. This region appears to offer no engineering difficulties to the construction of a railway from the sea-coast to the better country beyond. Some of the timber found in the country which sends its waters into James' Bay may prove to be of value for export. Among the kinds which it produces, may be mentioned white, red, and pitch pine, black and white spruce, balsam, larch, white cedar, and white birch. The numerous rivers which converge towards the head of James' Bay offer facilities for "driving" timber to points at which it may be shipped by sea-going vessels."

PROVINCE OF QUEBEC.

INTRODUCTORY NOTICE.

This Province is separated from Ontario on the west by almost the entire course of the Ottawa River, and by a line drawn directly north from Lake Temiscamingue: but a small area on the Ontario side of the Ottawa at its junction with the St. Lawrence, although properly within the limits of the western province, is given to Quebec. This area includes the small counties of Vandreuil and Soulanges. On the south, the Province of Quebec is bounded by the northern portions of the States of New York, Vermont, New Hampshire, and Maine; and by the New Brunswick counties of Victoria and Restigouche, and the Bay of Chaleurs. The St. Lawrence Gulf, Labrador, and Hudson's Bay bound it on the east and north, but its limits in the latter direction are still practically undefined. Its area is approximately stated at 190,000 square miles.

The river St. Lawrence, widening into its majestic Gulf, traverses the province from south-west to north-east, and receives, as principal affluents on its northern bank, the Ottawa, the Assomption, the St. Maurice, the St. Anne, the Montmorency, and the Saguenay; and on its southern bank, the Richelieu, the St. Francis, the Chaudière, the Du Loup, Metis and other rivers. Two mountain ranges follow roughly the course of the St. Lawrence: and to these, in conjunction with the great river and its affluents, the physical character of the province is chiefly subordinate. The range on the north (with an average elevation of 1500 feet above the sea, rising in places to over 2,000 feet) constitutes the Laurentide Mountains, and runs roughly parallel with the Ottawa until within about 30 miles of its mouth, when the range curves towards the east, and skirting the St. Lawrence a short distance inland, strikes the river at Cape Tourmente a little below the city of Quebec. From this point it follows the north shore of the river and Gulf to beyond the province boundary in Labrador. This range and the country which it traverses consist of greatly corrugated archæan gneiss, broken through by various dioritic and feldspathic rocks. The southern range is properly a continuation of the great Appalachian chain of the United States. It is known in Canada as the Notre Dame and Shickshock Mountain Range. It traverses the Eastern Townships, and gradually approaching the St. Lawrence, runs along the south shore, but at a distance of from 30 to about 10 or 12 miles inland, until it terminates in the high table-land of Gaspé at the extremity of the Province. This range, consisting of several roughly parallel lines of mountainous country, presents one or two points of nearly 4000 feet in altitude, and in the Gaspé peninsular the elevation averages 1,500 feet. It is made up largely of crystalline, magnesian rocks, including talcose and chloritic schists and beds of serpentine, associated in places with micaceous and gneissoid rocks and other crystalline representatives, the true age of which is still more or less uncertain.

On passing from the Laurentide Mountains southwards to the St. Lawrence, the gneissoid Archæan rocks become overlaid unconformably by Cambrian and Lower Silurian strata. Where the river enters the province, and for some distance eastward, the latter occur on both shores, and they continue along the north shore to beyond the city of Quebec. They reappear further east in the Mingan Islands, and also, with accompanying Upper Silurian strata, in the Island of Anticosti. On the south side of the river, Cambrian slates and other strata have been brought up by a great fault into a position apparently higher than that of the Hudson River beds. This fault, first indicated and traced out by Sir William Logan, extends along the bed of the gulf and river to the immediate vicinity of Quebec, and then turns inland towards the south-west, and continues in that direction to the head of Lake Champlain, beyond the province boundary. All the strata to the south and east of this fault have been greatly disturbed and broken up, and have been more or less altered by metamorphic agencies. Crystalline, gneissoid and magnesian rocks appear in part to underlie them, and partly to be mixed up with them in intricate foldings, by which their stratigraphic relations become greatly obscured. In many places also they are broken through by trachytic and granitic masses.

As regards its geology therefore, the Province admits of a subdivision into three natural areas: the Archæan area of the north; the typical Palæozoic area; and the disturbed Appalachian district including the Eastern Townships and the Gaspé peninsula. For convenience of description, however, the Island of Anticosti and the Mingan Islands may be separated from the second of the above areas, and regarded as forming a distinct palæozoic district.

The positions of these areas are shewn roughly in the annexed sketch-map :

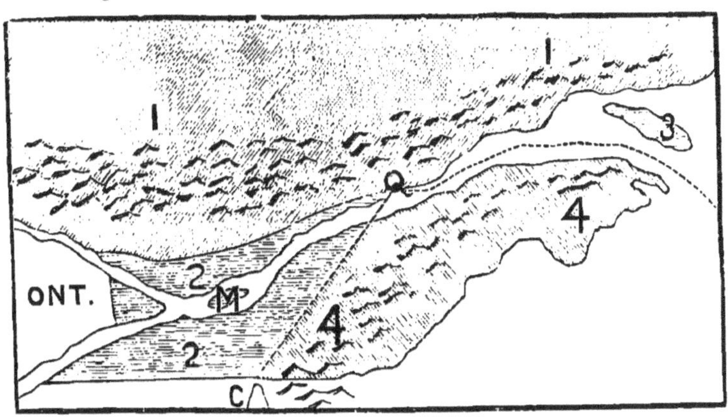

Fig. 246.

Sketch-Map of the Province of Quebec, shewing geological areas.

1. Northern Archæan district—shewn in part, only.
2. Palæozoic district of the Upper St. Lawrence.
3. Palæozoic district of Anticosti and the Gulf.
4. Appalachian and Gaspé district.

Q—Quebec; M—Montreal; C—Lake Champlain (Northern portion). The dotted line indicates the direction of the great Fault.

NORTHERN ARCHÆAN DISTRICT.

This is essentially a region of ancient crystalline strata—rocky and mountainous in character : an eastward extension of the Laurentian district of Ontario, but with certain special features of its own.   It comprises the wide expanse of territory lying between the Ottawa River and Labrador, with the exception of a comparatively narrow strip of country (occupied by Lower Palæozoic formations), extending along the St. Lawrence from near the junction of the two great rivers to a point a short distance below the city of Quebec.   It is traversed by the Laurentide Mountains, proper, which form within it several broken ranges curving roughly parallel with the course of the St. Lawrence.   The more southern of these gradually approach the river, and run closely adjacent to it along the lower part of its course.   The average height of the Laurentides generally, is from about 1,200 to 1,500 feet, but at one or two points they reach an altitude of over 2,000 feet above the sea.   Numerous rivers rise among them.   Some of the more important comprise : the Rivière du Moine, the Gattineau, the Rivière du Lièvre, the Rivière Rouge, and the Rivière du Nord, flowing into the Ottawa ; and the l'Assomption, Chicot, St. Maurice, Batiscan, Ste. Anne (Portneuf), Jacques Cartier, Montmorenci, Ste. Anne (Montmorenci), Murray, Saguenay, Moisie, and other eastern rivers, flowing into the St. Lawrence. The rocks of this Archæan area consist for the greater part of typical Laurentian gneiss (composed of quartz and orthoclase, with associated mica or hornblende) interbedded with quartzites and bands of crystalline limestone and iron ore.   These Laurentian strata, as a rule, are more or less strongly tilted and corrugated, or otherwise disturbed.   They are also traversed very generally by granitic or syenitic veins, and are broken through in places (more especially in Wentworth, Chatham, and Grenville on the Lower Ottawa) by enormous masses of eruptive syenite and greenstone.   The crystalline limestones very commonly contain diopside or light-coloured pyroxene, phlogopite, zircon, sphene, and other crystallized silicates, with scales of graphite, and grains and crystals of fluor-apatite.   The latter mineral also occurs in workable quantities, associated mostly with pyroxene, phlogopite mica, calcite, and scapolite, in broad veins which cut the gneissoid strata transversely—especially in the townships of Buckingham, Templeton, Lochaber, and Grenville, on the Ottawa,* and throughout that section of country, generally. About 20,000 tons were obtained from these deposits in 1886, and they are still being largely worked. Veins and large lenticular masses of graphite occur also in the rocks of this district, and mica of good quality has been obtained from localities in Grenville, Templeton, and adjacent townships. Iron ore in workable quantity occurs also in Hull, and elsewhere in the same district.

The orthoclase gneiss-rocks which form the prevailing strata of this archæan region north of the St. Lawrence, are overlaid in some few localities by comparatively limited areas of feldspathic rock composed of labradorite or other triclinic lime-feldspars; and here and there these labradoritic rocks or "anorthosites," see page 181, are apparently interstratified with the upper beds of the ordinary othoclase gneisses. They were at one time, and are still by some observers, regarded as indicating a newer or higher series of Laurentian strata, and were known as the Upper Laurentian, Labrador, or Norian formation, But in the main mass of these anorthosites there is no apparant stratification, and they are now regarded by Dr. Selwyn as essentially eruptive rocks of Laurentian age. This view, although not absolutely free from doubt, will probably meet with general acceptance. A large area of these anorthosic rocks occurs in the counties of Argenteuil, Terrebonne, Montcalm, and Joliette; and another, equally large, lies around the north-east and south sides of Lake St. John, at the head of the Saguenay River. Smaller areas occur on the north shore of the St. Lawrence in Montmorenci and Charlevoix, and a large exposure has been recognised on the branches of the river Moisie, off the Gulf. These feldspathic rocks present generally light shades of colour, and weather dull white. In most examples, some of the cleavage planes shew a blue or green opalescent play of colour, as in typical examples of labradorite; and occasionally, as at Château Richer and elsewhere, they contain scales and foliæ of bronze-coloured or green hypersthene or bronzite. Garnets, also, are frequently present in them; and they are associated in many localities with titaniferous iron ore. A very large deposit of the latter mineral occurs in these rocks near Baie St. Paul; and

---

* The distinctive characters etc., of this valuable mineral (known commercially as "phosphate") are given on page 135.

22

titaniferous iron sands are abundant around the mouth of the Moisie river on the Gulf.   Attempts to utilize these titaniferous ores have been made, but hitherto without success.*

In addition to the gneissic and other crystalline formations of this northern archæan region, a few outlying patches of Lower Silurian strata, consisting mostly of Trenton limestone and Utica shales, occur here and there within its area.   The largest of these Silurian outliers is seen around Lake St. John on the Upper Saguenay.   As remarked by Sir William Logan, the limestones at this locality shew characteristic Black River fossils associated with those of the Trenton formation proper,† and comprise principally: *Maclurea Logani* (fig. 222); *Stromatopora rugosa* (fig. 127) *Orthoceras Bigsbyi; Leptæna sericea* (196); *Strophomena alternata* (fig. 194); *Murchisonia gracilis* (fig. 220); *Calymene Blumenbachii* (fig. 179), etc.—with, near the base of the series, *Halysites catenulatus* (fig. 139), typically, an Upper Silurian form.   The fossils of the overlying Utica shales include various graptolites, broken crinoidal stems, species of *discina,*&c., and the characteristic Utica trilobite *Triarthrus Beckii* (fig. 177).   A small exposure of Hudson River beds is seen also on Snake Island in the lake.

Glacial boulders, clays, and gravels, with Post-Glacial sands and other superficial deposits, are distributed, as in other parts of Canada, more or less generally throughout this region; and many of the harder rocks shew glacial striæ.   These run most commonly either towards the south-east or south-west; but at some spots their direction is nearly north and south; and, in others, not far removed from east and west.

DISTRICT OF THE UPPER ST. LAWRENCE.

This is essentially a Palæozoic area, occupied—apart from some isolated eruptive-masses—by sandstones, limestones, and other strata,

---

* The presence of titanium in an iron ore is objectionable chiefly for the following reasons : (1) It diminishes, of course, the percentage of iron in the ore ; (2) it renders the ore 'very refractory ; and (3) it scours the ore, carrying off a large amount of iron in the form of slag.   To prevent the latter effect, a comparatively large amount of lime is required, and this fills up the furnace with an unproductive burden.   Very little titanium goes into the pig metal, even when highly titaniferous ores are used.

† A similar association of Black River and Trenton types was pointed out by the author, many years ago, as occurring at Shannonville in Ontario.   The two so-called formations cannot in fact be properly disunited.

which retain their original sedimentary aspect, and occur, for the greater part, in undisturbed beds. It extends along both sides of the St. Lawrence from the western boundary of the Province to the neighbourhood of Quebec. In the west, it includes the counties of Vaudreuil and Soulanges, lying in the point of the triangular space immediately west of the junction of the Ottawa and St. Lawrence Rivers. From the county of Vaudreuil, its northern boundary crosses the Ottawa, and then, keeping entirely on the north side of the St. Lawrence, runs along the southern edge of the Laurentide district already described, and gradually approaching the river, strikes it a short distance below Quebec. Its southern limit runs from the south-west corner of Huntingdon (south of the St. Lawrence), along the boundary-line between the Province and the State of New York, to a little beyond the River Richelieu at the northern extremity of Lake Champlain; and east of this, the district is bounded by the disturded and metamorphic area of the Eastern Townships—its actual limits in this direction being a remarkable line of dislocation, with accompanying fault, running (as first traced out by Sir William Logan) from near the north-east end of Lake Champlain to the vicinity of Point Lévis, and from thence, by the City of Quebec, along the north side of the Island of Orleans, and down the river and Gulf, between the Island of Anticosti and the Gaspé shore.

The rock-formations of the district belong to three distinct series, namely: stratified Palæozoic formations; eruptive rocks; and Glacial and other Post-Cainozoic deposits. The stratified rocks, proper, consist of representatives of the Potsdam, Calciferous, Chazy, Black River and Trenton, Utica, and Hudson River formations—with some small exposures, south of the St. Lawrence, of strata referred to the Medina group; and a few outlying patches of Upper Silurian strata (belonging to the Lower Helderberg formation) in the vicinity of Montreal. These formations are broken through in places by large eruptive masses of trachytic and trappean rock, forming a series of picturesque mountains, which rise abruptly from the generally level surface of the district in the more southern and western portions of its area; and in addition to these Palæozoic and Eruptive rocks, Glacial and Post-Glacial accumulations, with deposits of comparatively modern origin, occur throughout the district generally.

The Potsdam beds consist of coarse conglomerates and fine-grained siliceous sandstones—the latter in many localities sufficiently pure for glass-manufacture and for the hearths of furnaces. The formation is largely displayed in Hemmingford Mountain, and over large portions of Huntingdon, Chateauguay, and Beauharnois, from whence it crosses the St. Lawrence, and spreads over a large part of Soulanges and Vaudreuil; and from thence, passing across the western end of the Island of Montreal and Isle Bizard, it wraps around a large outlying mass of Laurentian gneiss (forming Mont Calvaire on the north shore), and continues uninterruptedly along the edge of the Laurentide district as far east as the River Chicot, where the continuity of the strata is broken by a fault, and limestones of the Trenton formation are let down against the Potsdam beds. East of this point, the formation only appears at one or two places—notably on the St. Maurice, where it exhibits a slight thickness of nearly horizontal beds of conglomerate and sandstone, resting upon gneiss. Throughout its range, as far east as the Chicot, it is accompanied by sandy and dolomitic limestones of the Calciferous formation, and these cover large areas south of the St. Lawrence, and in the country around the junction of the St. Lawrence and Ottawa. East of this formation, on the south side of the St. Lawrence, limestones of the Chazy and Trenton series, and dark bituminous shales of the Utica formation, with succeeding sandstones and arenaceous shales of the Hudson River formation, largely prevail—the latter, especially, east of Riche-lieu River. These formations cross the St. Lawrence, and range in regular sequence along the north shore between the Calciferous out-crop and the river bank. The intervening Island of Montreal, Isle Jésus, Isle Bizard, etc., consist essentially of Chazy, Trenton, and Utica strata—the Hudson River beds coming up farther east. The Chazy limestones of Caughnawaga and St. Doménique on the south shore, those of Ste. Geneviève on the Island of Montreal, of Isle Bizard, and of St. Lin on the north shore, yield marbles (red-spotted or uniformly red) of good quality. East of the River Chicot, which enters the St. Lawrence on the north shore, near the upper or western extremity of the expansion known as Lake St. Peter, the comparatively narrow strip of country between the Laurentian gneissoid rocks and the river margin is occupied almost entirely by Trenton, Utica, and Hudson River strata—one or two small exposures of the Potsdam formation on the St. Maurice and at St. Ambroise alone representing the lower beds as seen west of the Chicot fault. In this eastern portion of the district, the strata are tilted in many places at considerable angles, as near Pointe aux Trembles, Montmorenci Falls, etc., and their continuity at these spots is more or less disturbed by minor faults.*

As stated above, the Lower Silurian strata of the more southern and western portions of the Upper St. Lawrence district are broken through in places by trachytic and trappean masses, forming a series of isolated mountains which rise above the generally level surface of the country to elevations of from 600 to 800 feet. Most of these occur apparently upon a single line of fissure traversing the district in a general south-easterly direction. They comprise: (1) the Mountain of Rigaud in Vaudreuil, composed partly of a purely feldspathic, and partly of a dioritic or hornblendic trachyte, porphyritic in places; (2) the Montreal Mountain, composed essentially of augitic trap or dolerite, but traversed by dykes of compact and granitic trachyte; (3) Montarville or Boucherville Mountain, also essentially trappean in composition; (4) Belœil, a dioritic and micaceous trachyte; (5) Monnoir or Mount Johnson (south of Belœil), of the same mineral character; (6) Rougemont, in Rouville County, a trappean mass like that of Montreal in general composition; and (7), the Yamaska Mountain, essentially a micaceous trachyte. The Mountains of Brome and Shefford belong to the same eruptive series, but lie within the crystalline district to the east. In addition to these principal masses, many dykes of similar character traverse the surrounding strata; and some of these in the neighbourhood of Montreal and Lachine are intercalated with the soft shales of the Utica series, which have become more or less worn away, leaving the associated trap bands in the form of projecting ledges. Most of the rapids in this part of the St. Lawrence have been thus produced.

The superficial deposits of the district comprise Glacial boulders and related clays and gravels, with Post-Glacial and recent accumulations. Drift or Glacial deposits, proper, are of general distribution; and in some places, as on the Rigaud Mountain, the boulders form roughly parallel ridges, several feet in height. The Glacial striæ of the country have two prevailing directions—south-west and south-

* The fossils in these various Lower Silurian formations are practically identical with those of the same formations in Ontario: see *ante*, pages 315 to 317.

east respectively.   The Post-Glacial deposits belong chiefly to two series, as first determined by Sir J. W. Dawson of Montreal: a lower deep-sea formation, known as the "Leda Clay;" and a succeeding deposit, apparently a shallow-sea or shore-line accumulation, known as "Saxicava Sand."   These occur widely within the district, and at various elevations.*   On Montreal Mountain, beds of Saxicava sand, for example, form a series of terraces, one of which is at an altitude of nearly 500 feet above the present sea level.   Beauport, below Quebec, is another locality at which these deposits are well exposed ; but they occur also, and over large areas, around Murray Bay, as well as on the Lower St. Maurice, and elsewhere.   The more recent formations of the district comprise, principally, the bog iron ore and ochres of the St. Maurice and other localities on the north shore of the St. Lawrence ; the great peat-beds of Lanoraye, Lavaltrie, St. Sulpice, &c., on the same side of the river ; and those of Sherrington, Longueuil, and St. Doménique, with others on the south shore.   Most of these peat beds overlie deposits of shell marl.

THE ANTICOSTI AND MINGAN DISTRICT.

This division belongs strictly to the palæozoic area of the upper St. Lawrence, of which it forms a detached, eastern portion.   It includes the large island of Anticosti in the St. Lawrence Gulf, and the group of the Mingan Islands on the opposite northern shore, together with a narrow strip of the latter east of the Mingan River.   Its strata consist of Upper Cambrian and Lower and Upper Silurian formations.   On the Mingan coast and islands, and throughout Anticosti, these strata are practically undisturbed.   They dip, at slight inclinations, in regular sequence towards the Gulf.   The great line of dislocation, referred to on page 335 as passing from the head of Lake Champlain to the St. Lawrence in the vicinity of Quebec, and from thence eastwards along the river and the Gulf, runs between Anticosti and the Gaspé coast ; and appears to have produced

---

* The leda clay formation is characterised by the presence more especially of the lamellibranchiate, *leda truncata* (fig. 211, page 277); whilst the characteristic fossils of the higher deposit comprise : *saxicava rugosa* (fig. 216), *mya truncata* (fig. 214), and *buccinum undatum* (226).

(as in other places) much disturbance and contortion in the strata of the latter between the mouth of the Marsouin River and Cape Rosier at the extremity of the Gaspé peninsula.

The Island of Anticosti extends in a general north-west and south-east direction, with a length of about 150 miles, and a breadth in its broadest part, of about 35 miles, gradually tapering at the extremities. The northern coast presents bold ranges of cliffs, from 200 to 400 feet in height, cut through in places by deep water-courses. The interior of the island is thickly wooded, but is destitute of lakes and important streams. It appears to consist of a series of plateaux or broad terraces, gradually descending to the south shore. The latter, although showing in places high cliffs of drift clay, is mostly of a low and swampy character, and this part of the island is especially characterized by the presence of extensive beds of peat.

The Mingan Coast consists of arenaceous limestones and dolomites of the Calciferous formation, and similar strata on the islands are succeeded by Chazy beds composed of reddish and pale-gray limestones, with interstratified arenaceous shales. On the principal island (Large Island) of the Mingan group, light-coloured limestones, holding characteristic Lower Trenton or Black River fossils, overlie the Chazy beds—the whole dipping, at a slight angle, southwards or towards the Gulf. The next exposure (in the regular sequence of Lower Silurian formations) occurs along the opposite north coast of Anticosti, and consists of grayish and other coloured limestones, with interstratified shales and conglomerates, having an inland or southerly dip of very slight amount. These beds belong to the upper part of the Hudson River formation, and it may thus be legitimately inferred that the intervening area of the Gulf is occupied uninterruptedly by other Hudson River beds, with Utica and Trenton strata cropping out to the north successively from beneath them. In some of these Hudson River strata, examples of the curious stem-like corals *(Bea ricea undata)*, resembling the petrified trunks of large trees, occur in considerable abundance. The succeeding area of the Island to the south, is occupied by argillaceous and other limestones, the equivalents apparently of the Medina, Clinton, and Niagara formations of the West; but characteristic Niagara fossils are associated in some of these strata with Lower Silurian types.

The other rock-formations of Anticosti and the northern islands of

the district consist of Post-Cainozoic deposits.   Raised beaches, in the
form of a series of terraces, extending to a height of about 100 feet
above the sea, occur on some of the Mingan Islands; and other
evidences of elevation are seen in the pillared rocks left here and
there upon the surface, at heights of fifty or sixty feet above the
present sea level.   Drift clays, holding limestone pebbles, overlie the
calcareous strata of some parts of Anticosti, especially on the south-
west coast, where they form cliffs of considerable height.   But the
more remarkable of the Post-Cainozoic formations of Anticosti are the
great peat-beds, which cover large areas on the southern part of the
Island.   One of these extends in a narrow band along the south-east
coast, between Heath Point and South Point, over a length of nearly
eighty miles.

THE APPALACHIAN OR EASTERN TOWNSHIPS' AND GASPÉ DISTRICT.

The term "Appalachian region" was first bestowed on this part
of Canada by Dr. Sterry Hunt.   The district forms, indeed, a pro-
longation into Eastern Canada of the Appalachian region of the United
States : the Appalachian chain, with its tilted, contorted, and in great
part metamorphosed series of rocks, extending into the Stoke Mountains and other elevations of the Eastern Townships, and appearing
further east in the Shickshock Mountains of Gaspé.   The district
comprises all that portion of the Province which lies east and south
of the great line dislocation referred to under the preceding district
as running north-easterly from Lake Champlain to Quebec, and from
thence along the bed of the river and the Gulf.   Whilst to the west
and north of this dislocation, the strata are practically undisturbed,
those which lie directly east and south of it have been greatly tilted
and uplifted, and generally reversed in dip; and towards the more
central parts of the district many of the beds have been altered or
rendered crystalline by metamorphic agencies, and have been folded
up with one another, and with an older system of crystalline rocks
forming the axes of the higher elevations so as to produce great com-
plications of structure.   As regards physical features, the district is
more or less throughout of a mountainous character, but also, as a
rule, of good fertility—differing remarkably in this respect from the

mountainous Laurentian region of the northern portion of the Province. The average elevation of the. Gaspé peninsula is about 1500 feet, and that of the other parts of the district about 800 to 1000 feet above the sea; but several peaks in the Shickshock ranges of Gaspé approach 4000 feet in height, and the summits of some of the mountains in the Eastern Townships are apparantly over 3000 feet. Many lakes, but none of large size, occur within the district. Among these, Lake St. Francis lies at an elevation of 890 feet, and Lake Memphramagog at an elevation of 760 feet above the sea. In Gaspé Lake Temiscouta and Lake Matapedia lie respectively at altitudes of 470 and 480 feet. The district is greatly intersected by streams and rivers. Some of the principal comprise : the Yamaska and the St. Francis, the Chaudiere (with its tributaries, the Famine, Des Plantes, etc.,) and the Etchemin, in the more western portion of the district ; and the Riviere du Loup, Trois Pistoles, Rimouski, Metis, Matanne, Chatte, St. Anne, Magdalen, Cascapediac, Matapediac, and other rivers of the Gaspé peninsula, most of which flow in deeply excavated channels.

The geology of this district, so far as regards the actual sequence and relations of the rock groups within its area, is still very imperfectly known, and much diversity of opinion prevails respecting the true age and position of some of these formations.

Viewed broadly, the district may be regarded as divisible into three sub-areas : (1) a comparatively narrow central zone of mountainous country composed of crystalline and essentially metalliferous rocks, ranging from the Dominion boundary near Lake Champlain, through Sutton, Brome, Stukely, and other " Eastern Townships," in a north-easterly direction across the Chaudière River as far as the County of Kamouraska, and re-appearing farther east in the Shickshock Mountains of Gaspé ; with (2), a somewhat broader band between it and the St. Lawrence, composed of Cambrian and Lower Silurian strata ; and (3), a still broader area, underlaid essentially by Upper Silurian and Devonian beds, along its southern border. These areas are broken through in places by eruptive dykes and mountain-masses ; and are overlaid very generally by Drift and other surface-deposits.

Four series of rock-formations are thus recognizable within the district. These comprise : (1), Crystalline and sub-crystalline forma-

tions ; (2), Fossiliferous Palæozoic formations ; (3), Granites, Trachytes, and Trappean masses ; and (4), Superficial Post-Cainozoic
deposits.

Crystalline and Sub-crystalline formations :—These as described
above, form the main axis of the Appalachian District, extending in
a belt of mountainous country from the Dominion boundary, a little
east of Lake Champlain, through the Eastern Townships, and north-
easterly, beyond the Chaudière River to the border of Kamouraska.
Sutton Mountain, the Owl's Head in Lake Memphramagog, Victoria,
Mountain in Oxford, Pinnacle Mountain in Shipton, Ham
Mountain, the Stoke Mountains, and other elevated points, lie
within this area, and are composed essentially of crystalline
schists or related feldspathic and quartzoze rocks. Still farther
towards the north-east, these rocks re-appear in the Shickshock
Mountains of the Gaspé peninsula. Good exposures are seen at
numerous spots in Sutton, Potton, Brome, Bolton, Shefford, Stukely,
Orford, Melbourne, Acton, Shipton, Stoke, Ham, Cleveland, and
throughout the Townships generally. Also around Mt. Albert in
Gaspé, and at many other spots along the Shickshock Range,
where serpentines especially predominate. Almost everywhere,
these crystalline strata are greatly tilted, folded, and otherwise
disturbed. They consist essentially of gneissoid and micaceous beds,
talcose and chloritic schists and potstones, serpentines, amphibolites
and epidosites, crystalline limestones and dolomites, graphitic
argillites, quartzites, specular-iron schists, and other rocks of related
character, holding in many places large masses or "stocks" of copper
ore or magnetite, or containing these ores in veins or in disseminated grains, together with chromic iron ore, gold (in quartz veins,)
galena, antimony ores and nickel sulphide. Many of the crystalline
dolomites are intermixed with serpentine, forming green, chocolate-
brown and other coloured "serpentine marbles."

These crystalline schists and their associated quartzites, serpentines,
dolomites, etc., are now regarded as mainly of Pre-Cambrian, probably Huronian, age ; but it may be inferred that portions of Trenton
and higher palæozoic strata are enclosed here and there in a
metamorphosed condition, among their foldings.*

* Until recently, these metamorphic or crystalline formations were regarded as consisting
essentially of altered portions of Sir William Logan's " Quebec Group." The latter represents
a series of strata of about the age of the Calciferous formation, but holding peculiar graptolites

The more important economic minerals belonging to these crystalline areas, are enumerated briefly in the following list : *Copper Ores*— These comprise chiefly the Yellow Copper Pyrites, Bornite or Horse flesh ore, and Copper Glance, occasionally mixed with small portions of native copper and native silver. They occur mostly in lenticular or irregular masses ("stocks") often of considerable size, or in thickly disseminated grains, in chloritic schists and other rocks of the country. Indications occur in almost the entire series of the Eastern Townships, but copper mining has been carried on at a few spots only : notably at the Harvey Hill mine in Leeds, the Huntingdon and Ives mines in Bolton, the Hepburn, Suffield and Ascot mines in Ascot, and at spots in Acton, Sutton and Brome. *Chromic Iron Ore*—in beds in serpentine, in the townships of Ham, Bolton, and Melbourne; and largely at Mt. Albert in Gaspé. *Magnetic Iron Ore*—chiefly in beds, in Brome, Sutton, Leeds, and other townships, but often titaniferous. *Antimony*—native, but associated commonly with Antimony Glance and Kermesite, the red oxy-sulphide (page 80), in Ham township. *Nickel*—as sulphide, but in small quantities only, with chrome garnet in calcite, Township of Orford. *Gold*—native, in quartz veins in Leeds, Garthby, etc., but apparently in little more than specks or traces. *Graphite*—in crystalline limestone, in Wakefield. *Serpentine* and *Serpentine-marble*—in Orford, Melbourne, Ham, Broughton, etc., and around Mt. Albert in Gaspé. "*Asbestos*" (*Chrysotile*, page 118)—in veins or bands varying from about half-an-inch to three or four inches in width : Townships of Thetford, Coleraine, Leeds, Melbourne, Bolton. *Potstone* and *Soapstone*—in beds in Sutton, Potton, Bolton,

and some other distinct fossils. It was thought by Sir William Logan to admit of a separation into three parts, named by him (in supposed ascending order) the Levis, Lauzun, and Sillery formations—the Lauzen formation, or sub-formation, being the especially metalliferous portion of the series. It is now pretty well established that these sub-divisions cannot be maintained. The so-called Levi strata are shewn to overlie the Sillery beds, if the two can properly be separated ; and although both may be much altered in places there is certainly no evidence to support the assumption that they have been altered into the gneissoid and micaceous schists, the serpentines and other crystalline rocks of the Eastern Townships and central Gaspé ; or that what were thought to be the lower beds of these Quebec strata pass under the crystalline series.

Sir William Logan's opinion was first objected to by Mr. Macfarlane (now of Ottawa), and was subsequently opposed very strongly by Dr. Sterry Hunt, who had originally endorsed Sir William's interpretation. But the now general acceptance of the essentially Pre-Cambrian age of these crystalline rocks, is mainly due to the present Director of the Geological Survey, Dr. A. R. C. Selwyn. Whether these rocks should be regarded as Huronian, or as constituting a somewhat higher, or "Montalban" series, distinguished especially by their predominating magnesian, chromiferous character, is still an open question.

Wolfestown, Stanstead, Leeds etc.   *Roofing Slate* (in part belonging to altered Silurian beds)—Melbourne, Orford, Cleveland, Richmond *Whetstones*—Stanstead, Bolton, Kinsey.

Fossiliferous Palæozoic Formations :—As recognised within the Appalachian district, these comprise, in ascending order, representatives of the Calciferous Formation, with Lower Silurian, Upper Silurian, Devonian and Lower Carboniferous strata.

The *Calciferous Formation*—or geological horizon corresponding practically to the higher part of this—is represented by the uncrystalline portions of Sir William Logan's " Quebec Group." This series of strata is made up essentially of grey and greenish (in places glauconitic) sandstones, many of which hold small pebbles of white quartz, interstratified with olive-green, red, purplish, and gray shales, forming the so-called " Sillery formation "; and of grey, red. and black graptolitic shales associated with limestone conglomerates and yellowish-grey dolomitic beds, making up the " Lévis formation." The subdivisions of Sillery and Lévis, however, should properly be abandoned, the two sets of strata forming practically but a single series of varied mineral composition. The sandstones along parts of the Gaspé coast —as between the St. Anne des Monts and Grand Michaud rivers, more especially—have been worn by denudation into more or less isolated pillars which form prominent objects as seen from the Gulf. Hence the name of " pillar sandstones," by which they were at one time known. The black and other coloured shales are especially characterised by the presenee of " compound graptolites," represented by *Loganograptus*, *Phyllograptus* and related types (*ante*, page 228); and the calcareous beds contain many peculiar trilobites—species of *Agnostus* (fig. 167 *bis*.) *Dikelocephalus* (fig. 171), *Bathyurus*, *Arionellus*, etc. The shales and limestones contain also, in many places, examples of characteristic brachiopods, as *Obolella pretiosa*, *Lingula Quebecencis* (fig. 204), and small species of *Orthis*, *Leptæna*, and *Camerella*.

As regards general distribution, these Calciferous (Quebec) strata form a practically continuous band along the western and northern edge of the central crystalline areas of the district, from the province line at the head of Lake Champlain to the extremity of Gaspé ; and they are well displayed at Sillery on the St. Lawrence, and along the southern and central portions of the Island of Orleans. Throughout the greater part of their course they are more or less intimately associated in complicated stratigraphic relations, with the Lower Silurian Utica shales, the latter appearing in many places to dip under them. They are distinguished however from these Utica beds by their peculiar graptolites. In Brome and Shefford the formation is broken through by trachytic mountain-masses which cover areas of many square miles in extent (see below). These trachytic areas lie between the central crystalline range and the great Champlain fault, and they belong to the same eruptive series as the trachytic and trappean masses of the upper St. Lawrence District, described on a preceding page. The formation appears also along the southern edge of the crystalline country west of Kamouraska in contact with altered Upper Silurian strata, as seen about St. Joseph on the Chaudière and in the townships of Wolfestown, Ham, Richmond and Sherbrook.

The *Lower Silurian formations* within the Appalachian District lie on the western and northern limit of the Calciferous (Quebec) series, and in many places are mixed up with these latter in complicated relations. They consist mostly of Hudson River and Utica strata, but some Trenton (and perhaps Chazy ?) beds have been recognized in Missisquoi, at the extreme south-western limit of the district, and the steeply-dipping black slates which occur on the St. Lawrence near Point Lévis, and which were originally thought to occupy a geological position below the Potsdam horizon, may belong to the same formation, if not to the succeeding Utica and Hudson River series. Representatives of these latter—consisting mostly of grey and black slates and greyish sandstones, whilst the associated Calciferous (Quebec) beds are chiefly in the form of red and greenish, or more rarely black, shales, with limestone conglomerates, yellowish dolomites, and greenish-grey sandstones—occur largely along the coast as far east as the Magdalen River ; and beyond that point, they occupy the entire-coast line to Cape Rosier, in the form of disturbed and strikingly contorted strata. At the Magdalen and elsewhere, according to Mr. R. W. Ells (Geological Survey Report 1882), the Utica beds appear to underlie the older Calciferous formations. No certain evidence has been obtained of the occurence of Lower Silurian strata in that portion of the District which lies south of the central crystalline range of country, all the strata being there regarded as of Upper Silurian and Devonian age ; but some of the argillaceous slates which occur there may perhaps be Lower Silurian.

The Upper Silurian strata of the Appalachian District, although originally regarded as forming the principal part of the "Gaspé Limestone series" of the late Sir William Logan, are but slightly developed in Gaspé proper, the supposed Upper Silurian beds of that region being now shewn by their fossils to be chiefly of Devonian age.   But Upper Silurian strata —if rightly determined—occur somewhat largely in the south-eastern part of the district, south of the central crystalline region.   In the Eastern townships of Orford, Melbourne, Westbury and adjacent sites, and in their extension across the Chaudière, they consist essentially of dark-gray or black slates, and are seen in places to overlie the Calciferous (Quebec) strata unconformably.   These slates have been largely quarried for roofing purposes in the township of Melbourne, and to some extent also in Cleveland and adjacent townships.   The Chaudière slates in the country around St. George, and elsewhere in Beauce County, exactly resemble the hard black slates on the St. Lawrence opposite Quebec, and shew no trace of fossils.   They form the broken floor or " bed-rock " on which the alluvial gold of this part of the Chaudière valley essentially rests ; and here and there they are traversed by quartz veins carrying small quantities of gold.   In Gaspé proper, the Upper Silurian strata are composed mainly of calcareous beds, consisting of grey limestones and shales ; but according to Mr. Ells, to whom our recent knowledge of the geology of the Gaspé coast is chiefly due, these strata occupy a much smaller area then that formerly assigned to them, most of the Gaspé limestones holding Devonian fossils.   Those to which an Upper Silurian age may be attributed, are exposed chiefly on the Grand River of the south coast, and along the shore between the Pabos River and Cape D'Espoir.   At these points they underlie unconformably the Lower Carboniferous Bonaventure beds.

The *Devonian formations* of the region now under review, besides comprising the great body of the " Gaspé limestones and sandstones," occur undoubtedly in the more southern of the Eastern Townships, but, as a rule, in a more or less altered condition, and not clearly separable from associated Silurian strata.   Many of the grey, black, and marbled limestones, however, of this area, as those of Weedon, Dudswell and Lake Memphremagog, as well as the micaceous limestones of Compton and adjacent townships, are apparently Devonian.

Some of the Dudswell beds form veined marbles of considerable beauty, black and yellow, and white and yellow, in colour; and these limestones in their less altered portions shew Devonian fossils, or mixtures of these with Upper Silurian types. In Gaspé, proper, Devonian rocks occupy wide areas, and consist of limestones, sandstones, and conglomerates. Where in contact with Upper Silurian strata (below), and with Lower Carboniferous conglomerates (above), as in Percé, etc., they are unconformable to both of these formations. The arenaceous beds constitute the more typical strata of this Devonian series, and are made up essentially of gray shales and gray, brown, and reddish sandstones, holding in places, more especially around Gaspé Bay and near Fleurant Point on the Bay of Chaleurs, numerous plant remains; and at the latter locality the remains also of ganoid fishes (see *ante*, page 291). The fossil plants comprise examples of *Psilophyton* (fig. 120), *Lepidodendron* (fig. 121) *Stigmaria, Cordaites* (fig. 122), *Calamites* (fig. 119), *Annularia, Prototaxites, etc.* A thin seam of shaly coal, a few inches thick, occurs also in these beds at Little Gaspé Cove; and petroleum springs ooze through the strata at Douglastown on the south shore of Gaspé Bay, but borings put down in the vicinity of these have failed to obtain a profitable supply. At Tar Point, a grey amygdaloidal dyke contains pitch-like petroleum in some of its cavities. Other eruptive dykes break through the Devonian strata on this part of the Gaspé coast and on the Bay of Chaleurs, as described below.

The *Carboniferous strata* of the district belong to the lower part of the Coal Measures, but are entirely destitute of coal. They form the "Bonaventure formation" of the late Sir William Logan, and consist essentially of conglomerates, associated with red and brown sandstones and shales, many of which contain fossilized vegetable remains or casts of these. The conglomerates are made up of rounded masses or pebbles of limestone, sandstone, syenite, agate, quartz and other rock-matters, held together by an arenaceo-calcareous cement. These beds rests unconformably on the Devonian and Silurian strata of the Gaspé coast. They occupy the entire area of Bonaventure Island, and occur in comparatively narrow strips along the Bonaventure and Percé shore of the mainland. They form also the upper portion of the Percé mountain, but the lower conglomerates formerly assigned to them, both at this spot and at localities along the Bay of Chaleurs, are now regarded as Devonian *

Eruptive Rocks :—The more important eruptive rocks occuring within the Appalachian and Gaspé District, comprise : (1) the tra-chytic mountains of Brome and Shefford ; (2), the granites of the Great and Little Megantic Mountains, and other granitic masses of the Eastern Townships ; and (3), the trappean dykes and hills of Eastern Gaspé and the Bay of Chaleurs.

The Brome and Shefford mountains consist essentially of granitoid trachytes, composed of greyish-white or light-coloured feldspar with scales or grains of brown or black mica and dark hornblende, and some minute crystals and specks of sphene and magnetite.   The feld-spar, according to Dr. Sterry Hunt's analyses, contains both potash and soda but shews the cleavage of orthoclase.   The Brome (and Gale) Mountain covers an area of about twenty square miles, and the closely adjacent Shefford Mountain, about nine square miles.   Both belong to the same eruptive series as the trachytic and trappean mountains (Yamaska, Chambly, Montreal, etc.) of the Upper St. Lawrence District described on a preceding page, all of which are apparently of Upper Silurian or Devonian age.

The granites of this region are also of Devonian or later age.   They present a very general resemblance in aspect and composition, consist-ing almost uniformly of white orthoclase-feldspar and colourless quartz, with small quantities of black or brown mica.   Large ex-posures occur in Stanstead, one alone covering five or six square miles ; also in the country immediately south-west of Lake St. Francis, and around Lake Megantic in Compton ; and smaller exposures are seen in Barnston, Burford, Hereford, and other townships.   In most cases the granite has penetrated the surrounding rocks in veins or dykes, and has greatly altered these : rendering the limestones and dolomites more or less crystalline, with partial or complete oblite-ration of fossils ; and converting the slates in some places into mica-ceous schists, or developing in them crystals of chiastolite (see page 97), garnet, hornblende, and small cubes of pyrites, as well as pro-ducing fractures and other signs of mechanical disturbance.

The eruptive dykes of Gaspé consist partly of grey dolerites (in places amygdaloidal), and in part of green diorite with porphyritic

---

* For structural and other details, see Report on the Gaspé Peninsula by R. W. Ells, 1882

crystals of greenish-white feldspar; or otherwise they present the form of a more or less compact or granular greenstone, properly so-called. Grey doleritic dykes occur around Gaspé Bay, as at Tar Point (referred to on a preceding page) and elsewhere; and green feldspatho-porphyritic dykes are exceedingly numerous on the south coast at New Carlisle, as described many years ago by the late Sir William Logan (Geological Reports: 1844, 1863). Trappean hills and dykes occur also at other points along or near the coast between the Cascapedia and the Restigouche at the head of the Bay of Chaleurs.

Glacial and other surface deposits :—These are spread very generally within this district, as elsewhere, over the various rock-formations of the country. They comprise, in ascending series: (1), Beds of auriferous gravel and magnetic sand; (2), Boulder-clays or Drift deposits, proper, consisting of accumulations of clay and gravel with boulders of crystalline and other rocks; (3), Beds of Leda Clay (a more or less deep-sea formation) and Saxicava Sand (a shore-line or shallow-sea deposit: page 342); and (4), sundry superficial deposits of comparatively recent origin. The Drift clays in many parts of the Eastern Townships and adjacent area, are underlaid by (and also partially mixed with) layers of gravel and black magnetic sand, containing, very generally, fine grains, and occasionally small nuggets of free gold. These auriferous deposits have been recognized in the beds of most of the streams and rivers which flow through this section of the Province, and especially in the St. Francis, Chaudière, Famine, Rivière des Plantes, Etchemin, Gilbert, Metgermet, and Rivière du Loup.* The slate rock which underlies most of the Chaudière country, although exposed in many places, is covered as a rule, from the surface downwards, by a deposit of alluvial earth, clay, boulder-clay, black magnetic sand, and coarse gravel, varying from a few feet to over 150 feet in thickness. There is hardly a single stream connected with the Chaudière valley that does not carry more or less gold, chiefly in the gravel which fills the cracks and crevices of the rock which forms its bed. But the greater portion of this surface gold has long since been extracted—at least, as regards the smaller streams and rivers where the bed-rock can be easily got

---

*Of Beauce County : not the Rivière du Loup of Kamouraska and Temiscouta on the St. Lawrence

23

at and explored. It has been ascertained, however, that, in many if not in all cases, these existing streams and rivers flowed at one time in older and lower channels beneath their present beds. The ancient and in general wider channels have been subsequently covered up and hidden by deposits of glacial and post-glacial age, consisting, as already stated, of gravels and black magnetic sand in immediate contact with the bed-rock, overlaid by clays and boulder-clays, other gravels (in some places auriferous) and vegetable soil. The clays and boulder-clays contain apparently no trace of gold; and it is only in the lowest layer of gravel immediately above the bed-rock, and in the cracks and hollows and behind projecting ridges of the bed-rock itself, that gold occurs in paying quantity. All underground workings therefore have to be carried down to the rock floor. When the gravel is washed, a certain amount of black magnetic sand is almost always found with the gold in the sluices, but this arises from the comparative density of the sand. The black sand, in itself, is no absolute indication of the presence of gold in alluvial gravels, as it occurs almost everywhere in the detritus of our crystalline rocks.

The Leda Clay and Saxicava Sand deposits (see page 342) are largely displayed on the Trois Pistoles, Cacouna, Rivière du Loup (Temiscouta), St. Anne, Matanne, Métis, and other rivers, at various elevations from three or four, to over two hundred feet above the present sea-level.

The more superficial deposits of the district include the bog-iron ores of Stanbridge, Farnham, Simpson, Ascot, Stanstead, Ireland, St. Lambert, St. Vallier, Vallery, Cacouna, and other sites; the ochres of Durham; the shell marls of Stanstead, New Carlisle, etc.; and the peat beds of the Rivière Ouelle, Rivière du Loup, Métis, Rimouski, and Madawaska—most of which are of great extent, and from five to ten or fifteen feet in depth.

APPENDIX.

Sequence of Rock Formations recognized within the Provinces of Ontario and Quebec.

POST-CAINOZOIC FORMATIONS :

    24. Deposits of Recent Origin.

    23. Post-Glacial Deposits.

    22. Drift or Glacial Deposits.

PALÆOZOIC FORMATIONS :

  Carboniferous :

    21. Bonaventure Fn.

  Devonian :

    20. Portage-Chemung Fn.

    19. Hamilton (or Lambton) Fn.

    18. Corniferous Fn.

    17. Oriskany Fn.

  Upper Silurian :

    16. Lower Helderberg or Eurypterus Fn.

    15. Onondaga or Gypsiferous Fn.

    14. Guelph Fn.

    13. Niagara Fn.

    12. Clinton Fn.

    11. Medina Fn.

  Lower (or Cambrio-) Silurian :

    10. Hudson River Fn.

    9. Utica Fn.

    8. Trenton (and Black River) Fn.

    7. Chazy Fn.

  Cambrian :

    6. Calciferous (including Quebec) Fn.

    5. Potsdam Fn.

    4. Keweenian Fn.

    3. Animikie Fn.

ARCHÆAN FORMATIONS :

    2. Huronian Fn.

    1. Laurentian Fn.

INDEX.

——o——

---

* The cut illustrating this instrument has been placed incorrectly on one of its sides. The lettering shews the proper position of the instrument.

24

www.ingramcontent.com/pod-product-compliance
Lightning Source LLC
Chambersburg PA
CBHW021708110726
47902CB00005B/1102